Lecture Notes in Control and Information Sciences

Edited by M. Thoma and A. Wyner

Vol. 117: K.J. Hunt
Stochastic Optimal Control Theory
with Application in Self-Tuning Control
X, 308 pages, 1989.

Vol. 118: L. Dai
Singular Control Systems
IX, 332 pages, 1989

Vol. 119: T. Başar, P. Bernhard
Differential Games and Applications
VII, 201 pages, 1989

Vol. 120: L. Trave, A. Titli, A. M. Tarras
Large Scale Systems:
Decentralization, Structure Constraints
and Fixed Modes
XIV, 384 pages, 1989

Vol. 121: A. Blaquière (Editor)
Modeling and Control of Systems
in Engineering, Quantum Mechanics,
Economics and Biosciences
Proceedings of the Bellman Continuum
Workshop 1988, June 13–14, Sophia Antipolis, France
XXVI, 519 pages, 1989

Vol. 122: J. Descusse, M. Fliess, A. Isidori,
D. Leborgne (Eds.)
New Trends in Nonlinear Control Theory
Proceedings of an International
Conference on Nonlinear Systems,
Nantes, France, June 13–17, 1988
VIII, 528 pages, 1989

Vol. 123: C. W. de Silva, A. G. J. MacFarlane
Knowledge-Based Control with
Application to Robots
X, 196 pages, 1989

Vol. 124: A. A. Bahnasawi, M. S. Mahmoud
Control of Partially-Known
Dynamical Systems
XI, 228 pages, 1989

Vol. 125: J. Simon (Ed.)
Control of Boundaries and Stabilization
Proceedings of the IFIP WG 7.2 Conference
Clermont Ferrand, France, June 20–23, 1988
IX, 266 pages, 1989

Vol. 126: N. Christopeit, K. Helmes
M. Kohlmann (Eds.)
Stochastic Differential Systems
Proceedings of the 4th Bad Honnef Conference
June 20–24, 1988
IX, 342 pages, 1989

Vol.127: C. Heij
Deterministic Identification
of Dynamical Systems
VI, 292 pages, 1989

Vol. 128: G. Einarsson, T. Ericson,
I. Ingemarsson, R. Johannesson,
K. Zigangirov, C.-E. Sundberg
Topics in Coding Theory
VII, 176 pages, 1989

Vol. 129: W. A. Porter, S. C. Kak (Eds.)
Advances in Communications and
Signal Processing
VI, 376 pages, 1989.

Vol. 130: W. A. Porter, S. C. Kak,
J. L. Aravena (Eds.)
Advances in Computing and Control
VI, 367 pages, 1989

Vol. 131: S. M. Joshi
Control of Large Flexible Space Structures
IX, 196 pages, 1989.

Vol. 132: W.-Y. Ng
Interactive Multi-Objective Programming
as a Framework for Computer-Aided Control
System Design
XV, 182 pages, 1989.

Vol. 133: R. P. Leland
Stochastic Models for Laser Propagation
in Atmospheric Turbulence
VII, 145 pages, 1989.

Vol. 134: X. J. Zhang
Auxiliary Signal Design in Fault
Detection and Diagnosis
XII, 213 pages, 1989

Vol. 135: H. Nijmeijer, J. M. Schumacher (Eds.)
Three Decades of Mathematical System Theory
A Collection of Surveys at the Occasion of the
50th Birthday of Jan C. Willems
VI, 562 pages, 1989

Vol. 136: J. Zabczyk (Ed.)
Stochastic Systems and Optimization
Proceedings of the 6th IFIP WG 7.1
Working Conference,
Warsaw, Poland, September 12–16, 1988
VI, 374 pages. 1989

For information about Vols. 1–116 please contact your bookseller or Springer-Verlag

Lecture Notes
in Control and Information Sciences 178

Editors: M. Thoma and W. Wyner

J.P. Zolésio (Ed.)

Boundary Control and Boundary Variation

Proceedings of IFIP WG 7.2 Conference
Sophia Antipolis, France,
October 15 - 17, 1990

Springer-Verlag Berlin Heidelberg GmbH

Editor

Jean Paul Zolésio
Institut Non Linéaire de Nice
Université de Nice Sophia Antipolis
Faculté des Sciences, B.P. No. 71
06108 Nice Cedex 2, France
and
Centre de Mathématiques Appliquées
Ecole des Mines, B.P. 207
06904 Sophia Antipolis, France

ISBN 978-3-540-55351-9 ISBN 978-3-540-47029-8 (eBook)
DOI 10.1007/978-3-540-47029-8

Originally published by International Fedration for Information Processing, Geneva, Switzerland in 1992.

Typesetting: Camera ready by authors

60/3020 5 4 3 2 1 0 Printed on acid-free paper

PREFACE

This volume comprises the proceedings of the Working Conference " Boundary Control and Boundary Variation " held in Sophia-Antipolis (France), october 1990 .

The conference was organized for the Working Group 7.2 (Computational methods for control systems described by partial differential equations) of Technical Committe 7 (Modelling and Optimization Techniques) of the international Federation for Information Processing (IFIP).

That conference was the fourth one organized in south of France and sponsored by the following French institutions :

Centre National de la Recherche Scientifique (CNRS)
Centre de Mathématiques Appliquées (CMA) in Sophia Antipolis
Aérospatiale , Cannes-la-Bocca.

These conferences were:

- june 86 IFIP 7.2 in Nice , proc. "Boundary Variation and Boundary Control" , L.N.C.I.S. Springer Verlag, no 100, 1987.
- december 87 ,COMCON conference in Montpellier, proc. "Stabilization of Flexible Structures",Optimization SoftwareInc.,L.A.1988
-january 89 in Montpellier , proc. "Stabilization of Flexible Structure" ,L.N.C.I.S. Springer Verlag, no 147, 1989.
-october 90 IFIP 7.2 in Sophia-Antipolis , these proc.

The aim of these conferences is to stimulate exchange of ideas between the group working on Shape Optimization (including free boundary problems) and the group working on boundary control of hyperbolic systems (including Stabilization).

J was helped on the organizing committee by

J.P. Marmorat , CMA
J. Leblond , CMA & INRIA

My thanks go to the 35 authors of the contributions contained in this volume.

Sophia-Antipolis, july 1991 J.P. Zolésio

CONTENTS

CONTROL OF AN OVERHEAD CRANE : STABILIZATION OF FLEXIBILITIES *

B. d'Andrea-Novel, F. Boustany †
F. Conrad ‡

Abstract

This paper deals with the feedback stabilization of the cable of an overhead crane, by the means of the position of the platform. The well-posedness of the closed-loop PDE system with boundary control and homogeneous Neumann condition on part of the boundary is established, the asymptotic stabilization is proved by Lasalle's Invariance Principle for a class of simple feedbacks and decay estimates are given. Illustrative simulations are displayed.
Keywords : Asymptotic Stabilization, Boundary control, Distributed Parameter Systems, Flexibilities, Decay Estimates.

*Supported by a grant from EDF-DER, Chatou
†Centre d'Automatique, Ecole des Mines, 35, rue St Honoré 77305 Fontainebleau, Fax : 64 69 47 01
‡Département de Mathématiques, Université Nancy I, and U.A. CNRS 750, B.P. 239 54506 Vandoeuvre les Nancy

1 Statement of the problem.

We consider an overhead crane, consisting of a motorized platform moving along an horizontal beam [1] and equipped with a winch, around which a cable is enrolled, holding a load. (see Figure 1, and [3])
Several studies ([3, 5, 7, 8, 20]) dealt with the "Rigid Case", that is the case where the cable is supposed to be rigid (and generally with negligible mass). The system can then be seen as a pendulum with variable length and mobile basis. The aim is to stabilize the load as quickly as possible, or to make it follow a given trajectory as precisely as possible. The actuators are the force u_1 applied by the motor to the platform, and the torque u_2 applied on the winch. When using the non-linear approach, one must make the assumption that the whole state is observed, i.e. : the position of the platform, the length of the cable and the angle of the cable with the vertical axis, as well as their derivatives.

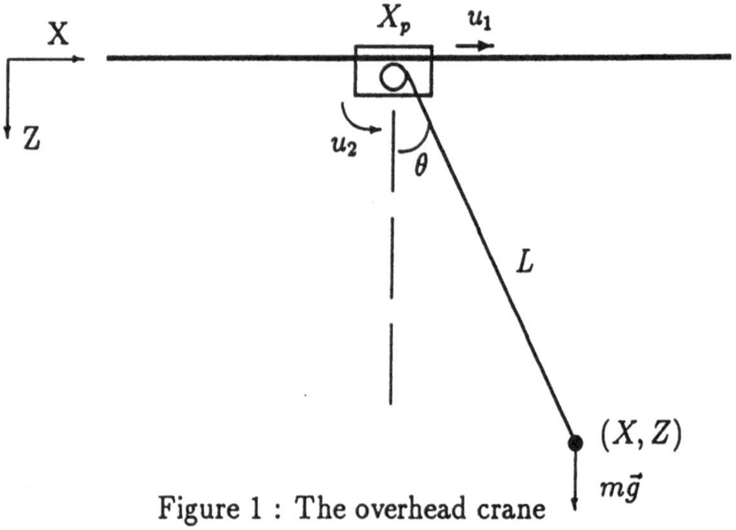

Figure 1 : The overhead crane

In [3, 5] it has been proved that this mechanical system, with less actuators than degrees of freedom, can be completely linearized by using **dynamic** non linear state feedback.

[1] This beam is assumed here to be fixed, so we are concerned with a two-dimensional problem

In this work, we focus on the cable. We take into account its mass and its flexibilities. We make the following assumptions :

Assumptions H1 :

- The cable is completely flexible and non-stretching.
- The length of the cable is constant.
- Displacements are small.
- The angle of the cable with the vertical axis is small everywhere along the cable.
- Dynamics of the platform is ignored, that is, the control is supposed to be the position or equivalently, the velocity of the platform, and not the force u_1.

Remark. For large displacements and length variation we can make use of the dynamic feedback linearization result for the rigid case [3, 5]. Then, close to the final desired position, and for the adequate length of the cable, assumptions H1 make sense and the objective is to stabilize the flexible modes of the cable.

The equation of heavy cables. We denote by s the curvilinear abscissa, and by $x(s,t)$ the position at time t of the point which has curvilinear abscissa s. (see Figure 2)

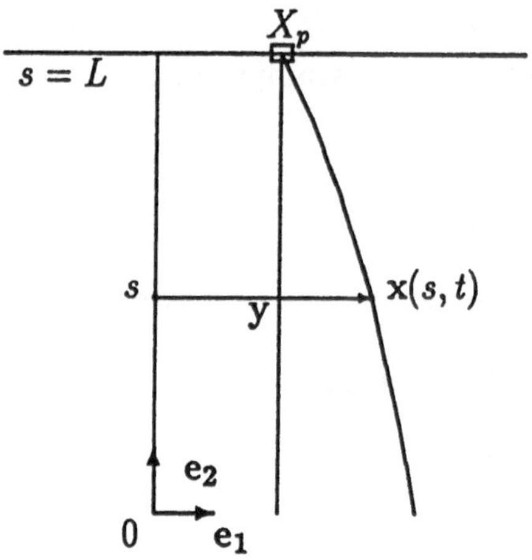

Figure 2 : Motion of the cable

Let $\tau(s,t)$ be the tension along the cable and ρ the lineic mass of the cable; the equation of heavy cables is the following ([17]) :

$$\rho x_{tt}(s,t) = \rho g + (\tau(s,t)x_s)_s \tag{1}$$

where x_s (resp. x_t) denotes the derivative of x with respect to s (resp. to t).
The non-stretching condition gives :

$$x_s(s,t).x_s(s,t) = 1 \tag{2}$$

Assumptions H1 enable us to study the tangent linearization of the system. To do this, we take the following equilibrium as reference :

$$\left\{ \begin{array}{lcl} x_{ref} & = & se_2 \\ \tau_{ref}(s) & = & (m + s\rho)g \end{array} \right. \tag{3}$$

and we set :

$$\left\{ \begin{array}{l} x(s,t) = x_{ref} + y \\ \tau(s,t) = \tau_{ref} + \delta \end{array} \right. \tag{4}$$

Replacing x and τ in (1) and (2), keeping only first-order terms and projecting upon the horizontal (e_1) axis (longitudinal vibrations are second-order quantities), we obtain the following system :

$$\begin{cases} y_{tt} - (ay_s)_s = 0 \\ y_s(0,t) = 0 \\ y(L,t) = X_p(t) \end{cases} \tag{5}$$

where

$$a(s) = g(\frac{m}{\rho} + s) \tag{6}$$

The first boundary condition expresses the verticality of the cable at its lowest end, due to the load; the second one means that the cable is, at its highest end, fixed to the platform.

Remark. Given X_p, the solution of system (5) can be explicitely written, using the eigenfunctions of the associated homogeneous problem, which are Bessel functions.

The equation describing the movement of the platform completes the dynamical model:

$$M\ddot{X}_p = (m + \rho L)g\theta + u_1 \tag{7}$$

where $(m + \rho L)g$ is an approximation of the tension, and θ is an approximation of $sin\theta$. System (5) with (7) leads to a hybrid PDE-ODE system (see [2]). Here (see Assumption H1), we study the simpler problem of stabilizing the cable, ignoring the dynamics of the platform.

Consider the following energy function :

$$E(t) = \frac{1}{2}\int_0^L (y_t^2 + ay_s^2)ds \ + \frac{k}{2}y(L,t)^2 \ , \ \ k > 0 \tag{8}$$

The integral is the internal energy of the cable, and the last factor could be replaced by $(y(L,t) - x_c)^2$, so that the platform could be steered to any equilibrium position x_c.

A formal computation of the time derivative of E gives :

$$\frac{dE}{dt} = y_t(L,t)\ [ay_s(L,t) + ky(L,t)] \tag{9}$$

which leads to the following dissipative boundary feedback laws, expressed in terms of boundary condition at $s = L$:

$$ay_s(L,t) + ky(L,t) = -g(y_t(L,t)) \tag{10}$$

where g is monotone (continuous), with $g(0) = 0$.

Consequently, the closed-loop system will be :

$$\begin{cases} y_{tt} - (ay_s)_s = 0 \\ y_s(0,t) = 0 \\ (ay_s)(L,t) + ky(L,t) = -g(y_t(L,t)) \end{cases} \tag{11}$$

Remark.

Problem (11) is similar to a model studied by Zuazua ([22, sect. 3.1]) concerning the wave equation in space dimension $n \leq 3$, in the case of active control on the whole boundary. The author also introduced the term $\frac{1}{2}y^2(L,t)$ in the energy, and obtained decay estimates for small k in the nonlinear case. In [23], it has been proved that this restriction on k is unnecessary when g is linear, and exponential uniform stability holds in that case. In our specific monodimensionnal problem, we improve the results of [22] in the sense that the coefficient of the PDE is not constant and we have an homogeneous Neumann condition on part of the boundary. These modifications are quite minor with respect to the work of [22]. More interesting is the fact that we obtain the estimates for any $k > 0$ even if g is nonlinear. Also, we handle both the superlinear case (as in [22]) and the sublinear case.

This paper is organized as follows :

- **Section 2:** we show that problem (11) is well-posed. This will give in particular a sense to the time-derivation (9).
- **Section 3 :** we prove that E decays to zero (strong asymptotic stabilization).
- **Section 4 :** we give estimates for this decay.
- **Section 5 :** we present some numerical results and we conclude.

2 Wellposedness

Though the results of this section hold for any $a \in L^\infty(0,L)$ satisfying :

$$a \geq \alpha > 0 \tag{12}$$

and for any maximal monotone graph g of $\mathbf{R}^2$ such that :

$$0 \in g(0) \tag{13}$$

we consider only the physical framework where a is affine and g is a continuous function.

Let H and V be the following Hilbert spaces :

$$\begin{array}{l} H = L^2(0,L) \\ V = H^1(0,L) \end{array} \tag{14}$$

Following the method of [9] it is possible to show that (11) defines a contraction semi-group on $V \times H$, associated with a nonlinear, maximal, monotone operator on $V \times H$, with dense domain. Instead, we apply an abstract result of [15].

Let $A \in \mathcal{L}(V, V')$ be defined by the following bilinear form :

$$< Au, v >_{V'V} = \int_0^L a u_s v_s ds + k u(L) v(L), \ \forall u, v \in V \tag{15}$$

Considering A as an unbounded operator on H, we have :

$$dom\ A = \{u \in V; (au_s)_s \in H; au_s(0) = 0; au_s(L) + ku(L) = 0\} \tag{16}$$

and, if u is in $dom\ A$:

$$Au = -(au_s)_s \tag{17}$$

A is a positive, self-adjoint, operator on H, and $V = dom\ A^{\frac{1}{2}}$. Moreover, $dom\ A$ is dense in V and in H. In the sequel, V will be equipped with the following scalar product :

$$(u, v)_V = \int_0^L a u_s v_s ds + k u(L) v(L) \tag{18}$$

defining on V a norm, equivalent to the usual norm. $V \times H$ is equipped with the scalar product :

$$((u,w),(v,z))_{V\times H} = \int_0^L a u_s v_s ds + ku(L)v(L) + \int_0^L wz ds \tag{19}$$

Next, we introduce :

$$Cv = v(L), \quad \forall v \in V \tag{20}$$

C belongs to $\mathcal{L}(V, \mathbf{R})$, and is surjective.

Multiplying the PDE in (11) by an element φ of V, and integrating by parts, we obtain the following abstract formulation :

$$y_{tt} + Ay + C^* g(Cy_t) = 0 \quad \text{in } V'; \; y \in V \tag{21}$$

where g, which is continuous and monotone, is the sub-differential of a convex, l.s.c (continuous) function on $\mathbf{R}$.

Hence, it is possible to use Theorem 2.2 from [15] : (21) written as a first-order system :

$$\begin{pmatrix} y \\ y_t \end{pmatrix}_t + \begin{pmatrix} 0 & -I \\ A & C^* g(C.) \end{pmatrix} \begin{pmatrix} y \\ y_t \end{pmatrix} = 0 \tag{22}$$

defines a nonlinear semigroup of contractions $S(t)$ on $V \times H$, associated with the following maximal monotone operator :

$$dom\, \mathcal{A} = \{ \begin{array}{l} (u,v) \in V \times V, (au_s)_s \in H, \\ au_s(0) = 0, au_s(L) + ku(L) = -g(v(L)) \} \end{array} \tag{23}$$

and, if $(u,v) \in dom\, \mathcal{A}$:

$$\mathcal{A}\begin{pmatrix} u \\ v \end{pmatrix} = \begin{pmatrix} -v \\ -(au_s)_s \end{pmatrix} \tag{24}$$

For initial data $(y(0), y_t(0))$ in $dom\, \mathcal{A}$,

$$(y(t), y_t(t)) = S(t)(y(0), y_t(0)) \tag{25}$$

is a strong solution of (22) i.e. $(y(t), y_t(t))$ is in *dom* $\mathcal{A}$ and $t \to (y(t), y_t(t))$ is Lipschitz continuous, almost everywhere differentiable, and (11) is satisfied almost everywhere for t positive (see [6]).

It is possible to extend the semi-group property to $V \times H$ by density of *dom* $\mathcal{A}$ in $V \times H$. Indeed, let :

$$E = dom\ A \times \{v \in V; v(L) = 0\} \tag{26}$$

On the one hand :

$$dom\ A = \{u \in V; (au_s)_s \in H; au_s(0) = 0; au_s(L) + ku(L) = 0\} \tag{27}$$

is dense in $V = dom\ A^{\frac{1}{2}}$. On the other hand, $\{v \in V; v(L) = 0\}$ is dense in $H = L^2$. Then E is dense in $V \times H$, and since E is a subset of *dom* $\mathcal{A}$ (because $g(0) = 0$), the result follows.

3 Strong asymptotic stabilization

We prove the strong asymptotic stability of system (11) in $V \times H$ by using *Lasalle's Invariance Principle* ([12]).

The results in this section can be established in the case when a is any function in $L^\infty(0, L)$ satisfying (12), and g is a maximal monotone graph, satisfying :

$$g(0) = \{0\}, \quad 0 \in g(\xi).\xi \Rightarrow \xi = 0 \tag{28}$$

Once again, we keep the physical framework.

Let (y, y_t) be a solution of (22). The energy associated with this solution is given by :

$$E(t) = \frac{1}{2}\int_0^L (y_t^2 + ay_s^2)ds \ + \frac{k}{2}y(L,t)^2 \tag{29}$$

Lemma 1
(i) E is nonincreasing w.r.t. t
(ii) if the initial condition is in dom $\mathcal{A}$ *then :*

$$\frac{dE}{dt} = -g(y_t(L,t))y_t(L,t) \leq 0 \tag{30}$$

Proof
Direct computation gives (ii). Then (i) is proved by density of *dom* $\mathcal{A}$ and contraction property of $S(t)$. □

E is then a Lyapunov Function. In order to use the Invariance Principle in infinite dimension, we need the following result :

Lemma 2 *The resolvent* $(I+\lambda\mathcal{A})^{-1}$ *is compact from* $V \times H$ *to* $V \times H$*, and* $0 \in R(\mathcal{A})$

Proof
We have immediatly :

$$\mathcal{A}(0) = 0 \tag{31}$$

The resolvent $(I + A)^{-1}$ is compact on H. Indeed, consider $f \in H$, $u = (I + A)^{-1}(f)$ satisfies the following system :

$$\begin{cases} u - (au_s)_s = f \\ au_s(0) = 0 \\ au_s(L) + ku(L) = 0 \end{cases} \tag{32}$$

Multiplying by u and integrating from 0 to L leads to the following estimate :

$$|u|_V^2 \leq Const.|f|_H^2 \tag{33}$$

ensuring the compactness. Since g is obviously compact from **R** to **R**, using [15, Th. 2.3] the resolvent $(I+\lambda\mathcal{A})^{-1}$ is then compact from $V \times H$ to $V \times H$. □

From Lemma 2 and classical results ([6, 12]) we have :

Proposition 1
For any $(y(0), y_t(0)) \in V \times H$*, the trajectories* $\{S(t)(y(0), y_t(0)), t \geq 0\}$ *are precompact in* $V \times H$*, and the* ω*-limit set* $\omega(y(0), y_t(0))$ *exists.*

Asymptotic stabilization can now be obtained, using another abstract result from [15], which is in fact the application of Lasalle's Invariance Principle [12].

Theorem 1 *Assume that $g(\xi) \neq 0$ for $\xi \neq 0$. Then for any $(y(0), y_t(0)) \in V \times H$, we have :*

$$(y(t), y_t(t)) \to 0, \quad in\ V \times H, \quad when\ t \to \infty \tag{34}$$

Proof

We show that $\omega(y(0), y_t(0)) = \{0\}$. In fact, it is enough to prove this for $(y(0), y_t(0)) \in Dom\ \mathcal{A}$. Let $\bar{w}_0$ be an element of $\omega(y(0), y_t(0))$, and let

$$\bar{w}(t) = (w, w_t)(t) = S(t)\bar{w}_0 \in \omega(y(0), y_t(0)) \tag{35}$$

First note that, by the Invariance Principle applied to $\bar{w}$, we have :

$$\frac{dE}{dt} = 0 \tag{36}$$

thus

$$g(w_t(L,t))w_t(L,t) = 0\ a.e. \tag{37}$$

We use the assumption on g to deduce

$$w_t(L,t) = (aw_s)(L,t) + kw(L,t) = 0\ a.e. \tag{38}$$

Proceeding as in [9] (see also [15, Thm 2.4 (ii)]), in order to show that $\omega(y(0), y_t(0)) = \{0\}$, we are led to prove the following uniqueness property :

Lemma 3 *Let $\bar{w} = (w, w_t)$ be a solution of the following overdetermined system :*

$$\begin{cases} w_{tt} - (aw_s)_s = 0 \\ (aw_s)(0,t) = 0 \\ (aw_s)(L,t) + kw(L,t) = 0 \\ w_t(L,t) = 0 \\ \bar{w}_0 \in \omega(y(0), y_t(0)) \subset Dom\ \mathcal{A} \end{cases} \tag{39}$$

Then

$$\bar{w} = 0 \tag{40}$$

Proof

We prove this result using Fourier series. We consider the eigenvalue problem :

$$\begin{cases} -(a\varphi_s)_s = \lambda\varphi \\ a\varphi_s(0) = 0 \\ a\varphi_s(L) + k\varphi(L) = 0 \end{cases} \tag{41}$$

Let (λ_n, φ_n) be the eigenvalues and eigenfunctions, with φ_n normalized in H. The solution of (39) for initial condition $\bar{w}_0 = (w(s,0), w_t(s,0))$ in *dom A* can be written as :

$$\begin{cases} w(t) = \sum_{i=1}^{\infty}(a_n cos\sqrt{\lambda_n}t + b_n sin\sqrt{\lambda_n}t)\frac{\varphi_n}{\sqrt{\lambda_n}} \\ \\ w_t(t) = \sum_{i=1}^{\infty}(b_n cos\sqrt{\lambda_n}t - a_n sin\sqrt{\lambda_n}t)\varphi_n \end{cases} \tag{42}$$

The first series converges in V and the second converges in H, with

$$\begin{cases} a_n = \int_0^L aw(s,0)_s\frac{\varphi_{n_s}}{\sqrt{\lambda_n}}ds + kw(L,0)\frac{\varphi_n(L)}{\sqrt{\lambda_n}} \\ \\ b_n = \int_0^L w_t(s,0)\varphi_n ds \end{cases} \tag{43}$$

As in [9], one can prove by a classical argument that $w(s,0)$ and $w_t(s,0)$ satisfy in fact the homogeneous boundary conditions, and consequently, integrating by parts (43):

$$\begin{cases} \sqrt{\lambda_n}a_n = \int_0^L (aw(s,0)_s)_s\varphi_n(s)ds \in l^2(\mathsf{N}) \\ \\ \sqrt{\lambda_n}b_n = \int_0^L aw_{ts}(s,0)\frac{\varphi_{n_s}(s)}{\sqrt{\lambda_n}}ds + kw_t(L,0)\frac{\varphi_n(L)}{\sqrt{\lambda_n}} \in l^2(\mathsf{N}) \end{cases} \tag{44}$$

since $w(s,0)$ and $w_t(s,0)$ lie in a suitable domain. Then, the second series in (42) converges in V and taking the value at L we have for almost every positive t :

$$0 = \sum_{i=1}^{\infty}(b_n cos\sqrt{\lambda_n}t - a_n sin\sqrt{\lambda_n}t)\varphi_n(L) \tag{45}$$

with an uniformly convergent series. As in [9], it is easy to see that (45) implies :

$$a_n\varphi_n(L) = b_n\varphi_n(L) = 0 \tag{46}$$

But we have $\varphi_n(L) \neq 0, \forall n \in \mathbb{N}$. Indeed, otherwise $\varphi = \varphi_n$ is solution of the following Cauchy problem :

$$\begin{cases} -(a\varphi_s)_s = \lambda\varphi \\ a\varphi_s(0) = 0; \;\; a\varphi_s(L) = 0 \\ \varphi(L) = 0 \end{cases} \Rightarrow \varphi = 0\,! \tag{47}$$

Thus (46) implies :

$$a_n = b_n = 0, \; \forall n \tag{48}$$

and consequently :

$$(w, w_t) = 0 \tag{49}$$

which completes the proofs of both Lemma 3 and Theorem 1. □

4 Decay estimates

Here, we assume a satisfies (6), in order to obtain the observability lemma.

Let $y(t)$ be a solution of (11), corresponding to initial conditions (y_0, y_1) in $V \times H$, and consider the associated energy :

$$E(t) = \frac{1}{2}\int_0^L (y_t^2 + ay_s^2)ds \;\; + \frac{k}{2}y(L,t)^2 \tag{50}$$

Theorem 2
(i) If

$$C_2|\xi| \leq |g(\xi)| \leq C_1|\xi|, \quad C_1, C_2 > 0 \tag{51}$$

then

$$E(t) \leq Me^{-\mu t}E(0) \tag{52}$$

where M, μ are positive constants.
(ii) Superlinear case : if

$$C_2\, min\{|\xi|, |\xi|^p\} \leq |g(\xi)| \leq C_1|\xi|, \quad C_1, C_2 > 0 \tag{53}$$

with $p > 1$, *then*

$$E(t) \leq M(\frac{1}{1+\mu t})^{\frac{2}{p-1}} E(0) \tag{54}$$

where M *is a positive constant, and* $\mu > 0$ *depends on* $E(0)$.
(iii) Sublinear case : if

$$C_2|\xi| \leq |g(\xi)| \leq C_1 \, max\{|\xi|, |\xi|^p\}, \quad C_1, C_2 > 0 \tag{55}$$

with $p < 1$, *then*

$$E(t) \leq M(\frac{1}{1+\mu t})^{\frac{2p}{p-1}} E(0) \tag{56}$$

where M *is a positive constant, and* $\mu > 0$ *depends on* $E(0)$.

Remarks

- Case (i) (in particular if g is linear) corresponds to uniform exponential stabilization.
- It would be of interest to study the decay rate with respect to k (at least numerically, and for linear g) .
- The proof of Theorem 2 is different from the one used in [22] or [23]. It is based upon an abstract result from [10], which adapts, to the nonlinear case and for boundary control, an idea of [13].

Proof.

First, we need the wellposedness of system (11), which has already been proved in Section 2.

Then, we need the following lemmas (the proofs will be given in the sequel) :

Lemma 4 *The following open-loop system :*

$$\begin{cases} z_{tt} - (az_s)_s = 0 \\ z_s(0,t) = 0 \\ (az_s)(L,t) + kz(L,t) = u(t) \end{cases} \tag{57}$$

is well posed, i.e. the mapping :

$$(z_0, z_1, u) \in V \times H \times L^2(0,T) \rightarrow (z, z_t, z_t(L,.)) \in L^2(0,T; V \times H) \times L^2(0,T) \tag{58}$$

is well defined and is continuous, $\forall u \in L^2(0,T)$, $\forall T > 0$, $\forall (z_0, z_1) \in V \times H$ $(H2)$.

Lemma 5 *The uncontrolled problem :*

$$\begin{cases} \varphi_{tt} - (a\varphi_s)_s = 0 \\ \varphi_s(0,t) = 0 \\ (a\varphi_s)(L,t) + k\varphi(L,t) = 0 \end{cases} \tag{59}$$

satisfies the **strong observability condition** *:*

$$\exists M_0 > 0, \exists T_0 > 0 \ s.t. \ |\varphi_0|_V^2 + |\varphi_1|_H^2 \leq M_0 \int_0^{T_0} \varphi_t^2(L,t)dt \ \ (H3) \tag{60}$$

We use then Theorem 1 from [10] : the wellposedness of system (11), (H2), (H3) and the strucural assumptions on g made in Theorem 2 ensure the estimates of this Theorem. □

We give now the proofs of Lemmas 4 and 5.

Proof of Lemma 4.
It all amounts to get the a priori estimate :
$\forall T > 0$, $\exists C_1(T), C_2(T)$ such that :

$$\int_0^T E(t)dt + \int_0^T z_t^2(L,t)dt \leq C_1(T)E(0) + C_2(T) \int_0^T u^2(t)dt \tag{61}$$

where E is given by (8).
To obtain (61) we first establish that :

$$E(t) \leq E(0) + \frac{\lambda}{2} \int_0^t z_t^2(L,\tau)d\tau + \frac{1}{2\lambda} \int_0^t u^2(\tau)d\tau, \ \forall \lambda > 0 \tag{62}$$

integrating the identity :

$$\frac{dE}{dt} = u(t)z_t(L,t) \tag{63}$$

and using Young's inequality.

Then we establish the following inequality :

$$\frac{1}{2}\int_0^t z_t^2(L,\tau)d\tau \leq C_1\int_0^t E(\tau)d\tau + C_2E(0) + C_2E(t) \tag{64}$$

by multiplication of the equation in (57) by saz_s and integration over space and time.

Plugging (64) into (62), applying Gronwall's lemma, and integrating from 0 to T leads to the following :

$$\int_0^T E(t)dt \leq C_1(T)E(0) + C_2(T)\int_0^T u^2(t)dt \tag{65}$$

From (64) and (65) it is easy to get (61).□

Proof of Lemma 5.
Condition (H3) can be proved by Fourier analysis, using an inequality established by Ingham and improved by Ball and Slemrod. The result, based on such an approach, can be stated as follows :
Consider again the following eigenvalue problem ($\omega > 0$) :

$$\begin{cases} -(a\varphi_s)_s = \omega^2\varphi \\ \varphi_s(0) = 0 \\ (a\varphi_s)(L) + k\varphi(L) = 0 \end{cases} \tag{66}$$

Let (ω_n^2) be the sequence of (simple) eigenvalues of (66), φ_n a sequence of associated eigenfunctions, normalized in $H = L^2(0,L)$. If :

$$\begin{array}{ll} (i) & |\varphi_n(L)| \geq \beta > 0, \quad \forall n \in \mathbb{N} \\ (ii) & \underline{lim}(\omega_{n+1} - \omega_n) = \alpha > 0, \quad n \to \infty \end{array} \tag{67}$$

then condition (H3) is true (for any $T_0 > \frac{2\pi}{\alpha}$)

We show now that (i) and (ii) are satisfied.

Proof of (i).
We multiply the equation in (66) by φ and integrate :

$$\int_0^L a\varphi_s^2 ds + k\varphi^2(L) = \omega^2 \tag{68}$$

Then, using a multiplier $\psi\varphi_s$, one gets

$$\int_0^L a(\frac{\psi}{a})_s a\varphi_s^2 ds + \omega^2 \int_0^L \psi_s \varphi^2 ds \leq [\omega^2 + \frac{k^2}{a}]\psi(L)\varphi^2(L) \tag{69}$$

Combining (69) with (68), and choosing $\psi \geq 0$ such that

$$a(\frac{\psi}{a})_s \geq 1 \text{ and } \psi_s \geq 1 \tag{70}$$

leads to :

$$\varphi^2(L) \geq \frac{2\omega^2}{(\omega^2 + \frac{k^2}{a})\psi(L) + k} > 0 \;\; \forall \omega = \omega_n. \tag{71}$$

When n tends to infinity, the fraction tends to 2. So (i) is true.

Proof of (ii).

We first define the spectral elements of problem (66) in a more precise way.

Let σ be defined by :

$$\sigma^2 = \frac{4\omega^2}{g}(s + \frac{m}{\rho}) \tag{72}$$

and let

$$\psi(\sigma) = \varphi(s) \tag{73}$$

Then, ψ is given by the following equation :

$$\psi''(\sigma) + \frac{\psi'(\sigma)}{\sigma} + \psi(\sigma) = 0 \tag{74}$$

The solutions of (74) can be written in the following form :

$$\psi(\sigma) = AJ_0(\sigma) + BY_0(\sigma) \tag{75}$$

where J_0 is the Bessel function of first kind, and Y_0 is the Neumann Bessel function of second kind (see [1]).

The boundary conditions give :

$$\begin{aligned} &AJ_0'(\sigma(0)) + BY_0'(\sigma(0)) = 0 \\ &a(L)\tfrac{d\sigma}{ds}(L)[AJ_0'(\sigma(L)) + BY_0'(\sigma(L))] + k[AJ_0(\sigma(L)) + BY_0(\sigma(L))] = 0 \end{aligned} \tag{76}$$

This system must have solutions other than $A = B = 0$. Thus, the ω_i are the zeros of the determinant :

$$\begin{aligned} F(\omega) &= a(L)\tfrac{d\sigma}{ds}[Y_0'(\sigma(L))J_0'(\sigma(0)) - Y_0'(\sigma(0))J_0'(\sigma(L))] \\ &+ k[J_0'(\sigma(0))Y_0(\sigma(L)) - J_0(\sigma(L))Y_0'(\sigma(0))] \end{aligned} \tag{77}$$

We use now the well-known asymptotic expansions of Bessel functions (see [19]) :

$$\begin{aligned} J_0(z) &= \sqrt{\frac{2}{\pi z}}cos(z - \frac{\pi}{4}) + O(\frac{1}{z^{\frac{3}{2}}}) \\ Y_0(z) &= \sqrt{\frac{2}{\pi z}}sin(z - \frac{\pi}{4}) + O(\frac{1}{z^{\frac{3}{2}}}) \\ J_0'(z) &= -\sqrt{\frac{2}{\pi z}}sin(z - \frac{\pi}{4}) + O(\frac{1}{z^{\frac{3}{2}}}) \\ Y_0'(z) &= \sqrt{\frac{2}{\pi z}}cos(z - \frac{\pi}{4}) + O(\frac{1}{z^{\frac{3}{2}}}) \end{aligned} \tag{78}$$

and we obtain :

$$\begin{aligned} F(\omega) &= F_0 sin(\sigma(L) - \sigma(0)) + O(\tfrac{1}{\omega}) \\ &= F_0 sin(\omega\tau) + O(\tfrac{1}{\omega}) \end{aligned} \tag{79}$$

where F_0 is a constant, and :

$$\tau = 2\sqrt{\frac{\frac{m}{\rho} + L}{g}} - 2\sqrt{\frac{m}{\rho g}} \tag{80}$$

In order to claim that the zeros of $F(\omega)$ are the zeros of its principal term, asymptotically, it is enough to prove that they are simple. We compute $F'(\omega)$ using (77), and eliminate the second order derivatives using (74). Using the asymptotic expansions (78), we finally obtain :

$$F'(\omega) = F_1.cos(\omega\tau) + O(\frac{1}{\omega}) \tag{81}$$

We denote by ω_i the strictly increasing sequence defined by (66) (i.e. of positive zeros of F), and by ω_i^0 the increasing sequence of positive zeros of $sin(\sigma(L) - \sigma(0))$:

$$\omega_i^0 = \frac{i\pi}{\tau} \tag{82}$$

Let ω_i and ω_{i+1} be two consecutive zeros of F. We have :

$$\begin{array}{l} \omega_i = \omega_j^0 + O(\frac{1}{\omega_i}) \\ \omega_{i+1} = \omega_k^0 + O(\frac{1}{\omega_i}) \end{array} \tag{83}$$

We use (81) to show that necessarily :

$$j \neq k \tag{84}$$

then :

$$\omega_{i+1} - \omega_i \geq |\omega_k^0 - \omega_j^0| - |O(\frac{1}{\omega_i})| = \frac{\pi}{\tau} - |O(\frac{1}{\omega_i})| \tag{85}$$

implying (ii), with :

$$\alpha = \frac{\pi}{\tau} \tag{86}$$

Finally, we have the observability condition (H3) for any T_0 such that :

$$T_0 > 2\tau \tag{87}$$

□

5 Numerical results

We have simulated the closed-loop system (11) in the case of a linear function $g(\xi) = \gamma\xi$:

$$\left\{ \begin{array}{l} y_{tt} - (ay_s)_s = 0 \\ y_s(0,t) = 0 \\ (ay_s)(L,t) + ky(L,t) = -\gamma y_t(L,t) \end{array} \right. \tag{88}$$

and we used a modal decomposition.

Modal study. Our goal here is to expand the solution of (88) in terms of the eigenfunctions of the open-loop system (see [21]). To do this, we have to consider homogeneous boundary conditions. Therefore we introduce a new function $z(s,t)$:

$$z(s,t) = y(s,t) - y(L,t) \tag{89}$$

which leads to the following PDE system :

$$\begin{cases} z_{tt} - (az_s)_s = -y_{tt}(L,t) = \dfrac{1}{\gamma}(az_{st}(L,t) + ky_t(L,t)) \\ z(L,t) = 0 \\ z_s(0,t) = 0 \\ I.C. \end{cases} \tag{90}$$

The eigenvalue problem associated with the open-loop system is the following :

$$\begin{cases} -(a\varphi_s)_s = \omega^2\varphi \\ a\varphi_s(0) = 0 \\ \varphi(L) = 0 \end{cases} \tag{91}$$

Let σ be defined by (72) and let ψ be defined from φ as in (73). Then, ψ satisfies (74) and can be written in the following form :

$$\psi(\sigma) = AJ_0(\sigma) + BY_0(\sigma) \tag{92}$$

where J_0 is the Bessel function of first kind, and Y_0 is the Neumann Bessel function of second kind (see [1]). To find the eigenfrequencies, as well as A and B, we write the boundary conditions :

$$\begin{cases} z(L,t) = 0 \;\Rightarrow\; AJ_0(\sigma(L)) + BY_0(\sigma(L)) = 0 \\ z_s(0,t) = 0 \;\Rightarrow\; AJ_0'(\sigma(0)) + BY_0'(\sigma(0)) = 0 \end{cases} \tag{93}$$

Thus, in order to obtain nonzero solutions we must have :

$$J_0(\sigma(L))Y_0'(\sigma(0)) - Y_0(\sigma(L))J_0'(\sigma(0)) = 0 \tag{94}$$

Using (72) and [1, §9.5.32], we find the solutions ω_i of this equation. The ω_i's are real, simple and increase indefinitely with i.
With every ω_i, we associate (see (72)) :

$$\sigma_i^2 = \frac{4\omega_i^2}{g}(s + \frac{m}{\rho}) \tag{95}$$

and the eigenfunction (see (73)) :

$$\varphi_i(s) = A_iJ_0(\sigma_i) + B_iY_0(\sigma_i) \tag{96}$$

where A_i and B_i are chosen to satisfy one of the boundary conditions and the normalization :

$$\int_0^L \varphi_i^2(s)ds = 1 \tag{97}$$

Then, we can write formally the solution of (90) in the following way :

$$z(s,t) = \sum_{i=1}^{\infty} \alpha_i(t)\varphi_i(s) \tag{98}$$

where the α_i satisfy :

$$\ddot{\alpha}_i + \omega_i^2 \alpha_i = \frac{K_i}{\gamma}(az_{st}(L,t) + ky_t(L,t)) \tag{99}$$

with

$$K_i = \int_0^L \varphi_i(s)ds \tag{100}$$

Noticing that the evolution of $y(L,t)$ is given by :

$$y_t(L,t) = -\frac{1}{\gamma}(az_s(L,t) + ky(L,t)) \tag{101}$$

(99) gives a linear infinite dimensional state formulation of the system. In order to perform numerical computations, we kept the first five modes only, obtaining the following system :

$$\left\{ \begin{array}{l} \ddot{\alpha}_i + \omega_i^2 \alpha_i = \frac{K_i}{\gamma}(az_{st}(L,t) + ky_t(L,t)) \ i = 1, \ldots, 5 \\ y_t(L,t) = -\frac{1}{\gamma}(az_s(L,t) + ky(L,t)) \end{array} \right. \tag{102}$$

or, equivalently :

$$\left\{ \begin{array}{l} \ddot{\alpha}_i + \omega_i^2 \alpha_i = \frac{K_i}{\gamma}(a \sum_{i=1}^{5} \dot{\alpha}_i(t) \frac{d\varphi_i}{ds}(L) + ky_t(L,t)), \quad i = 1, \ldots, 5 \\ y_t(L,t) = -\frac{1}{\gamma}(a \sum_{i=1}^{5} \alpha_i(t) \frac{d\varphi_i}{ds}(L) + ky(L,t)) \end{array} \right. \tag{103}$$

Remark : The angular variable θ (in the rigid case) is nothing but :

$$\theta = -y_s(L,t) = -z_s(L,t) = -\sum_{i=1}^{\infty} \frac{d\varphi_i}{ds}(L)\alpha_i(t) \tag{104}$$

We used Basile to simulate system (103), results are displayed in Figure 3.

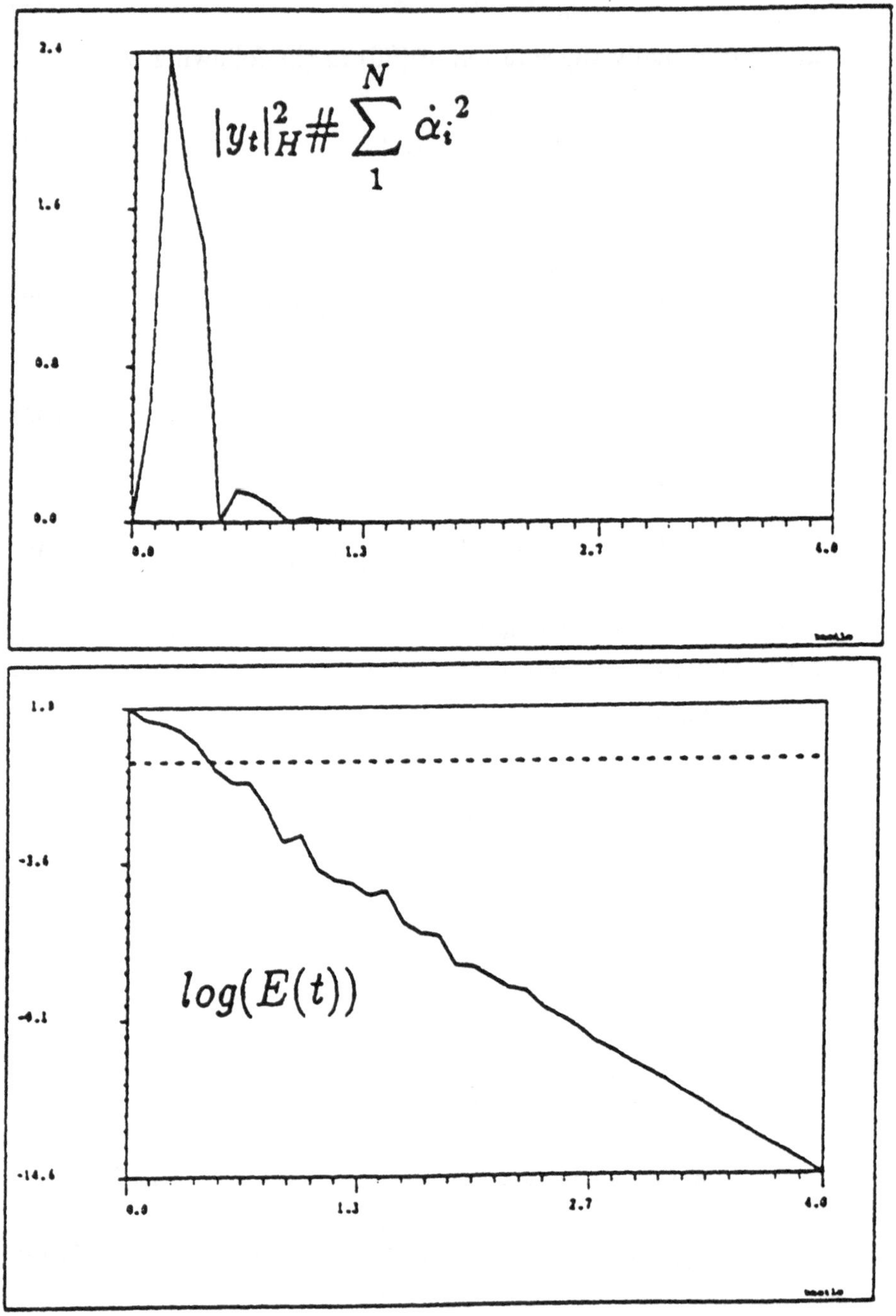

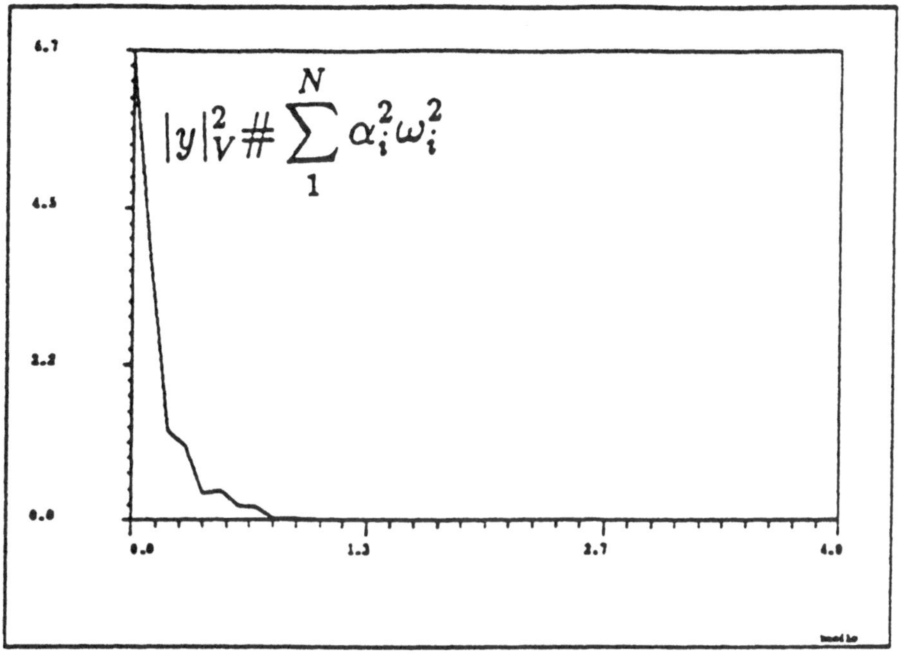

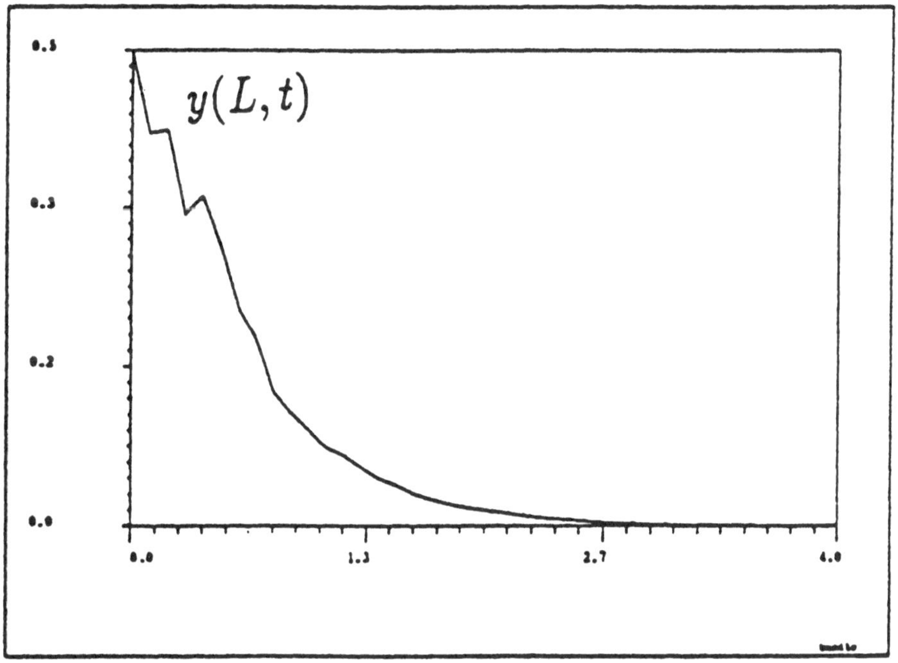

Figure 3 : Time evolution for a modal approximation with linear feedback.

As a conclusion we can stress the simplicity of the feedback. It involves only quantities usually observed in the rigid case. Practically, sensors are not able to take into account all the modes, very high frequencies are neglected. As a model case, simulations with a feedback not including all the modes (e.g. considering only the first three α_i) were performed. The residual modes (here fourth and fifth) were not stabilized, but they remained bounded. This fact is not surprising, and was already mentionned in several papers (see [4] for instance).

The complete "hybrid" PDE-ODE system, taking into account the dynamical equation (7) governing the platform, as well as the effect of observation delays on the performances of the feedback, is the subject of present investigations.

Acknowledgments The authors are indebted to Professor J.M. CORON for helpful suggestions.

References

[1] ABRAMOWITZ, STEGUN, *Handbook of Mathematical Functions,* Dover Publ., 1965.

[2] B. d'ANDREA-NOVEL, F. BOUSTANY, B.P. RAO, *Control of an Overhead Crane : Feedback Stabilization of an Hybrid PDE-ODE System,* submitted to the European Control Conference, July 1991, Grenoble, France.

[3] B. d'ANDREA-NOVEL, J. LEVINE, *Modelling and Nonlinear Control of a Overhead Crane,* MTNS Amsterdam, 1989.

[4] M. BALAS, *Modal Control of Certain Flexible Dynamic Systems,* SIAM J. Control and Optimization, 16, 3, 1978, pp. 450-462.

[5] F. BOUSTANY, *Commande d'un Pont Roulant*, Rapport de DEA, Ecole des Mines de Paris - Université Paris IX Dauphine, 1989.

[6] H. BREZIS, *Opérateurs Maximaux Monotones et Semi-Groupes de Contraction dans les espaces de Hilbert,* North Holland, 1973.

[7] B. CARON, *Etude de la Structure de Commande d'un Pont Roulant,* GRECO, Pôle SARTA, Grenoble, juin 1990.

[8] T. CHOROT, *Commande non linéaire de systèmes mécaniques hamiltoniens. Application à un Pont Roulant,* IASTED, Control'90, Switzerland.

[9] F. CONRAD, M. PIERRE, *Stabilization of Euler-Bernouilly Beams by Nonlinear Boundary Feedback,* INRIA report # 1235, 1990.

[10] F. CONRAD, J. LEBLOND, J.P. MARMORAT, *Stabilization of Second Order Evolution Equations by Unbounded Nonlinear Feedback,* IFAC Symp. Control of Distributed Parameter Systems, Perpignan, 1989.

[11] R. COURANT, D. HILBERT *Methods of Mathematical Physics,* Vol II, Interscience Publishers, 1962.

[12] C.M. DAFERMOS, M. SLEMROD, *Asymptotic behaviour of nonlinear contraction semi-groups,* J. Funct. Anal. 13, 1973, pp. 97-106.

[13] A. HARAUX, *Une Remarque sur la Stabilisation de certains Systèmes du Deuxième Ordre en Temps,* Port. Math. 56, 3, 1989, pp. 245-258.

[14] V. KOMORNIK, *Stabilisation Frontière Rapide de l'Equation des Ondes,* C. R. Acad. Sci. Paris Sér. I Math. 807, 1987, pp. 397-401.

[15] I. LASIECKA, *Stabilization of Wave and Plate like Equations with Nonlinear Dissipation on the Boundary,* J. Diff. Eq. 79, 2, 1989, pp. 340-381.

[16] A. PAZY,*Initial value probles for nonlinear differential equations in Banach spaces,* proc. Sem. Collège de France 84,5, 1982, pp. 154-172.

[17] M. REEKEN, *The Equation of Motion of a Chain,* Mat. Zeitschrift 155, 1977.

[18] F. RIESZ, B. SZ. NAGY, *Leçons d'analyse fonctionnelle,* Gauthiers-Villars, 1972.

[19] L. SCHWARTZ, *Méthodes Mathématiques pour la Physique,* Masson.

R. SEPULCHRE, *Commande Par Bouclage Dynamique Non Linéaire de Systèmes Mécaniques de type Pont Roulant,* Mémoire de fin d'études ERASMUS, Louvain-la-Neuve, 1990.

SNEDDON, *Fourier Transforms,* McGraw-Hill, 1951.

E. ZUAZUA, *Uniform Stabilization of the Wave Equation by Nonlinear Boundary Feedback,* SIAM J. Control and Optimization, 28, 2, 1990, pp. 466-477.

E. ZUAZUA, *Some Remarks on the Boundary Stabilization of the Wave Equation,* IFIP Conf. Control of Boundaries and Stabilization, Clermont Ferrand, June 1988, Lect. Notes in Control and Inf. Sciences # 125, pp. 251-266.

EXACT CONTROLLABILITY FOR LINEAR DYNAMIC SYSTEMS WITH SKEW SYMMETRIC OPERATORS

A. Bensoussan[1]

Introduction

We present in this paper an algebraic characterization of exact controllability for infinite dimensional dynamic systems. This characterization yields all available results for systems governed by linear hyperbolic equations. The key property is the skew symmetry of the operator, which implies a purely imaginary point spectrum.

1. General formulation of linear dynamic systems and the problem of controllability

1.1. Definitions

Let us consider two Hilbert spaces H and U, where
H is the Hilbert space of states
U is the Hilbert space of controls.

Let us consider two linear operators A, B

$$A : D(A) \to H, D(A) \text{ dense in } H \text{ domain of } A$$

$$B \in L(U; H)$$

We assume that A is the infinitesimal generator of a strongly continuous semi-group e^{At} in H.

Let us consider the dynamic system

$$z' = Az + Bv, \quad v(.) \text{ locally } L^2$$

$$z(0) = 0$$

then the solution $z(t)$ is expressed by

$$z(t) = \int_0^t e^{A(t-s)} Bv(s)ds$$

[1] University of Paris Dauphine and INRIA, BP 105, 78153 Le Chesnay Cedex, France

Define

$$F_T = \text{ range of } z(T; v(.)),$$
$$v(.) \in L^2(0,T;U)$$

F_T can be structured as a Hilbert space with the bijection

$$F_T \overset{\Psi_T}{\Leftrightarrow} L^2(0,T;U)/_{N_T}$$

where

$$N_T = \{v(.) | z(T; v(.)) = 0\}$$

defined by

$$\Psi_T(\tilde{u}) = z(T; u).$$

Note that $F_T \subset H$ with continuous injection.

We introduce the

Definition The pair (A, B) is ***exactly*** controllable at T if $D(A) \subset F_T$ with dense and continuous injection.
The pair (A, B) is ***approximatively*** controllable at T if

$$F_T \text{ is dense in } H.$$

1.2. Controllability operator
Consider the operator

$$\Gamma_T = \int_0^T e^{tA} B B^* e^{tA^*} \, dt$$

then

$$\Gamma_T \in L(H; H).$$

Define π = injection of F_T in H, then we can split Γ_T as

$$\Gamma_T = \pi \Psi_T \Psi_T^* \pi^*.$$

If the pair (A, B) is approximatively controllable at T, then π^* is injective and $\pi^* H$ is identified with H.
In particular

$$F_T \subset H \subset F_T'$$

and

$$\Gamma_T = \Psi_T \Psi_T^* \in L(F_T'; F_T)$$

If the pair (A, B) is exactly controllable at T, then

$$D(A) \subset F_T \subset H \subset F_T' \subset D(A))'$$

each space beeing density and continuously embedded in the next one.

Proposition 1.1.
(A, B) approximately controllable $\Leftrightarrow$
Γ_T is a norm on $H \Leftrightarrow$
A^*, B^* is observable on $(0, T)$, which means

$$B^* e^{tA^*} h = 0, \forall t \in (0, T) \rightarrow h = 0$$

Observability is equivalent to a uniqueness property.
Uniqueness

$$z' = A^* z$$

$$B^* z = 0 \quad \text{on} \quad (0, T)$$

$$\Rightarrow z = 0$$

On the other hand exact controllability is obtained by estimates of the following type :
Exact controllability :

$$(\Gamma_T h, h) \geq C(T - T_o)\|h\|^2_{(D(A))} \Leftrightarrow \|h\|^2_{F'_T} \geq C(T - T_o)\|h\|^2_{(D(A))'}$$

Remark 1.1.
Exact controllability is constructive. Indeed, given ξ in F_T, we solve

$$\Gamma_T h = \xi \Rightarrow h \in F'_T \text{ unique}$$

Then

$$\Psi^*_T h \in L^2(0, T; U)/_{N_T}$$

is the equivalence class of controls which realize Ψ.

Remark 1.2.
If $h \in H$, then $\Psi^*_T h$ is the equivalence class of $B^* e^{(T-t)A^*} h$.

2. A general framework which yields to skew symmetric operators

2.1. Notation

Let L be a self adjoint operator in a Hilbert space H_L (identified with its dual), $D(L)$ domain of L.
Assume
$V_L \subset H_L \subset V'_L$ continuously and densily embedded

$$< Lz_1, z_2 >= ((z_1, z_2)) \forall z_1, z_2 \in V_L$$

Injection of V_L into H_L is compact.
Consider the set of eigenvalues,

$$Lw_j = \lambda_j w_j, \quad |w_j|_{H_L} = 1$$

$$0 < \lambda_1 \leq \lambda_2 \ldots \leq \lambda_j \ldots \uparrow \infty.$$

Define

$$\Delta_L = \{z \in V_L | Lz \in V_L\}$$

then one has

$$\Delta_L \subset D(L) \subset V_L \subset H_L \subset V_L' \subset (D(L))' \subset \Delta_L'$$

with dense and continuous injection of each space in the next one.

We define

$$H = H_L \times V_L'$$

$$A = \begin{pmatrix} 0 & I \\ -L & 0 \end{pmatrix} \qquad \begin{aligned} D(A) &= V_L \times H_L \\ (D(A))' &= V_L' \times (D(L))' \end{aligned}$$

$$U = U_1 \times U_2 \qquad \begin{aligned} q_1 &\in L(D(L); U_1) \\ q_2 &\in L(V_L; V_3) \end{aligned}$$

$$Bv = -\begin{pmatrix} L^{-1} q_1^* v_1 \\ q_2^* v_2 \end{pmatrix} \quad \Rightarrow B \in L(U; H)$$

$$B^* h = -\begin{pmatrix} q_1 L^{-1} h_1 \\ q_2 L^{-1} h_2 \end{pmatrix}$$

With these definitions, we can proceed with the interpretation of the differential operational equation

$$z' = Az + Bv \; z(0) = 0$$

$$z \in C([0,T]; H), z' \in L^2(0,T; (D(A))')$$

We get

$$z_1' = z_2 - L^{-1} q_1^* v_1 \qquad z_1(0) = 0$$

$$z_2' = -Lz_1 - q_2^* v_2 \qquad z_2(0) = 0$$

$$z_1 \in C([0,T]; H_L), z_2 \in C([0,T]; V_L')$$

$$z_1' \in L^2(0,T; V_L'), z_2' \in L^2(0,T; (D(L))')$$

set $\eta = z_2$ then we deduce that η is the solution of

$$\eta" + L\eta = q_1^* v_1 - q_2^* v_2'$$

$$\eta(0); \eta'(0) + q_2^* v_2(0) = 0$$

$$\eta \in C([0,T]; V_L'), \eta' + q_2^* v_2 \in C([0,T]; (D(L))')$$

$$\eta' \in L^2(0,T; (D(L))')$$

$$\eta" \in C([0,T]; \Delta_L') \oplus L^2(0,T; (D(L))') \oplus H^{-1}(0,T; V_L').$$

It is important to notice that $\eta'(t)$ and $v_2(t)$ are defined a.e.t, but the quantity $\eta'(t) + q_2^* v_2(t)$ is defined at any time t.

We can use the transposition method, alternatively. Let ψ be the solution of

$$\psi'' + L\psi = 0$$

$$\psi(T) = \psi_o, \psi'(T) = \psi_1$$

$$\psi_o \in D(L), \psi_1 \in V_L$$

$$\psi \in C([0,T]; D(L)), \psi' \in C([0,T]; V_L)$$

then the pair $\eta(T), \eta'(T) + q_2^* v_2(T)$is defined by

$$< \eta'(T) + q_2^* v_2(T), \psi_o > - < \eta(T), \psi_1 >=$$
$$\int_0^T [(v_1, q_1\psi) + (v_2, q_2\psi')]dt.$$

Therefore the problem of exact controllability at T is formulated as follows :

Find $v_1(.), v_2(.)$ so that $\eta(T)$ and $\eta'(T) + q_2^* v_2(T)$ take *prescribed values in* H_L *and* V_L'.

There is another and mnemonic way to write the equation.

Consider the operator

$$J_o \in L(H^1(0,T;U_2); L^2(0,T;U_2))$$

defined by

$$J_o u = u'$$

and consider its transpose

$$J_o^* \in L(L^2(0,T;U_2); (H^1(0,T;U_2))')$$

Remark 2.1.

If $u \in H_0^1(0,T;U_2)$, then $J_0^* = -u'$.

In general, $J_0^* \neq -u' \in H^{-1}(0,T;U_2)$.

We write the equation of η as follows.

$$\eta'' + L\eta = q_1^* v_1 + q_2^* J_o^* v_2$$

$$\eta \in C([0,T]; V_L') \quad \eta' \in L^2(0,T; (D(L))')$$

$$\eta'' \in C([0,T]; \Delta_L') \oplus L^2(0,T; (D(L))') \oplus (H^1(0,T; V_L))'$$

$$\eta(0) = 0$$

Remark 2.2.

One has

$$q_2 \in L(H^1(0,T;V_L); H^1(0,T;U_2))$$

and

$$q_2^* J_0^* \in L(L^2(0,T;U_2); (H^1(0,T;V_L))').$$

In the previous notation η'' is slightly misleading.

What it means is the following. Let $\psi \in H^1(0,T;D(L)) \cap L^2(0,T;\Delta_L)$ with $\psi(T) = 0$, then one has

$$-\int_0^T (\eta', \psi')dt + \int_0^T (\eta, L\psi)dt = \int_0^T [(v_1, q_1\psi) + (v_2, q_2\psi')]dt$$

where all terms make sense. Now if we do not assume $\psi(T) = 0$, then we should write the above as

$$< \eta'(T), \psi(T) > - \int_0^T (\eta', \psi')dt + \int_0^T (\eta, L\psi)dt = \int_0^T [(v_1, q_1\psi) + (v_2, q_2\psi')]dt$$

but $\eta'(T)$ is not the value of the derivative η' at time T, in spite of the notation. It is a short for $\eta'(T) + q_2^* v_2(T)$. Considering ψ as in the transposition method above, we deduce again

$$< \eta'(T), \psi_o > - < \eta(T), \psi_1 >= \int_0^T [(v_1, q_1\psi) + (v_2, q_2\psi')]dt$$

this expression serves as the definition of a pair $\eta(T) \in V_L', \eta'(T) \in (D(L))'$ (η' is not a derivative ; again a short for $\eta' + q_2^* v_2$).

With this notation, the controllability operator writes

$$\Gamma_T h = -\begin{pmatrix} L^{-1}\eta'(T) \\ -\eta(T) \end{pmatrix}$$

The interest of the notation η' for $\eta' + q_2^* v_2$ is that we can write $\eta'(0) = 0$.

2.2. Exact controllability control

Recall that $D(A) \subset F_T \subset H \subset F_T' \subset (D(A))'$.

Take $\Gamma_T h$ given in $D(A) \Rightarrow h \in F_T'$

hence, we can fix the values $\eta(T) = y_o \in H_L, \eta'(T) = y_1 \in V_L'$ (where again $\eta'(T)$ is short for $\eta'(T) + q_2^* v_2(T)$). The system

$$\eta" + L\eta = -q_1^* q_1 \varphi + q_2^* J_o^* q_2 \varphi'$$

$$\varphi" + L\varphi = 0, \quad \eta(0) = 0, \quad \eta'(0) = 0$$

has a solution

$$\varphi \in C([0,T]; V_L)$$
$$\varphi' \in C([0,T]; H_L)$$
$$q_1\varphi \in L^2(0,T;U_1), \quad q_2\varphi' \in L^2(0,T;U_2)$$
$$\eta \in C([0,T]; V_L'), \quad \eta' \in L^2(0,T;(D(L))')$$

(here the derivative).

We can also claim that the following equation in φ has a solution

$$< y_1, \psi_o > - < y_o, \psi_1 >= \int_0^T [-(q_1\varphi, q_1\psi) + (q_2\varphi', q_2\psi')]dt$$

$$\forall\psi_o \in V_L, \psi_1 \in H_L, \psi'' + L\psi = 0, \psi(T) = \psi_o\, \psi'(T) = \psi_1$$

such that

$$q_1\psi \in L^2(0,T;U_1)$$

$$q_2\psi' \in L^2(0,T;U_2)$$

3. Theorem leading to exact controllability

3.1. Assumptions and statement

There exist operators Π_1, Π_2 and a form $b(\xi,\xi')$ such that

$$\Pi_1 \in L(D(L);(D(L))'), \Pi_2 \in L(V_L;V_L')$$

b is bilinear continuous on $H_L \times V_L$

$$|q_1\zeta|^2 \geq (\Pi_1\zeta,\zeta), \forall\zeta \in D(L)$$

$$|q_2\zeta|^2 \geq (\Pi_2\zeta,\zeta), \forall\zeta \in V_L$$

$$(\Pi_1 w_j, w_k) = \lambda_j b(w_j,w_k) + \lambda_k b(w_k,w_j), \forall j \neq k \geq N+1$$

$$b(w_j,w_k) + b(w_k,w_j) + (\Pi_2 w_j, w_k) = 0, \forall j \neq k \geq N+1$$

$$\frac{1}{\lambda_j}(\Pi_1 w_j, w_j) + (\Pi_2 w_j, w_j) \geq 2c_o, \forall j \geq N+1$$

$$(*) \qquad \left|\frac{(\Pi_1 w_j, w_j)}{\lambda_j^{\frac{3}{2}}} - \frac{(\Pi_2 w_j, w_j)}{\lambda_j^{\frac{1}{2}}} - 4\frac{b(w_j,w_j)}{\lambda_j^{\frac{1}{2}}}\right| \leq 2K, \forall j \geq N+1$$

$\lambda_{N+1} > \lambda_N$; for $j \geq N$, we consider only the λ_j having different values and call $P_j(\bar{P}_j)$ the projector on the finite dimensional eigensubspace of H corresponding to the eigenvalue $i\sqrt{\lambda_j}(-i\sqrt{\lambda_j})$ then

$$|B^* P_j z|^2 \geq c_1 \frac{|P_j z|^2}{\lambda_j}, \forall z \in H, j \leq N$$

(N can be 0).

We can state the following

Theorem 3.1.

Under the above assumptions, there is exact controllability for T sufficiently large.

3.2. Exact controllability for T arbitrarily small

We make the following modifications

Assumptions : replace (*) by

$$\left|\frac{(\Pi_1 w_j, w_j)}{\lambda_j} - (\Pi_2 w_j, w_j) - 4b(w_j,w_j)\right| \leq 2K, \forall j \geq N+1$$

Assume also that there exists a Hilbert space W_L such that $V_L \subset W_L \subset H_L$, the injection of V_L in W_L is compact.
b is bilinear continuous on $H_L \times W_L$.

Theorem 3.2.

With the above two modifications, there is exact controllability for arbitrary $T > 0$.

4. Examples

4.1. Example 1

Wave equation with boundary control

We take

$$L = -\Delta, \quad H_L = L^2(\Omega), \quad V_L = H_0^1(\Omega)$$

$$D(L) = H^2 \cap H_0^1, \quad \Delta_L = \{z \in D(L) | \Delta z \in H_0^1\}$$

$$U_1 = U_2 = L^2(\Gamma), \quad q_1 = \frac{\partial}{\partial \nu}, \quad q_2 = 0$$

then the state equation writes

$$\eta" - \Delta\eta = (\frac{\partial}{\partial \nu})^* v$$

$$\eta \in C([0,T]; H^{-1}), \eta' \in L^2(O,T;(D(L))')$$

(in fact here $\eta' \in C([0,T];(D(L))')$

$$\eta" \in C[O,T]; \Delta_L') \oplus L^2(O,T;(D(L))').$$

Using the transposition method, we get

$$< \eta'(T), \psi_o > - < \eta(T), \psi_1 >= \int_0^T \int_\Gamma v \frac{\partial}{\partial \nu} \psi dt d\Gamma$$

where

$$\psi" - \Delta\psi = 0, \quad \psi(T) = \psi_o, \quad \psi'(T) = \psi_1$$

$$\psi_o \in D(L), \quad \psi_1 \in H_0^1$$

$$\psi \in C([0,T]; D(L)), \quad \psi' \in C([0,T]; V_L)$$

A formal writing of the state equation is

$$\eta" - \Delta\eta = 0 \quad \eta|_\Gamma = -v$$

$$\eta(0) = \eta'(0) = 0$$

We can check the conditions of exact controllability.
Let $m(x) = x - x_o, R(x_o) = sup|m(x)|, x \in \Gamma$

We take

$$(\Pi_1\zeta,\zeta') = \frac{1}{R(x_o)}\int_\Omega \Sigma_\alpha m_\alpha z \frac{\partial\zeta}{\partial x_\alpha} dx$$

Index $N = 0$
Eigenvectors $-\Delta w_j = \lambda_j w_j, w_j|_\Gamma = 0$

Verification of conditions : it is a consequence of the relation

$$\int_\Gamma m.\nu \frac{\partial w_j}{\partial\nu}\frac{\partial w_k}{\partial\nu} d\Gamma = (2-n)\sqrt{\lambda_j\lambda_k}\delta_{jk} - \int_\Omega \sum_\alpha m_\alpha(\lambda_j w_j \frac{\partial w_k}{\partial x_\alpha} + \lambda_k w_k \frac{\partial w_j}{\partial x_\alpha})dx$$

Exact controllability

We can state that the system

$$\eta" - \Delta\eta = 0 \quad \varphi" - \Delta\varphi = 0$$

$$\eta|_\Gamma = \frac{\partial\varphi}{\partial\nu} \quad \varphi|_\Gamma = 0$$

$$\eta(0) = \eta'(0) = 0, \quad \eta(T) = y_o, \quad \eta'(T) = y_1$$

has a solution for any $y_o \in L^2, y_1 \in H^{-1}$ with

$$\varphi \in C([0,T];H_0^1) \quad \varphi' \in C([0,T];L^2)$$

$$\frac{\partial\varphi}{\partial\nu} \in L^2(0,T;L^2(\Gamma))$$

$$\eta \in C([0,T];H^{-1}) \quad \eta' \in C([0,T];(D(L))')$$

4.2. Example 2 :

Plate equation with Neumann Control

We define

$$H_L = L^2(\Omega), V_L = H_0^2(\Omega), D(L) = H^4(\Omega) \cap H_0^2$$

$$L = \Delta^2, \Delta(L) = \{z \in H_0^2 | \Delta^2 z \in H_0^2\}$$

Eigenvalues

$$\Delta^2 w_j = \lambda_j w_j$$

$$w_j|_\Gamma = \frac{\partial w_j}{\partial\nu}|_\Gamma = 0, |w_j| = 1$$

Note : Use multiplier $m(x) = x - x_o$
We have the relation

$$\int_\Gamma m.\nu\Delta w_j \Delta w_k d\Gamma = (4-n)\sqrt{\lambda}_j\lambda_k\delta_{jk}$$
$$- \int_\Omega \Sigma_\alpha m_\alpha(\lambda_j w_j \frac{\partial w_k}{\partial x_\alpha} + \lambda_k w_k \frac{\partial w_j}{\partial x_\alpha})dx$$

We take

$$U_1 = U_2 = L^2(\Gamma_o) \text{ where } \Gamma_o = \{x \in \Gamma | m.\nu \geq 0\}$$

then

$$q_1 = -\gamma_o \Delta \quad q_2 = 0.$$

All assumptions on exact controllability at arbitrary small $T > 0$ are verified, taking

$$(\Pi_1 \zeta, \zeta') = \frac{1}{R(x_o)} \int_\Gamma m.\nu \Delta\zeta \Delta\zeta' d\Gamma$$

$$\Pi_2 = 0$$

$$b(z, \zeta) = -\frac{1}{R(x_o)} \int_\Omega m_\alpha z \frac{\partial \zeta}{\partial x_\alpha} dx$$

$$N = 0; \quad W_L = H_0^1$$

The exact controllability property is stated as follows :
Given $y_o \in L^2(\Omega), y_1 \in H^{-2}(\Omega)$

$$\eta" + \Delta^2 \eta = 0, \quad \varphi" + \Delta^2 \varphi = 0$$

$$\eta(0) = \eta'(0) = 0$$

$$\eta(T) = y_o, \eta'(T) = y_1$$

$$\frac{\partial \eta}{\partial \nu}|_{\Gamma_o} = \Delta\varphi|_{\Gamma_o}, \frac{\partial \eta}{\partial \nu}|_{\Gamma_1} = 0, \eta|_\Gamma = 0, \varphi|_\Gamma = \frac{\partial \varphi}{\partial \nu}|_\Gamma = 0$$

$$\eta \in C([0,T]; V_L') \quad \eta' \in C([0,T]; (D(L))')$$

$$\varphi \in C([0,T]; H_0^2) \quad \varphi' \in C([0,T]; L^2)$$

$$\Delta\varphi|_{\Gamma_o} \in L^2(0,T; L^2(\Gamma_o))$$

has a solution for any $T > 0$.

References

[1] C. BARDOS, G. LEBEAU, T. RAUCH, Microlocal ideas in control and stabilization, Proc. Clermont-Ferrand Colloquium, June 1988, Springer-Verlag.

[2] A. BENSOUSSAN, On the General Theory of Exact Controllability for Skew symmetric Operators, Acta Applicandae Mathematicae, 1990, 197-229.

[3] J.E. LAGNESE, J.L. LIONS, Modelling Analysis and Control of thin Plates, Masson, RMA vol. 6, 1988.

[4] J.L. LIONS, Contrôlabilité exacte, perturbations et stabilisation de systèmes distribués, Vol. 1 and 2, Masson, Paris, 1988.

[5] R. TRIGGIANI, Exact boundary controllability on $L^2(\Omega) \times H^{-1}(\Omega)$ of the wave equation with Dirichlet boundary control acting on a portion of the boundary $\partial\Omega$ and related problems, Appl. Math. Opt. 18 (1988), 241-277.

BOUNDARY HOMOGENIZATION AND BOUNDARY SHAPE OPTIMIZATION FOR A CYLINDER

Alain BRILLARD

Faculté des Sciences et Techniques
4 Rue des Frères Lumière, F-68093 MULHOUSE Cedex

INTRODUCTION

Let Ω be the cylinder $\{(x,y,z) \in \mathbb{R}^3 \,/\, x^2+y^2 < 1 \,,\, 0 < z < h\}$. The lower (resp. upper) face of Ω is denoted by $\partial\Omega^-$ (resp. $\partial\Omega^+$) :

$$\partial\Omega^- = \{(x,y,z) \in \mathbb{R}^3 \,/\, x^2+y^2 < 1 \,;\, z = 0\},$$

$$\partial\Omega^+ = \{(x,y,z) \in \mathbb{R}^3 \,/\, x^2+y^2 < 1 \,;\, z = h\}.$$

Σ is the lateral boundary of Ω :

$$\Sigma = \{(x,y,z) \in \mathbb{R}^3 \,/\, x^2 + y^2 = 1 \,;\, 0 < z < h\}.$$

For every measurable and open subset D of Σ, we consider the solution u_D of :

$$\left| \begin{array}{ll} -\Delta u_D = f & \text{in } \Omega \quad (f \in L^2(\Omega)), \\ u_{D|\partial\Omega^-} = 0, & \\ \dfrac{\partial u_D}{\partial z} = 0 & \text{on } \partial\Omega^+, \\ u_{D|\Sigma} = 0 & \text{on } D, \\ \dfrac{\partial u_D}{\partial \nu_\Sigma} = 0 & \text{on } \Sigma\setminus D, \ (\nu_\Sigma \text{ is the outer normal to } \Sigma). \end{array} \right. \tag{1}$$

If $\delta_{V(D)}$ denotes the indicator function of the closed subspace V(D) of $H^1(\Omega)$:

$$V(D) = \{u \in H^1(\Omega) \,/\, u_{|\partial\Omega^-} = 0 \,;\, u_{|\Sigma} = 0 \text{ on } D\}, \tag{2}$$

corresponding to the homogeneous Dirichlet boundary conditions appearing in (1), then u_D is the solution of the minimization problem :

$$\min_{u \in H^1(\Omega)} F(u,D),$$

where F(u,D) is equal to $\displaystyle \frac{1}{2} \int_\Omega |\nabla u|^2 \, dx + \delta_{V(D)}(u) - \int_\Omega f.u \, dx.$ (3)

u_D is also the solution of :

$$\int_\Omega \nabla u_D . \nabla u \, dx = \int_\Omega f.u \, dx, \text{ for every u in } V(D). \tag{4}$$

The first part of this work is devoted to the study of an homogenization phenomena, by considering the domain D_ε equal to the union of identical rings of width r_ε, ε-periodically disposed on Σ ($0 < r_\varepsilon < \varepsilon$). We completely describe the limit problem, when the parameter ε goes to 0, taking its values in a sequence decreasing to 0. Then, we present the limit problem associated to any sequence $(D_n)_n$ of measurable and open subsets of Σ.

In the second paragraph, we suppose that we are given a cost function $J(D) = J(u_D)$ and a class $\mathcal{U}_{ad}$ of admissible domains. D represents the control and u_D the corresponding state. We study, on three examples, the existence of an optimal domain, that is a solution D_{opt} of the minimization problem :

$$\operatorname*{Inf}_{D\in\mathcal{U}_{ad}} J(D). \tag{5}$$

The classical approach consists introducing a minimizing sequence $(D_n)_n$ for the cost function J :

$$\lim_{n\to+\infty} J(D_n) = \operatorname*{Inf}_{D\in\mathcal{U}_{ad}} J(D),$$

and the corresponding sequence $(u_{D_n})_n$ of states. Then, one needs assumptions on $\mathcal{U}_{ad}$ and on J, in order to obtain the convergence of $(D_n)_n$ to some D_{opt}, belonging to $\mathcal{U}_{ad}$, and the convergence of $(u_{D_n})_n$ to $u_{D_{opt}}$, in such a way that :

$$\liminf_{n\to+\infty} J(D_n) \geq J(D_{opt}).$$

Hence, D_{opt} is a solution for the minimization problem (5).

These assumptions have to eliminate too large variations of the boundary of D. Indeed, we prove in the first example that, without such assumptions, the homogenization phenomena described in the first paragraph may occur. In this situation, we prove that there does not exist an optimal solution. The "optimal domain" has to be searched in the relaxed setting presented in Theorem 9.

In the second example, we suppose that the cost function J involves the gradient of u_D. Thus, the variations of u_D and therefore of D are limited. We prove the existence of an optimal domain.

In the last example, we assume that the boundary of D is sufficiently smooth and that D contains a fixed non-empty subset D_o of Σ. We also prove the existence of an optimal domain.

I - BOUNDARY HOMOGENIZATION

A) PRESENTATION

We denote by D_ε the union $\bigcup_{k=1}^{K(\varepsilon)} A_{k\varepsilon}$ of identical rings of width r_ε, ε-periodically distributed on Σ, with $0 < r_\varepsilon < \varepsilon$, according to fig. 1 below.

The purpose of this paragraph is to compute the limit of $(u_{D_\varepsilon})_\varepsilon$, when the parameter ε goes to 0. The subset D_ε becomes more and more fragmented : the number K(ε) of the rings increases as h/ε, while their common width r_ε decreases to 0.

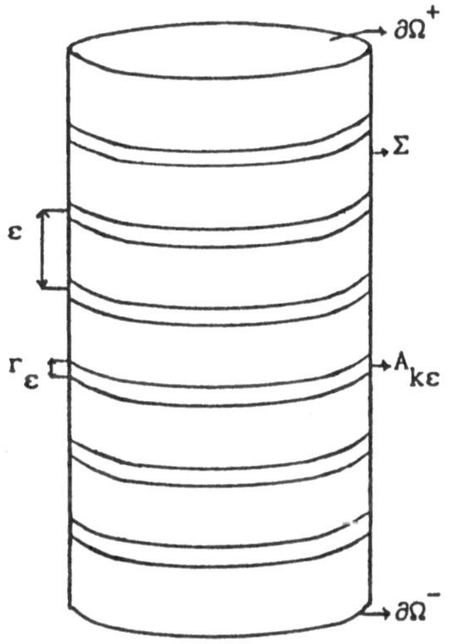

fig. 1 : ε-periodic repartition of identical rings of width r_ε

B) CASE : $\lim_{\varepsilon\to 0} r_\varepsilon/\varepsilon = 0.$

THEOREM 1 [5]

1) $(u_{D_\varepsilon})_\varepsilon$ converges in the weak topology of $H^1(\Omega)$ to the solution u_o of :

$$-\Delta u_o = f \qquad \text{in } \Omega, \tag{6.1}$$

$$u_o|_{\partial\Omega^-} = 0, \tag{6.2}$$

$$\frac{\partial u_o}{\partial z} = 0 \qquad \text{on } \partial\Omega^+, \tag{6.3}$$

$$\left|\;\begin{aligned} &\frac{\partial u_o}{\partial \nu_\Sigma} + C\, u_{o|\Sigma} = 0 \qquad \text{on } \Sigma, \quad \text{if } C = \lim_{\varepsilon\to 0} \frac{\pi}{\varepsilon|\ln r_\varepsilon|} < +\infty, && (6.4)\\ &u_{o|\Sigma} = 0 \qquad \text{on } \Sigma, \quad \text{if } C = +\infty, && (6.5)\end{aligned}\right.$$

that is, the solution of the minimization problem :

$$\left|\;\begin{aligned} &\underset{H^1(\Omega)}{\mathrm{Min}}\ \Big\{\frac{1}{2}\int_\Omega |\nabla u|^2\,dx + \frac{C}{2}\int_\Sigma (u_{|\Sigma})^2\,d\sigma + \delta_V(u) - \int_\Omega f.u\,dx\Big\}, \\ &\qquad\qquad\qquad\qquad\qquad\qquad\qquad\qquad \text{if } C < +\infty, \\ &\underset{H^1(\Omega)}{\mathrm{Min}}\ \Big\{\frac{1}{2}\int_\Omega |\nabla u|^2\,dx + \delta_{V(\Sigma)}(u) - \int_\Omega f.u\,dx\Big\}, \qquad \text{if } C = +\infty,\end{aligned}\right. \qquad (7)$$

where V is the closed subspace of $H^1(\Omega)$ defined by :

$$V = \{u \in H^1(\Omega) \;/\; u_{|\partial\Omega^-} = 0\} \qquad (8)$$

and $V(\Sigma)$ is defined by (2) for $D = \Sigma$.

2) $\left(\int_\Omega |\nabla u_{D_\varepsilon}|^2\,dx\right)_\varepsilon$ *converges to* $\int_\Omega |\nabla u_o|^2\,dx + C\int_\Sigma (u_{o|\Sigma})^2\,d\sigma.$

Sketch of the proof

We consider the function : $w_{\varepsilon k}(x,y,z) = \dfrac{\ln((1-\rho)^2 + (z-\varepsilon k)^2) - \ln(r_\varepsilon^2/4)}{2\ln(r_\varepsilon/\varepsilon)}$,

(where ρ is equal to $(x^2 + y^2)^{1/2}$), defined in $T_{\varepsilon k}$:

$$T_{\varepsilon k} = \{(x,y,z) \in \Omega \;/\; r_\varepsilon^2/4 < (1-\rho)^2 + (z-\varepsilon k)^2 < \varepsilon^2/4\}$$

and let w_ε denote the function equal to $w_{\varepsilon k}$ in $T_{\varepsilon k}$ $(1 \le k \le K(\varepsilon))$ and extended by 1 in $\Omega \setminus \bigcup_k^{K(\varepsilon)} T_{\varepsilon k}$. One can prove (see [5] for the details), that, whenever C is finite, $(w_\varepsilon)_\varepsilon$ is bounded in $H^1(\Omega)$ and converges to 1, in the weak topology of this space. Then, for every function u in $C^1(\bar{\Omega})$, such that $u_{|\partial\Omega^-}$ is equal to 0, the sequence $(uw_\varepsilon)_\varepsilon$ converges to u in the weak topology of $H^1(\Omega)$. Moreover, if u_o denotes any limit point of $(u_{D_\varepsilon})_\varepsilon$, in the weak topology of $H^1(\Omega)$, one establishes the convergence :

$$\int_\Omega \nabla u_{D_\varepsilon}.\nabla(uw_\varepsilon)\,dx \xrightarrow[\varepsilon\to+\infty]{} \int_\Omega \nabla u_o.\nabla u\,dx + C\int_\Sigma u_{o|\Sigma}.u_{|\Sigma}\,d\sigma,$$

by integrating by parts the first expression and making use of the properties of w_ε. This convergence implies that u_o satisfies (6.1),...,(6.4). Hence u_o is unique and the whole sequence $(u_{D_\varepsilon})_\varepsilon$ converges to u_o.

The case C equal to +∞ and the limit of $(\int_\Omega |\nabla u_{D_\varepsilon}|^2\, dx)_\varepsilon$ can be studied by proving the epi-convergence, in the weak topology of $H^1(\Omega)$, of $(F(.,D_\varepsilon))_\varepsilon$ to the functional F^{hom} associated to the homogenized problem (6.1),..., (6.4) or (6.5) (see (7)). ∎

REMARK 2

1) For the definition and the properties of the epi-convergence, we refer to [1] (see also the references therein).

2) From the definition of the constant C (see (6.4)), one infers the existence of a critical size r_ε^c of the rings : $r_\varepsilon^c = \exp(-\frac{a}{\varepsilon})$ $(a > 0)$, such that, if $\lim_{\varepsilon\to 0} \frac{r_\varepsilon}{r_\varepsilon^c}$ is equal to 0 (resp. +∞), then C is equal to 0 (resp. +∞) and the limit boundary condition on Σ is an homogeneous Neumann (resp. Dirichlet) one.

3) Notice that in the homogenized problem (6.1),..., (6.4) or (6.5), the shape of the rings has no influence on the limit boundary condition on Σ. Therefore the results exposed in Theorem 1 are also valid in the two following situations :

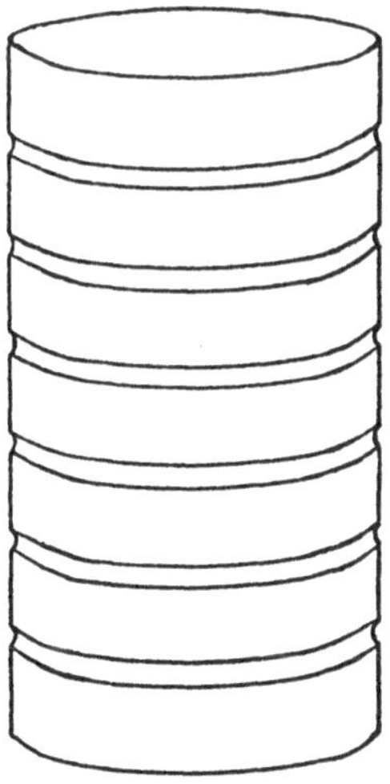

fig.2 : in-going semi-tores

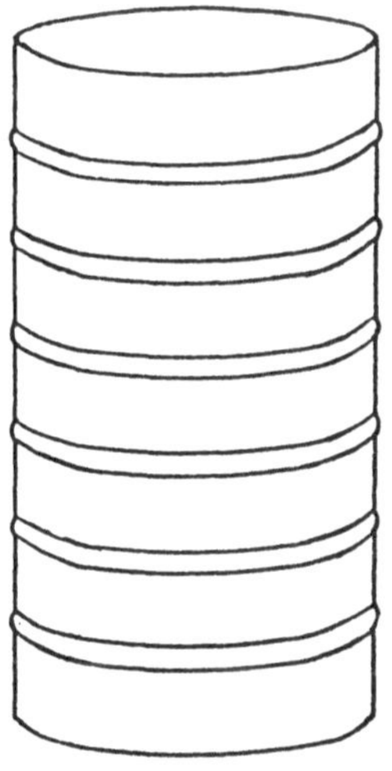

fig.3 : out-going semi-tores

C) **CASE** : $r_\varepsilon = \alpha\varepsilon$ ($\alpha > 0$).

From Theorem 1 (cf. Remark 2, 2)), one deduces that, in this case, $(u_{D_\varepsilon})_\varepsilon$ converges, in the weak topology of $H^1(\Omega)$, to the solution u_o of (6.1), (6.2), (6.3) and (6.5). Therefore, $(u_{D_\varepsilon|\Sigma})_\varepsilon$ converges to 0 in the strong topology of $L^2(\Sigma)$. In the following theorem, the rate of convergence is described :

THEOREM 3 [5]

Assume that Ω is smooth and $u_{D_\varepsilon|\partial\Omega^+} = 0$, hence $(u_{D_\varepsilon})_\varepsilon$ converges, in the weak topology of $H^1(\Omega)$, to the solution u_o^* of :

$$\left|\begin{array}{l} -\Delta u_o^* = f \quad \text{in } \Omega, \\ u_{o|\partial\Omega^-}^* = 0, \\ u_{o|\partial\Omega^+}^* = 0, \\ u_{o|\Sigma}^* = 0. \end{array}\right.$$

Then,

1) $(\frac{1}{\varepsilon} u_{D_\varepsilon|\Sigma})_\varepsilon$ *is bounded in* $L^2(\Sigma)$.

2) $(\frac{1}{\sqrt{\varepsilon}} (u_{D_\varepsilon} - u_o^*))_\varepsilon$ *converges to 0, in the weak topology of* $H^1(\Omega)$.

3) *For every strictly positive* η, $(\frac{1}{\sqrt{\varepsilon}} (u_{D_\varepsilon} - u_o^*))_\varepsilon$ *converges to 0, in the strong topology of* $H^1(\Omega_\eta)$, *where* $\Omega_\eta = \{(x,y,z) \in \Omega \,/\, d(x,\Sigma) > \eta\}$.

In order to conclude the study of this homogenization problem, we determine the limit of $(\frac{1}{\varepsilon} u_{D_\varepsilon|\Sigma})_\varepsilon$, in the weak topology of $L^2(\Sigma)$. We introduce the strip $S = \{(a,b) \in \mathbb{R}^2 /\ a < 0 ;\ |b| < \frac{1}{2}\}$ and the solution W of :

$$\left|\begin{array}{ll} -(\frac{\partial^2 W}{\partial a^2} + \frac{\partial^2 W}{\partial b^2})(a,b) = 0 & \text{in } S, \\ W(0,b) = 0 & \text{for every } b :\ |b| < \frac{\alpha}{2}, \\ \frac{\partial W}{\partial a}(0,b) = 1 & \text{for every } b :\ \frac{\alpha}{2} < |b| < \frac{1}{2}, \\ W(a,\frac{1}{2}) = W(a,-\frac{1}{2}) & \text{for every } a :\ a < 0. \end{array}\right.$$

THEOREM 4 [5]

Assume that Ω *is smooth and* $u_{D_\varepsilon}|_{\partial\Omega^+} = 0$. *Then,* $(\frac{1}{\varepsilon} u_{D_\varepsilon}|_\Sigma)_\varepsilon$ *converges to :*

$$- \frac{\partial u_o^*}{\partial \nu_\Sigma} \cdot \int_{-1/2}^{1/2} W(0,b)\, db,$$

in the weak topology of $L^2(\Sigma)$.

Sketch of the proof

We consider the function W_ε defined by $W_\varepsilon(x,y,z) = W(\frac{1}{\varepsilon}(\rho-1), \frac{1}{\varepsilon}(z-\varepsilon k))$, for z satisfying $|z-\varepsilon k| \le \varepsilon/2$. From the results of [8], one derives that the sequence $(W_\varepsilon)_\varepsilon$ (resp. $(\sqrt{\varepsilon}\ \nabla W_\varepsilon)_\varepsilon$) is bounded in $L^2(\Omega)$ (resp. in $(L^2(\Omega)^3)$). Then, we compute the difference :

$$\int_\Omega \nabla(u_{D_\varepsilon} - u_o^*).\nabla(W_\varepsilon.\psi)\, dx - \int_\Omega \nabla((u_{D_\varepsilon} - u_o^*).\psi).\nabla W_\varepsilon\, dx, \qquad (9)$$

for every smooth cut-off function ψ satisfying :

$$\psi|_{\partial\Omega^+\cup\partial\Omega^-} = 0 \ ; \ \psi \equiv 0 \text{ in } \{(x,y,z) \in \Omega \ / \ x^2 + y^2 < 1/4\}.$$

The above-indicated estimates on W_ε and the estimates on u_{D_ε}, established in Theorem 3, imply that the expression (9) converges to 0 when ε goes to 0. An integration by parts proves that this quantity (9) is equal to :

$$\int_\Sigma \frac{1}{\varepsilon} u_{D_\varepsilon}|_\Sigma.\psi\, d\sigma + \int_\Sigma \frac{\partial u_o^*}{\partial \nu_\Sigma}.W_\varepsilon.\psi\, d\sigma + o_\varepsilon , \quad (\text{with } \lim_{\varepsilon\to 0} o_\varepsilon = 0).$$

Then, the proof of Theorem 4 is obtained by passing to the limit in (9) :

$$0 = \int_\Sigma (\lim_{\varepsilon\to 0} \frac{1}{\varepsilon} u_{D_\varepsilon}|_\Sigma).\psi\, d\sigma + \int_\Sigma \frac{\partial u_o^*}{\partial \nu_\Sigma}.\psi.(\int_{-1/2}^{1/2} W(0,b)\, db)\, d\sigma. \qquad \blacksquare$$

REMARK 5

One can notice that the limit of $(\frac{1}{\varepsilon} u_{D_\varepsilon}|_\Sigma)_\varepsilon$ does not depend on the shape of the rings, because neither u_o^* nor W depend on this shape.

Let us end this paragraph giving the general limit problem associated to any sequence $(D_n)_n$ of measurable and open subsets of Σ.

THEOREM 6 [3]

For every sequence $(D_n)_n$ *of measurable and open subsets of* Σ, *there exists :*

- *a subsequence* $(n_k)_k$,

- *a measure* μ *in* $H^{-1/2}(\Sigma)^{+}$, *a* μ*-measurable function h with values in* $\overline{\mathbb{R}}^{+}$,
- *a rich family* $\mathcal{R}$ *of Borel subsets of* Σ, *(we recall that a family* $\mathcal{R}$ *of subsets of* Σ *is called rich if, for every family* $(D_t)_{t\in[0,1]}$ *of Borel subsets of* Σ, *satisfying :* $\overline{D}_s \subset \dot{D}_t$, *whenever* $s < t$, *then the set :* $\{t \in [0,1] \ / \ D_t \notin \mathcal{R}\}$ *is at most countable),*

such that, for every D in $\mathcal{R}$, *the solution* $u_{D_{n_k} \cap D}$ *of (1) converges in the weak topology of* $H^1(\Omega)$ *to the solution of :*

$$\underset{u\in H^1(\Omega)}{Min} \left\{\frac{1}{2}\int_\Omega |\nabla u|^2\,dx + \frac{1}{2}\int_D h(\sigma).(\tilde{u}_{|\Sigma})^2\,d\mu(\sigma) - \int_\Omega f.u\,dx\right\},$$

where $\tilde{u}$ *denotes the quasi-continuous representation of u (*$\tilde{u}$ *is continuous on* Ω *up to a set of null* $H^1(\Omega)$*-capacity), and conversely.*

REMARK 7

In the homogenization situation studied in Theorem 1, and when D_ε is the union of identical rings of width $r_\varepsilon = \exp(-\frac{\pi}{C\varepsilon})$, for a finite and strictly positive constant C :

- the rich family $\mathcal{R}$ is equal to the whole set $\mathcal{B}(\Sigma)$ of Borel subsets of Σ,
- (μ,h) is equal to $(d\sigma, C)$.

II - SHAPE OPTIMIZATION

In this paragraph, we study, on three examples, the existence of a solution D_{opt} of the minimization problem (5).

A) EXAMPLE 1

$$\mathcal{U}_{ad,1} = \{D \text{ measurable and open subset of } \Sigma \ / \ meas(D) > 0\},$$

$$J_1(D) = \int_\Omega |u_D - u_o|^2\,dx \quad \text{or} \quad \int_\Sigma |u_{D|\Sigma} - u_{o|\Sigma}|^2\,d\sigma,$$

where u_o is the solution of (6.1)...(6.4), for a finite and strictly positive constant C, and represents the desired state.

THEOREM 8 [4]

There exists no optimal domain D_{opt}.

<u>Proof</u>

Consider $D_\varepsilon = \bigcup_k^{K(\varepsilon)} A_{k\varepsilon}$, where the width r_ε of the rings is equal to

$\exp(-\pi/C\varepsilon)$. From Theorem 1, one deduces that the infimum : $\inf_{D\in U_{ad,1}} J_1(D)$ is equal to 0. Since u_o satisfies the boundary condition (6.4) on Σ, the minimum of J_1 on $U_{ad,1}$ cannot be achieved. ∎

In this special case, the "optimal domain" has to be searched in the following relaxed setting :

THEOREM 9 [3]

Consider :

$$\tilde{U}_{ad,1} = \{(\mu,h) \;/\; \mu \in H^{-1/2}(\Sigma)^+ \;;\; h \;\mu\text{-measurable}\;,\; h \geq 0\},$$

$$\tilde{J}_1((\mu,h)) = \int_\Omega |u_{(\mu,h)} - u_o|^2 \, dx \text{ or } \tilde{J}_1((\mu,h)) = \int_\Sigma |u_{(\mu,h)|\Sigma} - u_{o|\Sigma}|^2 \, d\sigma,$$

where $u_{(\mu,h)}$ is the solution of the minimization problem :

$$\min_{u\in V} \{\frac{1}{2}\int_\Omega |\nabla u|^2 \, dx + \frac{1}{2}\int_\Sigma h(\sigma).(\tilde{u}_{|\Sigma})^2(\sigma) \, d\mu(\sigma) - \int_\Omega f.u \, dx\},$$

that is the solution of :

$$\int_\Omega \nabla u_{(\mu,h)}.\nabla u \, dx + \int_\Sigma h.\tilde{u}_{(\mu,h)|\Sigma}.\tilde{u}_{|\Sigma} \, d\mu = \int_\Omega f.u \, dx, \quad \forall\, u \in V,$$

where V is the subspace defined by (8).

Then,

$$\inf_{D\in U_{ad,1}} J_1(D) = \min_{(\mu,h)\in\tilde{U}_{ad,1}} \tilde{J}_1((\mu,h)).$$

Proof

Let $M = \inf_{D\in U_{ad,1}} J_1(D)$ and $\tilde{M} = \inf_{(\mu,h)\in\tilde{U}_{ad,1}} \tilde{J}_1((\mu,h))$.

We first prove that M is equal to $\tilde{M}$.

Choosing $\mu = d\sigma$ and $h = \begin{cases} +\infty & \text{on } D, \\ 0 & \text{on } \Sigma\setminus D \end{cases}$ (we note : $h = \infty_D$), one observes that $J_1(D)$ is equal to $\tilde{J}_1((d\sigma,\infty_D))$, hence : $M \geq \tilde{M}$.

In order to prove the converse inequality, we use Theorem 6. For every element (μ,h) of $\tilde{U}_{ad,1}$, Theorem 6 implies the existence of a sequence $(D_n)_n$ of measurable subsets of Σ, such that, for every D in $\mathcal{R}$:

$$J_1(D_n \cap D) = \tilde{J}_1((d\sigma,\infty_{D_n\cap D})) \xrightarrow[n\to+\infty]{} \tilde{J}_1((\mu,h.\chi_D)).$$

Hence, using the properties of $\mathcal{R}$ and Theorem 2.40 of [1], concerning the epi-convergence of increasing sequences of functionals, one obtains : $M \leq \tilde{M}$. Therefore, M is equal to $\tilde{M}$.

Let us now prove that the infimum $\tilde{M}$ of $\tilde{J}_1$ is achieved on $\tilde{U}_{ad,1}$. This is still a consequence of Theorem 6. If $(D_n)_n$ denotes any minimizing sequence for J_1, one infers, from Theorem 6, the existence of (μ_o,h_o) in $\tilde{U}_{ad,1}$, such that $(u_{D_n\cap D})_n$ converges in the weak topology of $H^1(\Omega)$ to $u_{(\mu_o,h_o.\chi_D)}$, for every element D of a rich family $\mathcal{R}_o$ of Borel subsets of Σ. Hence, $(J_1(D_n\cap D))_n$ converges to $\tilde{J}_1((\mu_o,h_o.\chi_D))$, for every D in $\mathcal{R}_o$. Considering the properties of $\mathcal{R}_o$ and of the epi-convergence, one proves that the infimum M of J_1 on $U_{ad,1}$ is equal to $\tilde{J}_1((\mu_o,h_o))$. This ends the proof of Theorem 9. ∎

REMARK 10

Theorem 9 has to be compared to the general result presented in [6], in a different situation. See also [2] for an analogous result, in a quite similar case.

By means of an elementary approximation argument, one proves :

COROLLARY 11 [3]

$$\underset{D\in U_{ad,1}}{Inf} J_1(D) = \underset{\theta\in L^\infty(\Sigma)^+}{Inf} \tilde{J}_1((d\sigma,\theta)).$$

Proof

Every measure $h.d\mu$, where (μ,h) belongs to $\tilde{U}_{ad,1}$, can be approximated by an increasing sequence of measures $(\theta_n.d\sigma)_n$, with θ_n in $L^\infty(\Sigma)^+$. Then, we use Theorem 2.40 of [1], in order to prove that $(u_{(d\sigma,\theta_n)})_n$ converges to $u_{(\mu,h)}$, in the weak topology of $H^1(\Omega)$. Hence, the sequence $(\tilde{J}_1((d\sigma,\theta_n)))_n$ converges to $\tilde{J}_1((\mu,h))$. ∎

B) EXAMPLE 2

$U_{ad,2}$ = {D measurable and open subset of Σ / meas(D) = m > 0},

$$J_2(D) = \frac{1}{2}\int_\Omega |\nabla u_D|^2 \, dx - \int_\Omega f.u_D \, dx.$$

REMARK 12

Notice that J_2 is exactly equal to the functional $F(.,D)$ defined in (3).

THEOREM 13 [4]

An optimal domain exists.

Proof

Let $(D_n)_n$ be any minimizing sequence for J_2. Then, $(u_{D_n})_n$ is bounded in $H^1(\Omega)$ and let u^* (resp. θ^*) be any limit point of $(u_{D_n})_n$, in the weak topology of $H^1(\Omega)$ (resp. $(\chi_{D_n})_n$, in the topology $\sigma(L^\infty(\Sigma), L^1(\Sigma))$). Since :

$$\int_\Sigma (u_{D_n|\Sigma})^2 . \chi_{D_n} \, d\sigma = 0,$$

one proves that :

$$\int_\Sigma (u^*_{|\Sigma})^2 . \theta^* \, d\sigma = 0.$$

Moreover, θ^* satisfies :

$$0 \le \theta^* \le 1 \quad ; \quad \int_\Sigma \theta^* \, d\sigma = m \quad ; \quad \text{meas}\{x \in \Sigma \ / \ \theta^* > 0\} \ge m.$$

If D_{opt} denotes any measurable subset of the set $\{x \in \Sigma \ / \ \theta^* > 0\}$ satisfying : $\text{meas}(D_{opt}) = m$, then, $u^*_{|\Sigma}$ is equal to 0 on D_{opt}. From the lower-semicontinuity of J_2, with respect to the weak topology of $H^1(\Omega)$, one derives :

$$\liminf_{n \to +\infty} J_2(D_n) \ge \frac{1}{2} \int_\Omega |\nabla u^*|^2 \, dx - \int_\Omega f.u^* \, dx.$$

Since $u_{D_{opt}}$ is the solution of (1), the preceding inequality implies that :

$$\liminf_{n \to +\infty} J_2(D_n) \ge \frac{1}{2} \int_\Omega |\nabla u_{D_{opt}}|^2 \, dx - \int_\Omega f.u_{D_{opt}} \, dx = J_2(D_{opt}),$$

(see Remark 12). Hence, D_{opt} is an optimal domain. ■

C) **EXAMPLE 3**

DEFINITION 14

A subset D of Σ satisfies the ζ-projected cone property, with ζ in $]0,\pi/2[$, if, for every x of the boundary ∂D, there exists a unit vector ξ_x in the tangent plane T_x at x, such that, for every y in $B(x,\zeta/2)$, the projection on Σ of the portion of cone $C(proj_{T_x}(y), \xi_x, \zeta)$ defined by :

$$C(proj_{T_x}(y),\xi_x,\zeta) = \{z \in T_x \;/\; |z-proj_{T_x}(y)| \le \frac{\zeta}{2},$$
$$\langle z-proj_{T_x}(y),\xi_x\rangle \ge |z-proj_{T_x}(y)|.\cos(\zeta)\},$$

is included in D.

REMARK 15

The notion of ζ-projected cone property is derived from the notion of ζ-cone property studied in [7] and is equivalent to some $k(\zeta)$-lipschitz continuity of ∂D (see [7]).

Then, for every open, measurable and non empty subset D_o of Σ, we consider the class of admissible domains given by :

$$\mathbb{U}_{ad,3} = \mathbb{U}_{ad,3}(D_o,\zeta) = \{D \text{ measurable subset of } \Sigma \;/\; D \supset D_o,\ D \text{ satisfies the } \zeta\text{-projected cone property}\}$$

and the cost function :

$$\mathbb{J}_3(D) = \int_\Omega |u_D - u_d|^2\, dx \quad \text{or} \quad \int_\Sigma |u_{D|\Sigma} - u_d|^2\, d\sigma,$$

where u_d belongs to $L^2(\Omega)$ or $L^2(\Sigma)$ and represents the desired state.

THEOREM 16 [4]

There exists an optimal domain.

Proof

Let $(D_n)_n$ be any minimizing sequence for $\mathbb{J}_3$: there exists D_{opt} in $\mathbb{U}_{ad,3}$ such that $(\bar{D}_n)_n$ and $(\overline{\Sigma\backslash D_n})_n$ converge respectively (up to subsequences) to $\bar{D}_{opt}$ and $\overline{\Sigma\backslash D_{opt}}$, for the Hausdorff convergence, see [4], [7], [9]. Moreover, $(\chi_{D_n})_n$ converges to $\chi_{D_{opt}}$ for the topology $\sigma(L^\infty(\Sigma),L^1(\Sigma))$ and almost everywhere. Then, one derives from (1) that $(u_{D_n})_n$ is bounded in $H^1(\Omega)$ and let $\bar{u}$ be any limit point of this sequence, in the weak topology of $H^1(\Omega)$. These convergences imply that $\bar{u}_{|\Sigma}$ is equal to 0 on D_{opt}. In order to prove that $\bar{u}$ is equal to $u_{D_{opt}}$, we use the following :

LEMMA 17 [4]

With the above-mentionned notations, for every v in $V(D_{opt})$, there exists a sequence $(v_n)_n$, with v_n in $V(D_n)$, which converges to v in the strong topology of $H^1(\Omega)$.

Proof of Lemma 17

From the construction of D_{opt}, one infers the existence of a function φ_n in $H^1(\Omega)$ satisfying :

$$\varphi_{n|\partial\Omega^-} = 0 \; ; \; \varphi_{n|\Sigma} = 1 \text{ on } (D_n \setminus \bar{D}_{opt} \cup D_{opt} \setminus \bar{D}_n) \; ; \; \lim_{n\to+\infty} \|\varphi_n\|_{H^1(\Omega)} = 0$$

which means that the $H^1(\Omega)$-capacity of the set $D_n \setminus \bar{D}_{opt} \cup D_{opt} \setminus \bar{D}_n$ converges to 0.

Suppose first that v belongs to $V(D_{opt}) \cap C^1(\bar{\Omega})$. The function v_n equal to $v - \varphi_n v$ satisfies the property announced in Lemma 17. This proves that the closed set $\liminf_{n\to+\infty} V(D_n)$, in Kuratowski's sense and for the strong topology of $H^1(\Omega)$, contains $V(D_{opt}) \cap C^1(\bar{\Omega})$. Hence, the proof of Lemma 17 is complete. ■

Let us end the proof of Theorem 16. Lemma 17 implies that, for every v in $V(D_{opt})$:

$$\lim_{n\to+\infty} \int_\Omega \nabla u_{D_n} . \nabla v_n \, dx = \int_\Omega \nabla \bar{u} . \nabla v \, dx.$$

Therefore, $\bar{u}$ satisfies (4) for D equal to D_{opt}. Hence, $\bar{u}$ is equal to $u_{D_{opt}}$ and the whole sequence $(u_{D_n})_n$ converges to $u_{D_{opt}}$, in the weak topology of $H^1(\Omega)$. ■

REFERENCES

[1] ATTOUCH H. Variational convergence for functions and operators. Pitman (London) 1984.

[2] ATTOUCH H., PICARD C. On the control of transmission problems across perforated walls. IFAC Perpignan (1987)

[3] BRILLARD A. Prépublication Mulhouse N° 54 (1990)

[4] BRILLARD A. Thèse d'Etat. Montpellier II (1990)

[5] BRILLARD A., PEREZ M. Comportement asymptotique d'un corps homogène isotrope gelé sur une partie de sa frontière. Prépublication Mulhouse N° 47 (1988)

[6] BUTTAZZO G. These proceedings.

[7] CHESNAIS D. On the existence of a solution in a domain identification problem. J. Maths. Anal. Applic. 52 p. 189-289 (1975)

[8] OLEINIK O.A., YOSIF'IAN G.A. On the behaviour at infinity of solutions of second order elliptic equations in domains with noncompact boundary. Math. USSR Sbornik 40 N° 4 p. 527-548 (1981)

[9] PIRONNEAU O. Optimal shape design for elliptic systems. Springer Verlag, Berlin (1983).

Relaxed Formulation for a Class of Shape Optimization Problems

Giuseppe BUTTAZZO

Istituto di Matematiche Applicate "Ulisse Dini"
Via Bonanno, 25/B
56100 PISA (ITALY)

1. Introduction

A shape optimization problem can be considered as a minimization problem of the form

$$\min\left\{J(A)\ :\ A\in\mathcal{A}\right\} \tag{1.1}$$

where J is a cost functional which has to be minimized over a class $\mathcal{A}$ of admissible domains. Problems of this kind arise in various questions of mechanics and structural engineering, where several "physical constraints" on the admissible domain are also imposed.

The main difficulty in proving the existence of a solution for problems of the form (1.1) is that the family $\mathcal{A}$ does not have any vector structure or convexity properties, so that the usual methods of functional analysis and of convex analysis fail. Moreover, the direct method of the calculus of variations fails too, because the usual topologies on families of sets do not provide coerciveness and lower semicontinuity for the cost functional we have in mind. This is the main reason for the lack of existence in several shape optimization problems (see Examples 2.5 and 3.1), which justifies the introduction of relaxed solutions.

In many situations the cost functional J is given by

$$J(A) = F(u_A)$$

where u_A is the solution of some differential equation in the set A

$$E_A(u) = 0,$$

and F is a functional defined on a space $\mathcal{U}$ of functions. In this way, problem (1.1) becomes an optimal control problem of the form

$$(1.2) \qquad \min\left\{F(u) \; : \; E_A(u) = 0,\; A \in \mathcal{A},\; u \in \mathcal{U}\right\}$$

where u is the state variable and A is the control.

Here we consider a model problem in shape optimization for the domain of an elliptic equation with Dirichlet boundary conditions. More precisely, given a bounded open subset Ω of $\mathbf{R}^n$ ($n \geq 2$), a function $f \in L^2(\Omega)$, and a function $g : \Omega \times \mathbf{R} \to \mathbf{R}$, we denote by $\mathcal{A}(\Omega)$ the family of all open subsets of Ω, and for every $A \in \mathcal{A}(\Omega)$ we set

$$J(A) = \int_\Omega g\big(x, u_A(x)\big)\, dx$$

where u_A is the solution of the Dirichlet problem

$$(1.3) \qquad \begin{cases} -\Delta u = f & \text{in } A \\ u \in H_0^1(A) \end{cases}$$

extended by 0 to $\Omega \setminus A$. Therefore, the minimization problem (1.1) becomes

$$(1.4) \qquad \min\left\{\int_\Omega g\big(x, u(x)\big)\, dx \; : \; -\Delta u = f \text{ in } A,\; u \in H_0^1(A),\; A \in \mathcal{A}(\Omega)\right\}.$$

In the form (1.2) the space $\mathcal{U}$ is the Sobolev space $H_0^1(\Omega)$, the functional F is given by

$$(1.5) \qquad F(u) = \int_\Omega g\big(x, u(x)\big)\, dx,$$

and the differential problem $E_A(u) = 0$ is given by (1.3).

We shall see (Examples 2.5 and 3.1) that, in general, problem (1.4) has no solution; the reason is that, although the solution u_{A_h} of (1.3) corresponding to a minimizing sequence (A_h) of (1.4) always admit a limit point u in the weak topology of $H_0^1(\Omega)$, we can not find, in general, an open subset A of Ω such that $u = u_A$. On the contrary, it can be proved that the limit function u is the solution of a relaxed Dirichlet problem of the form

$$(1.6) \qquad \begin{cases} -\Delta u + u\mu = f & \text{in } \Omega \\ u \in H_0^1(\Omega) \cap L_\mu^2(\Omega) \end{cases}$$

where μ is a suitable nonnegative Borel measure on Ω which vanishes on all sets of harmonic capacity 0, but may take the value $+\infty$ on some subsets of Ω.

This suggests to introduce the following relaxed formulation for the shape optimization problem (1.4):

$$(1.7) \quad \min\left\{F(u) \; : \; -\Delta u + u\mu = f \text{ in } \Omega,\; u \in H_0^1(A) \cap L_\mu^2(\Omega),\; \mu \in \mathcal{M}(\Omega)\right\}.$$

where $\mathcal{M}(\Omega)$ is the class of all measures allowed in (1.6) and, for every $\mu \in \mathcal{M}(\Omega)$, u_μ denotes the corresponding solution of (1.6).

We shall describe the relationship between problems (1.4) and (1.7); moreover we shall give some necessary conditions of optimality for the solutions of the relaxed problem (1.7) (which always exist).

Relaxed formulations for different classes of shape optimization problems have been considered in the literature by several authors (see for instance Kohn & Strang [7], Kohn & Vogelius [8], Murat & Tartar [10]).

2. The Relaxed Problem

In this section we describe, with the notation introduced in Section 1, the relaxed problem (1.7) and its relationship with the original problem (1.4).

In order to give the precise meaning of problem (1.6), and the definition of the space $\mathcal{M}(\Omega)$, we recall that the *capacity* of a Borel subset B of $\mathbf{R}^n$ is defined by

$$\operatorname{cap}(E) = \inf\left\{\int_{\mathbf{R}^n} (|Du|^2 + |u|^2)\, dx \;:\; u \in \mathcal{U}_B\right\}$$

where $\mathcal{U}_B$ is the set of all functions $u \in H^1(\mathbf{R}^n)$ such that $u \geq 1$ almost everywhere in a neighborhood of B (depending on u). Moreover, if a property $P(x)$ holds for all $x \in B$, except for a set $A \subset B$ with $\operatorname{cap}(A) = 0$, then we say that $P(x)$ holds *quasi everywhere* on B (shortly q.e. on B).

We denote by $\mathcal{M}(\Omega)$ the set of all nonnegative Borel measures μ on Ω such that $\mu(B) = 0$ for every Borel subset B of Ω wih $\operatorname{cap}(B) = 0$. It is possible to see that if $n-2 < \alpha \leq n$ the α dimensional Hausdorff measure H^α belongs to $\mathcal{M}(\Omega)$; in particular, the n-dimensional Lebesgue measure $\mathcal{L}^n$ and the $n-1$ dimensional Hausdorff measure $\mathcal{H}^{n-1}$ belong to $\mathcal{M}(\Omega)$. Another example of measure of the class $\mathcal{M}(\Omega)$, which plays an important role in our problem, is, for every $S \subset \Omega$, the Borel measure ∞_S defined for every Borel set $B \subset \Omega$ by

$$\infty_S(B) = \begin{cases} 0 & \text{if } \operatorname{cap}(B \cap S) = 0 \\ +\infty & \text{if } \operatorname{cap}(B \cap S) > 0. \end{cases} \tag{2.1}$$

We may now give the precise meaning of problem (1.6). Let $\mu \in \mathcal{M}(\Omega)$ and $f \in L^2(\Omega)$ be fixed; following Dal Maso & Mosco [5] we say that $u \in H_0^1(\Omega) \cap L^2_\mu(\Omega)$ is a solution of problem (1.6) if and only if

$$\int_\Omega DuDv\, dx + \int_\Omega uv\, d\mu = \int_\Omega fv\, dx \tag{2.2}$$

for every $v \in H_0^1(\Omega) \cap L^2_\mu(\Omega)$. It is possible to prove that in this way the solution of (1.6) exists and is unique; it will be denoted by u_μ. Therefore, the relaxed minimization problem (1.7) can be written as

$$\min \{F(u_\mu) \ : \ \mu \in \mathcal{M}(\Omega)\}. \tag{2.3}$$

where F is given by (1.5).

Remark 2.1. Let $A \in \mathcal{A}(\Omega)$ and let $\mu = \infty_{\Omega \setminus A}$. It is easy to see that u is a solution of (1.6) if and only if it is a solution of (1.3), that is $u = u_A$. In other words, the new family of relaxed Dirichlet problems of the form (1.6) with $\mu \in \mathcal{M}(\Omega)$ includes all classical Dirichlet problems.

The main feature of the space $\mathcal{M}(\Omega)$ is that it is possible to define a convergence on it, which has suitable compactness and density properties. More precisely we say that a sequence (μ_h) in $\mathcal{M}(\Omega)$ γ-converges to a measure $\mu \in \mathcal{M}(\Omega)$ if and only if for every $f \in L^2(\Omega)$ we have

$$u_{\mu_h} \to u_\mu \qquad \text{strongly in } f \in L^2(\Omega)$$

where u_{μ_h} and u_μ denote the solutions of (1.6) with μ_h and μ respectively. The name γ-convergence comes from the fact that this notion can be defined in an equivalent way in terms of the Γ-convergence of the functionals

$$\int_\Omega |Du|^2 \, dx + \int_\Omega u^2 \, d\mu_h.$$

The main compactness and density properties of γ-convergence can be summarized in the following theorem (see Dal Maso & Mosco [5]).

Theorem 2.2. *The γ-convergence on $\mathcal{M}(\Omega)$ has the following properties:*

(i) there exists a metric on $\mathcal{M}(\Omega)$ inducing the γ-convergence;

(ii) for every sequence (μ_h) in $\mathcal{M}(\Omega)$ there exists a subsequence (μ_{h_k}) which γ-converges to a measure $\mu \in \mathcal{M}(\Omega)$;

(iii) for every $\mu \in \mathcal{M}(\Omega)$ there exists a sequence (a_h) of nonnegative functions in $C_c^\infty(\Omega)$ such that the measures $a_h \mathcal{L}^n$ γ-converge to μ;

(iv) for every $\mu \in \mathcal{M}(\Omega)$ there exists a sequence (S_h) of smooth compact subsets of Ω such that the measures ∞_{S_h} γ-converge to μ.

In other words, Theorem 2.2 states that $\mathcal{M}(\Omega)$ endowed with the γ-convergence can be viewed as a compact metric space in which both $C_c^\infty(\Omega)$ and $\mathcal{A}(\Omega)$ are dense. In the following, when no confusion is possible, we often identify a domain A with the associated measure $\infty_{\Omega \setminus A}$.

We are now in a position to state the relaxation result.

Theorem 2.3. *Let $f \in L^2(\Omega)$ and let $g : \Omega \times \mathbf{R} \to \mathbf{R}$ be a function satisfying the following conditions:*

(2.4) $g(\cdot, s)$ *is* $\mathcal{L}^n$*-measurable in* Ω *for every* $s \in R^n$*;*
(2.5) $g(\cdot, s)$ *is continuous in* $\mathbf{R}$ *for a.e.* $x \in \Omega$*;*
(2.6) there exists $a \in L^1(\Omega)$ *and* $b \in \mathbf{R}$ *such that for a.e.* $x \in \Omega$ *and for every* $s \in \mathbf{R}$

$$|g(x,s)| \leq a(x) + b|s|^2.$$

Then problem (1.7) admits a solution and

$$\min\left\{F(u_\mu)\ :\ \mu \in \mathcal{M}(\Omega)\right\} = \inf\left\{F(u_A)\ :\ A \in \mathcal{A}(\Omega)\right\}.$$

Moreover we have:

(i) for every minimizing sequence (A_h) *of problem (1.4) there exists a subsequence* (A_{h_k}) γ*-converging to a solution* μ *of problem (1.7);*

(ii) if the original problem (1.4) admits a solution $A \in \mathcal{A}(\Omega)$*, then the measure* $\infty_{\Omega \setminus A}$ *is a solution of the relaxed problem (1.7).*

Remark 2.4. By using the Sobolev imbedding theorem it is possible to replace the inequality in (2.6) by the weaker condition

$$|g(x,s)| \leq a(x) + b|s|^p.$$

with $0 \leq p < 2n/(n-2)$.

Example 2.5. Assume $f(x) > 0$ for a.e. $x \in \Omega$, let w be the solution of the problem

$$\begin{cases} -\Delta w = f & \text{in } \Omega \\ w \in H_0^1(\Omega), \end{cases}$$

and let g be the function

$$g(x,s) = |2s - w(x)|^2.$$

It is easy to see that the relaxed problem (1.7) attains its minimum value 0 at the measure

$$\mu = \frac{f}{w}\mathcal{L}^n$$

which corresponds to $u_\mu = w/2$. On the other hand, by the maximum principle we have $w > 0$ in Ω and so, if for a domain $A \in \mathcal{A}(\Omega)$ it is

$$g\big(x, u_A(x)\big) = 0 \qquad \text{for a.e. } x \in \Omega,$$

we must necessarily have $A = \Omega$, which is impossible because $u_\Omega = w$ and $g\big(x, w(x)\big) = |w(x)|^2$.

The fact that the dimension n is greater or equal to 2 is crucial in Example 2.5; indeed, in the one-dimensional case the following result holds.

Proposition 2.6. *Assume $n = 1$; then for every $f \in L^2(\Omega)$ and $g : \Omega \times \mathbf{R} \to \mathbf{R}$ satisfying (2.4), (2.5), (2.6), there exists $A \in \mathcal{A}(\Omega)$ solution of problem (1.4).*

Proof. Let (A_h) be a minimizing sequence for problem (1.4) and let u_{A_h} be the corresponding solutions of problems (1.3). It is easy to see that the sequence (u_{A_h}) is bounded in $H_0^1(\Omega)$; hence, possibly passing to a subsequence, we may assume that (u_{A_h}) converges weakly in $H_0^1(\Omega)$ and uniformly in Ω to a function $u \in H_0^1(\Omega)$. Set

$$A = \{x \in \Omega \ : \ u(x) \neq 0\};$$

since u is continuous, the set A is open and $u \in H_0^1(A)$. Moreover, if $\varphi \in C_c^\infty(A)$ and U is a neighbourhood of the support of φ relatively compact in A, it is $u_h \neq 0$ on U for h large enough. Hence

$$\int_A u_h' \varphi' \, dx = \int_A f \varphi \, dx,$$

and passing to the limit as $h \to +\infty$ we get

$$-u'' = f \qquad \text{in } A.$$

Therefore $u = u_A$ and

$$F(u_A) = \lim_{h \to +\infty} F(u_{A_h}),$$

which proves that A is a solution of (1.4). ■

Remark 2.7. Proposition 2.6 still holds if the assumptions on g are weakened by requiring only it is a normal integrand such that for every $r > 0$ the function

$$\alpha_r(x) = \inf \{g(x, s) \ : \ |s| \leq r\}$$

is in $L^1(\Omega)$.

3. Some Optimality Conditions

In this section we show some optimality conditions for a solution $\mu \in \mathcal{M}(\Omega)$ of the relaxed shape optimization problem (1.7). The general method to prove these results consists in computing the limit

$$\lim_{\varepsilon \to 0^+} \frac{1}{\varepsilon} \big[F(u_{\mu_\varepsilon}) - F(u_\mu)\big] \tag{3.1}$$

for suitable families $(\mu_\varepsilon)_{\varepsilon > 0}$ in $\mathcal{M}(\Omega)$; the optimality conditions will be obtained from the fact that the limit above is nonnegative.

Let us fix a function $f \in L^2(\Omega)$ and a function $g : \Omega \times \mathbf{R} \to \mathbf{R}$ satisfying conditions (2.4), (2.5), (2.6). We assume, in addition, that

(3.2) for a.e. $x \in \Omega$ the function $g(x, \cdot)$ is differentiable, and its derivative $g_s(x, \cdot)$ is continuous on $\mathbf{R}$;

(3.3) for every $s \in \mathbf{R}$ the function $g_s(\cdot, s)$ is $\mathcal{L}^n$-measurable on Ω;

(3.4) there exist $a_1 \in L^2(\Omega)$ and $b_1 \in \mathbf{R}$ such that for a.e. $x \in \Omega$ and for every $s \in \mathbf{R}$

$$|g_s(x, s)| \leq a_1(x) + b_1|s|.$$

Let us denote by μ a solution of the relaxed problem (1.7), by u the solution of the problem

$$\begin{cases} -\Delta u + u\mu = f & \text{in } \Omega \\ u \in H_0^1(\Omega) \cap L_\mu^2(\Omega), \end{cases}$$

and by v the solution of the adjoint problem

$$\begin{cases} -\Delta v + v\mu = g_s(x, u) & \text{in } \Omega \\ v \in H_0^1(\Omega) \cap L_\mu^2(\Omega). \end{cases}$$

Taking a function $\varphi \in L^\infty(\Omega)$ with $\varphi \geq 0$ a.e. in Ω, and taking for every $\varepsilon > 0$

$$\mu_\varepsilon = \mu + \varepsilon\varphi\mathcal{L}^n,$$

it is possible to prove that the limit in (3.1) is equal to

$$-\int_\Omega uv\varphi\, dx.$$

Since the limit in (3.1) is nonnegative, and since φ is arbitrary, we obtain the first optimality condition:

$$uv \leq 0 \qquad \text{a.e. in } \Omega.$$

which can be refined into

(3.5) $$uv \leq 0 \qquad \text{q.e. in } \Omega.$$

Taking for every $\varepsilon \in]0, 1[$

$$\mu_\varepsilon = (1 - \varepsilon)\mu$$

it is possible to prove that the limit in (3.1) is equal to

$$\int_\Omega uv\, d\mu$$

which, taking (3.5) into account, gives the second optimality condition:

(3.6) $$uv = 0 \qquad \mu\text{-a.e. in } \Omega.$$

The statements of the next two optimality conditions are much more complicated in the general case of $\mu \in \mathcal{M}(\Omega)$; indeed, they require a very weak definition of normal derivative on boundaries of Borel sets, and several tools from potential theory. For this reason we refer the reader interested in the general case to Buttazzo & Dal Maso [2] and [3], and we consider here the simpler case when μ is of the form

$$\mu = \infty_{\Omega \setminus A} + a\mathcal{L}^n \tag{3.7}$$

where $a \in L^\infty(\Omega)$ with $a(x) \geq 0$ a.e. on Ω, and A is an open subset of Ω with a boundary ∂A of class C^2.

Let $\varphi, \psi \in C(\overline{\Omega})$ with

$$\inf\left\{\varphi(x)\ :\ x \in \Omega\right\} > 0, \qquad \inf\left\{\psi(x)\ :\ x \in \Omega\right\} > 0.$$

Taking for every $\varepsilon > 0$

$$\mu_\varepsilon = \mu\big|_A + \frac{1}{\varepsilon}\left(\frac{1}{\varphi}\mathcal{L}^n\big|_A + \frac{1}{\psi}\mathcal{H}^{n-1}\big|_{\partial A}\right)$$

it is possible to prove that the limit in (3.1) is equal to

$$\int_{\Omega \setminus A} f g_s(\cdot, 0)\varphi\,dx + \int_{\Omega \cap \partial A} \frac{\partial u}{\partial \nu}\frac{\partial v}{\partial \nu}\psi\,d\mathcal{H}^{n-1}.$$

Since this limit is nonnegative and since φ and ψ are arbitrary, we obtain the third and fourth optimality conditions:

$$f(\cdot)g_s(\cdot, 0) \geq 0 \qquad \text{a.e. on } \Omega \setminus A; \tag{3.8}$$

$$\frac{\partial u}{\partial \nu}\frac{\partial v}{\partial \nu} \geq 0 \qquad \mathcal{H}^{n-1}\text{-a.e. on } \Omega \cap \partial A. \tag{3.9}$$

Taking (3.5) into account, one can easily show that (3.9) implies

$$\frac{\partial u}{\partial \nu}\frac{\partial v}{\partial \nu} = 0 \qquad \mathcal{H}^{n-1}\text{-a.e. on } \Omega \cap \partial A. \tag{3.10}$$

Specializing the optimality conditions (3.5), (3.6), (3.8), (3.10) to the particular case of (3.7) in which $a = 0$, we obtain that if the original shape optimization problem (1.4) admits a solution A with a boundary of class C^2, then (3.6) is trivial, whereas the remaining ones give

$$uv \leq 0 \qquad \text{q.e. on } A;$$

$$f(\cdot)g_s(\cdot, 0) \geq 0 \qquad \text{a.e. on } \Omega \setminus A;$$

$$\frac{\partial u}{\partial \nu}\frac{\partial v}{\partial \nu} = 0 \qquad \mathcal{H}^{n-1}\text{-a.e. on } \Omega \cap \partial A.$$

Example 3.1. Let f and u_0 be two functions in $L^2(\Omega)$ and let

$$g(x,s) = |s - u_0(x)|^2.$$

Assume that $f(x) > 0$ for a.e. $x \in \Omega$ and that the essential infimum of u_0 on every compact subset of Ω is positive. Then, if

$$\int_\Omega |u_0|^2\,dx$$

is sufficiently small, problem (1.4) has no solution. Indeed, assume by contradiction a solution A exists; by the optimality condition (3.8) we get

$$-f(x)u_0(x) \geq 0 \qquad \text{for a.e. } x \in \Omega \setminus A$$

which implies, for the assumptions made on f and u_0, that $A = \Omega$ up to a set of measure zero. A more careful inspection (see Chipot & Dal Maso [4]) shows that the set A should actually coincide with Ω. But this is impossible because the empty set $A = \emptyset$, which corresponds to $u = 0$, would improve the cost whenever

$$\int_\Omega |u_0|^2\,dx < \int_\Omega |u_\Omega - u_0|^2\,dx$$

being u_Ω the solution corresponding to $A = \Omega$. If u_0 is small enough (in L^2 norm) the inequality above is satisfied, and so problem (1.4) cannot have a solution.

References

[1] G. BUTTAZZO: *Semicontinuity, Relaxation and Integral Representation in the Calculus of Variations.* Pitman Res. Notes Math. Ser. **207**, Longman, Harlow (1989).

[2] G. BUTTAZZO & G. DAL MASO: *Shape optimization for Dirichlet problems: relaxed solutions and optimality conditions.* Bull. Amer. Math. Soc., **23** (1990), 531–535.

[3] G. BUTTAZZO & G. DAL MASO: *Shape optimization for Dirichlet problems: relaxed formulation and optimality conditions.* Appl. Math. Optim., **23** (1991), 17–49.

[4] M. CHIPOT & G. DAL MASO: *Relaxed shape optimization: the case of nonnegative data for the Dirichlet problem.* IMA preprint 635, Minneapolis (1990).

[5] G. DAL MASO & U. MOSCO: *Wiener's criterion and* Γ*-convergence.* Appl. Math. Optim., **15** (1987), 15–63.

[6] S. FINZI VITA: *Numerical shape optimization for relaxed Dirichlet problems.* Preprint Università di Roma "La Sapienza", Roma (1990).

[7] R. V. KOHN & G. STRANG: *Optimal design and relaxation of variational problems, I,II,III.* Comm. Pure Appl. Math., **39** (1986), 113–137, 139–182, 353–377.

[8] R. V. KOHN & M. VOGELIUS: *Relaxation of a variational method for impedance computed tomography.* Comm. Pure Appl. Math., **40** (1987), 745–777.

[9] F. MURAT & J. SIMON: *Sur le contrôle par un domaine géometrique.* Preprint 76015, Univ. Paris VI, (1976).

[10] F. MURAT & L. TARTAR: *Optimality conditions and homogenization.* Proceedings of "Nonlinear variational problems", Isola d'Elba 1983, Res. Notes in Math. **127**, Pitman, London, (1985), 1–8.

On the smoothness of the value function along optimal trajectories.

PIERMARCO CANNARSA *
FAUSTO GOZZI **

Abstract. We consider a finite horizon optimal control problem in Mayer form for a system governed by a semilinear state equation. We prove that, under suitable assumptions, the associated value function is differentiable along optimal trajectories. For this purpose we prove a backward uniqueness result for a class of abstract evolution equation of parabolic type.

Key words. Optimal control, value fuction, analytic semigroup, backward equation.

AMS (MOS) subject classification. 49c10, 49c20.

1. Introduction.

This paper is devoted to the study of differentiability properties for the value function of optimal control problems in Mayer form. Let us consider an optimal control problem of Mayer type:

$$(P) \qquad \textit{minimize} \quad g(x(T;t_0,x_0,u)) \quad \textit{over all measurable controls } u:[t_0,T]\to U$$

where $x(\cdot;t_0,x_0,u)$ is the solution of the infinite dimensional control system:

$$(1.1) \qquad \begin{cases} x'(t) = Ax(t) + f(t,x(t),u(t)) & t\in[t_0,T] \text{ and } u(t)\in U \ \forall t\in[t_0,T]; \\ x(t_0) = x_0 & x_0\in X,\ t_0>0. \end{cases}$$

Here X (the *state space*) is a real Hilbert space and U (the *control space*) is a metric space. Moreover $u:[t_0,T]\to U$ is measurable and A is a generator of a strongly continuous semigroup on X. A measurable function $u:[t_0,T]\to U$ at which the minimum is attained is called an optimal control for problem (P), and the associate solution of equation (1.1) is called the corresponding optimal trajectory of the system.

The value function of the problem is defined as:

$$(1.2) \qquad V(t_0,x_0) \stackrel{def}{=} \inf\{\, g(x(T;t_0,x_0,u)) \mid u:[t_0,T]\to U \text{ is measurable}\}$$

letting (t_0,x_0) range over $[0,T]\times X$.

* Dipartimento di Matematica, II Università di Roma "Tor Vergata" Via Fontanile di Carcaricola, 00133 Roma. Ph. 39.6.79794694; FAX 39.6.79794699; Email VAXTVM::CANNARSA.

** Dipartimento di Matematica, via F. Buonarroti 2, 56127 Pisa, Italy. Ph. 39.50.599542; FAX 39.50.599524; Email GOZZI@IPISNSIB (Bitnet), VAXSNS::GOZZI (Decnet).

Many results about the continuity and Lipschitz continuity of V have been proved (see the next section). What is more difficult to study is the differentiability of the value function V. In general V is not differentiable in any variable, but partial results in this direction can be proved. This is the subject of our work.

A differentiability result for the value function is proved in a recent paper by Cannarsa and Frankowska (see [7]). Under suitable assumptions they prove that the value function is differentiable with respect to x in the Gateaux sense, along optimal trajectories starting at a differentiability point. However they have to assume that the operator A generates a group (see[7] Theorem 5.1). For similar results in finite dimensions see [5], [6].

The purpose of this work is to show the differentiability of the value function along optimal trajectories starting at a differentiability point, under the assumption that A generates an analytic semigroup. Moreover we obtain Fréchet differentiability (see section 3, Theorem 3.4) for the function $V(t,\cdot)$.

Our method uses the basic fact, proved in [7], that, along an optimal trajectory $\overline{x}(\cdot)$, we have:

$$\bar{p}(t) + D_x^+ V(t, \bar{x}(t)) \subset (G(t,t_0)X)^{\perp}$$

where $\bar{p}(\cdot)$ is the co-state corresponding to $\overline{x}(\cdot)$ (see (5.4)) and $G(s,t)$ is the solution operator of the linear system:

$$\begin{cases} \dfrac{\partial G}{\partial s}(s,t) = \Big(A + \dfrac{\partial f}{\partial x}(s, \bar{x}(s), \bar{u}(s))\Big) G(s,t) & t \le s \le T \\ G(t,t) = \mathbf{I} \end{cases}$$

At this point, in order to conclude, we have to overcome two difficulties.

First we need to show that $G(t,t_0)X$ is dense in X. By duality, this result is equivalent to showing a backward uniqueness property for the (parabolic) semilinear evolution system which is solved by the adjoint of G (see the Appendix).

Second, to obtain the Fréchet differentiability of $V(t,\cdot)$, we use a regularity property of $V(t,\cdot)$ proved in [8], that is

$$|V(t,x_1) - V(t,x_0)| \le C|(-A)^{-\alpha}(x_1 - x_0)|$$

uniformly on all bounded subsets of $[0,T] \times X$. As a consequence of this fact it follows that the superdifferential $D_x^+ V(t,x)$ is contained in $D((-A)^{\alpha})$, for every $\alpha \in [0,1[$ and all $(t,x) \in [0,T[\times X$ (see Corollary 4.3). This compactness-like result and the semi-concavity of $V(t,\cdot)$ allow us to prove that $V(t,\cdot)$ is Fréchet Differentiable whenever $D_x^+ V(t,x)$ is a singleton.

We conclude this introduction with an application of the results contained in this paper.

First of all we recall that V is the unique viscosity solution of the Hamilton-Jacobi equation (see [12]):

$$(HJ) \qquad \begin{cases} -V_t(t,x) + H(t,x,-V_x(t,x)) - \langle V_x(t,x), Ax\rangle = 0 \\ \\ V(T,x) = 0 \qquad \forall x \in X \end{cases}$$

where the Hamiltonian H is defined as:

$$H(t,x,p) = \sup_{u\in U} \langle p, f(t,x,u)\rangle$$

Moreover, by dynamic programming arguments, it can be proved (see [6], [7]) that an optimal state, if it exists, satisfies in weak sense the closed loop equation:

$$(1.3) \qquad \begin{cases} x'(t) = Ax(t) + \dfrac{\partial H}{\partial p}(t, x(t), -\nabla_x V(t,x(t))) \\ \qquad \text{for } t \in [t_0, T] \text{ and } u(t) \in U \ \forall t \in [t_0, T] \\ x(t_0) = x_0 \qquad\qquad x_0 \in X,\ t_0 > 0; \end{cases}$$

Clearly, if we know that the function $V(t,\cdot)$ is Fréchet Differentiable at x_0 and we use the above differentiability result, (assuming the regularity of the Hamiltonian H), then we are able to consider the closed loop equation (1.3) in classical sense.

2. Preliminaries.

Let X a real Hilbert space with norm $|\cdot|$ and scalar product $\langle\cdot,\cdot\rangle$, and U a metric space. With obvious modifications all the results of this paper can be adapted to complex Hilbert spaces.

Let $A : D(A) \subset X \to X$ be the infinitesimal generator of an analytic semigroup, $e^{tA} (t \geq 0)$, in X. Then it is well known that there exist constants M_0, $M_1 > 0$ and $\omega \in \mathbf{R}$ such that:

$$(2.1) \qquad \begin{cases} (i) \qquad |e^{tA}x| \leq M_0 e^{\omega t}|x| \\ \\ (ii) \qquad |Ae^{tA}x| \leq \left(\omega M_0 + \dfrac{M_1}{t}\right) e^{\omega t}|x| \end{cases}$$

for all $x \in X$ and $t > 0$ (see e.g. [20], p. 60).

Suppose now that $\omega < 0$ and 0 belongs to the resolvent set of A, $\rho(A)$. We denote by $(-A)^\alpha$, $\alpha \in \mathbf{R}$, the fractional powers of $-A$ with domain $D((-A)^\alpha)$ (see [20], p. 69) and set:

$$(2.2) \qquad |x|_\alpha = |(-A)^\alpha x|$$

for all $x \in D((-A)^\alpha)$. Estimate (2.1) (ii) has the following version for fractional powers:

$$|(-A)^\alpha e^{tA} x| \le \frac{M_\alpha}{t^\alpha}|x| \tag{2.3}$$

for all $x \in X$, $t > 0$ and some constants $M_\alpha > 0$ (see [20], p. 74).

Let $T > 0$, $x_0 \in X$, $f \in L^p(0,T;X)$, $p > 1$. Then the Cauchy problem:

$$\begin{cases} x'(t) = Ax(t) + f(t) & \text{on } [0,T]; \\ x(0) = x_0 & x_0 \in X; \end{cases}$$

has a unique strong solution $u \in C([0,T];X)$ given by the formula

$$x(t) = e^{tA}x_0 + \int_0^t e^{(t-\tau)A} f(\tau)d\tau\ , \qquad t \in [0,T]. \tag{2.4}$$

Assume further that $f \in L^\infty(0,T;X)$. Then it is well known that $x(t) \in D((-A)^\alpha)$ for any $\alpha \in [0,1[$ and $t > 0$. In fact, estimate (2.3) and formula (2.4) yield

$$|x(t)|_\alpha \le M_\alpha \left(t^{-\alpha}|x_0| + \frac{t^{1-\alpha}}{1-\alpha}\|f\|_{L^\infty(0,T;X)} \right) \tag{2.5}$$

for all $t > 0$ and $\alpha \in]0,1[$.

At this point we recall some generalization of the notion of gradient for non-smooth functions. Let Ω be an open subset of X and $\psi : \Omega \to \mathbf{R}$. For any $x_0 \in \Omega$, the *semi-differentials* $D^+\psi(x_0)$, $D^-\psi(x_0)$ are defined as follows (see e.g. [12]):

$$D^+\psi(x_0) = \left\{ p \in X \,|\, \limsup_{x \to x_0} \frac{\psi(x) - \psi(x_0) - \langle p, x - x_0 \rangle}{|x - x_0|} \le 0 \right\}$$

$$D^-\psi(x_0) = \left\{ p \in X \,|\, \liminf_{x \to x_0} \frac{\psi(x) - \psi(x_0) - \langle p, x - x_0 \rangle}{|x - x_0|} \ge 0 \right\}$$

The function ψ is Fréchet differentiable at x_0, if and only if $D^+\psi(x_0)$ and $D^-\psi(x_0)$ are both non-empty. Moreover in this case:

$$D^+\psi(x_0) = D^-\psi(x_0) = \{D\psi(x_0)\}$$

where $D\psi$ denotes the Fréchet derivative.

We denote by $D^*\psi(x_0)$ the set of all $p \in X$ for which there exists a sequence $\{x_n\}_{n \in N}$ of points of Ω such that

$$\begin{cases} \text{(i)} & x_n \text{ converges to } x_0 \text{ as } n \to \infty \\ \text{(ii)} & \psi \text{ is Fréchet differentiable at } x_n \text{ for all } n \in N \\ \text{(iii)} & \nabla\psi(x_n) \text{ weakly converges to } p \text{ as } n \to \infty \end{cases}$$

In general, $D^*\psi(x_0)$ may be empty. However, if ψ is Lipschitz continuous in a neighborood of x_0, then a result proved (under more general assumptions) by Preiss (see [22]) states that ψ is Fréchet differentiable on a dense subset of Ω. Hence, in this case we conclude that $D^*\psi(x_0) \neq \emptyset$.

Let us now introduce the notion of semi-concavity. For $x_0 \in X$ we denote by $B_r(x_0), r > 0$, the closed ball in X with radius r and center at x_0. Let $\Omega \subset X$ be convex and $\psi : \Omega \to \mathbf{R}$. We say that ψ is *semi-concave* if there exists a function

$$\text{(2.6)} \qquad \omega : \mathbf{R}_+ \times \mathbf{R}_+ \to \mathbf{R}_+ \quad \textit{satisfying} \begin{cases} \forall r \leq R, s \leq S, \quad \omega(r,s) \leq \omega(R,S) \\ \forall r > 0 \quad \lim_{s \downarrow 0} \omega(r,s) = 0 \end{cases}$$

such that, for every $r > 0$, $\lambda \in [0,1]$ and $x, y \in \Omega \cap B_r(0)$,

$$\text{(2.7)} \qquad \lambda\psi(x) + (1-\lambda)\psi(y) - \psi(\lambda x + [1-\lambda]y) \leq \lambda(1-\lambda)|x-y|\omega(r, |x-y|).$$

For instance, any continuously differentiable function is semi-concave provided the derivative is bounded on bounded sets.

Superdifferentials of semi-concave functions enjoy several regularity properties. We collect some of them in the following proposition:

PROPOSITION 2.1. *If ψ is semi-concave and Lipschitz in $B_r(x_0)$ for some $r > 0$ then*

i)
$$D^+\psi(x_0) = \overline{co} D^*\psi(x_0)$$

where $\overline{co}$ denotes the closed convex hull. In particular this implies that $D^+\psi(x_0) \neq \emptyset$.

ii) For every $p \in D^+\psi(x_0)$

$$\psi(x) - \psi(x_0) - \langle p, x - x_0 \rangle \leq |x - x_0|\omega(r, |x - x_0|)$$

for all $x \in B_r(x_0)$.

iii) The set $D^+\psi(x_0)$ is a singleton if and only if ψ is Gateaux Differentiable at x_0.

iv) If $D^+\psi(x_0)$ is a singleton and $D^+\psi(x)$ is contained in some compact subset of X for all $x \in B_r(x_0)$,
then ψ is Fréchet differentiable at x_0.

PROOF. We sketch the proof for reader's convenience.

i) and ii) The proof of the first statement i) is given in [4], Corollary 4.7. The second statement is proved in [4], section 4. These two results are proved under a slightly different definition of semi-concavity. However, the proofs of [4] can be adapted to the situation at hand with obvious modifications. For this reason they are omitted.

iii) First we use the fact that $D^+\psi(x)$ coincides with the Clarke's generalized gradient ([11]) if ψ is semiconcave (see Proposition 4.8 in [4]). Now, if the Clarke's generalized gradient at x_0 is a singleton, then ψ is strictly differentiable at x_0 (see [11] p.33). This implies, in particular, that ψ is Gateaux Differentiable at x_0.
Conversely, assume that ψ is Gateaux Differentiable at x_0. By part i) we know that $D^+\psi(x_0)$ is nonempty. Moreover, if $p \in D_x^+V(t_0,x_0)$, then

$$\langle \nabla_x\psi(x_0) - p, x\rangle = \lim_{h\downarrow 0} \frac{\psi(x_0+hx) - \psi(x_0) - h\langle p,x\rangle}{h} \leq 0$$

for all $x \in X$. So, $D_x^+\psi(x_0) = \{\nabla_x\psi(x_0)\}$.

iv) The last statement can be obtained by adapting the proof of corollary 4.12 in [9] as follows. Let $\{p_0\} = D^+\psi(x_0)$. Then, by definition of superdifferential:

$$\limsup_{x\to x_0} \frac{\psi(x) - \psi(x_0) - \langle p_0, x - x_0\rangle}{|x - x_0|} \leq 0$$

Equivalently, for any strongly converging sequence $\{x_n\}$ with limit x_0, we have:

$$\limsup_{n\to+\infty} \frac{\psi(x_n) - \psi(x_0) - \langle p_0, x_n - x_0\rangle}{|x_n - x_0|} \leq 0 \tag{2.8}$$

Now by part i) the superdifferential $D^+\psi(x_n)$ is nonempty. Let $p_n \in D^+\psi(x_n)$. Then, due to the compactness assumption in iv), there exists a subsequence (still denoted by p_n) which strongly converges to an element $p_\infty \in X$. In view of ii), $p_\infty \in D^+\psi(x_0)$ and therefore:

$$p_n \xrightarrow{strongly} p_0 = p_\infty \ . \tag{2.9}$$

Again by ii) it follows that:

$$\psi(x_0) - \psi(x_n) - \langle p_n, x_0 - x_n\rangle \leq |x_0 - x_n|\omega(r, |x_0 - x_n|)$$

for all $x \in B_r(x_0)$. Now, by the above formula we have:

$$\begin{aligned} &\frac{\psi(x_n) - \psi(x_0) - \langle p_0, x_n - x_0\rangle}{|x_n - x_0|} = \\ &= \frac{\psi(x_n) - \psi(x_0) - \langle p_n, x_n - x_0\rangle}{|x_n - x_0|} + \frac{\langle p_n - p_0, x_n - x_0\rangle}{|x_n - x_0|} \geq \\ &\geq -\omega(r, |x_n - x_0|) + \langle p_n - p_0, \frac{x_n - x_0}{|x_n - x_0|}\rangle \end{aligned}$$

Then, taking the $\liminf_{n\to+\infty}$ and using (2.9) we obtain:

$$\liminf_{n\to+\infty} \frac{\psi(x_n) - \psi(x_0) - \langle p_0, x_n - x_0\rangle}{|x_n - x_0|} \geq 0$$

which gives:

$$\lim_{n\to+\infty} \frac{\psi(x_n) - \psi(x_0) - \langle p_0, x_n - x_0\rangle}{|x_n - x_0|} = 0$$

For a subsequence of the original sequence $\{x_n\}$. The claim easily follows.

We conclude this section with the following lemma which is a simple consequence of Gronwall's inequality.

LEMMA 2.3. *Let $\psi : [a,b] \to \mathbf{R}$ be an integrable function such that:*

$$\psi(t) \le A + \frac{B}{(t-a)^\alpha} + L\int_a^t \psi(s)ds \tag{2.10}$$

for almost everywhere $t \in [a,b]$ and some constants $L, A, B > 0$, $\alpha \in]0,1[$. Then, for a.e. $t \in [a,b]$:

$$\psi(t) \le \left[1 + (b-a)e^{L(b-a)}\right] A + \left[\frac{e^{L(b-a)}}{1-\alpha}(b-a)^{1-\alpha} + \frac{1}{(t-a)^\alpha}\right] B \tag{2.11}$$

PROOF. The proof of Lemma 2.3, given in [8], can be easily recovered applying Gronwall's inequality to the function:

$$\Psi(r) = \int_0^r \psi(t)dt.$$

Q.E.D.

REMARK 2.4. If $\alpha = 1$, then the statement of Lemmma 2.3 is not true. However it is possible to prove that, if a function ψ satisfies the inequality:

$$\psi(t) \le \frac{B}{t} + L\int_0^t \psi(s)ds \tag{2.12}$$

a.e. in $[0,T]$ for some constants B and L, then there exists a constant K, *independent of* B, such that:

$$\psi(t) \le K\left[1 + \frac{B|\log B|}{t}\right]$$

This estimate does not hold true if we substitute $B|\log B|$ by B, i.e. the dependence on B cannot be linear.

To prove this fact it is sufficient to take $B \in (0, \frac{1}{2})$ and:

$$\psi(t) = \begin{cases} 1 & 0 \le t < B \\ \frac{B}{t} + B\log\frac{t}{B} & B \le t \le \frac{1}{2} \end{cases}$$

For this function we have by easy computations:

$$\psi(t) \leq \frac{B}{t} + \int_0^t \psi(s)ds$$

However, it is not possible to find $K > 0$, independent of B, such that:

$$\psi(t) \leq K\frac{B}{t}$$

3. The optimal control problem: assumptions and statement of the main result.

Let X be a real Hilbert space and U a complete separable metric space. Fix $T > 0$ and let $(t_0, x_0) \in [0, T] \times X$.

We will consider an infinite dimensional control system governed by the following semilinear state equation:

$$\text{(3.1)} \qquad \begin{cases} x'(t) = Ax(t) + f(t, x(t), u(t)) & \text{on } [t_0, T]; \\ x(t_0) = x_0 & x_0 \in X,\ t_0 > 0; \end{cases}$$

where $u : [t_0, T] \to U$ is measurable and:

$$\text{(3.2)} \qquad \begin{cases} i) \ A : D(A) \subset X \to X \text{ is the infinitesimal generator} \\ \quad \text{of an analytic semigroup} \\ \quad \text{of compact operators } e^{tA}, t > 0 \\ \quad \text{and, } \forall x \in D(A): \\ \qquad\qquad \langle Ax, x\rangle \leq 0 \\ ii) \ f : [0, T] \times X \times U \to \mathbf{R} \text{ is continuous in both variables and:} \\ \qquad |f(t, x, u)| \leq C_0(1 + |x|) \quad \forall x \in X \\ \qquad |f(t, x, u) - f(t, y, u)| \leq C_0|x - y| \quad \forall x, y \in X \\ \quad \text{for some constant } C_0 > 0. \end{cases}$$

It is well known that hypotheses i) and ii) guarantee existence and uniqueness of a mild solution for the equation (3.1) satisfying:

$$\text{(3.3)} \qquad x(t) = e^{(t-t_0)A}x_0 + \int_0^t e^{(t-s)A} f(s, x(s), u(s))ds$$

(see for instance [20] and [15]). We will denote such a solution by $x(\cdot; t_0, x_0, u)$.

Moreover by (3.2), (3.3) and the Gronwall Lemma it follows:

$$\text{(3.4)} \qquad |x(t)| \leq e^{C_0T}(|x_0| + C_0T), \qquad \forall t \in [t_0, T].$$

Let $g : X \to \mathbf{R}$ be a Lipschitz continuous function on all bounded subsets of X. We are interested in the Mayer problem:

$$\text{(3.5)} \qquad \textit{minimize} \quad g(x(T;x_0,t_0,u)) \quad \textit{over all measurable controls } u : [t_0,T] \to U$$

REMARK 3.1. As well known, assumption (3.2) i)) implies that A generates a contraction semigroup, that is (2.1) i) holds with $M_0 = 1$ and $\omega = 0$. While the latter assumption implies no loss of generality (as one can see replacing the state $x(t)$ by $e^{-\omega(t-t_0)}x(t)$), the fact that $M_0 = 1$ is essential in this paper just for the backward uniqueness result proved in the Appendix.

Now we define the value function of the problem (3.5) letting (t_0, x_0) range in $[t_0,T]\times X$ as follows:

$$\text{(3.7)} \qquad V(t_0,x_0) \stackrel{def}{=} \inf\{\, g(x(T;t_0,x_0,u)) \mid u : [t_0,T] \to U \text{ is measurable}\}$$

A measurable function $u : [t_0,T] \to U$ at which the minimum is attained is called an optimal control for problem (3.5), and the associate solution of equation (3.1) is called the corresponding optimal trajectory of the system.

We recall a classical regularity result for the value functions of this kind of problems (see for instance [2]):

PROPOSITION 3.2. *Under assumptions (3.2) V is continuous in $[0,T]\times X$ and bounded on all bounded subsets of $[0,T]\times X$. Moreover, $x \to V(t,x)$ is Lipschitz on all bounded subsets of X, uniformly in $t \in [0,T]$ and $t \to V(t,x)$ is Lipschitz continuous on $[0,T]$ for every $x \in D(A)$.*

Furthermore, for every measurable $u : [0,T] \to U$, the function $V(\cdot, x(\cdot;t_0,x_0,u))$ is non-decreasing in $[t_0,T]$ and, for all $t \in [t_0,T]$ and $x_0 \in X$ we have:

$$\text{(3.8)} \qquad V(t_0,x_0) = \inf\{V(t,x(t;t_0,x_0,u)) | u : [t_0,T] \to U \textit{ is measurable}\}$$

Finally $V(\cdot, x(\cdot;t_0,x_0,u))$ is constant if and only if u is optimal for problem (3.5).

Now, we recall a semi-concavity result for the value function V (see [7] and also [4] for a related result concerning the Bolza problem).

THEOREM 3.3. - *Assume (3.2) and let g be semi-concave. Suppose further that f is Fréchet differentiable with respect to x and there exists a function ω satisfying (2.6) such that*

$$\|\frac{\partial f}{\partial x}(t,x,u) - \frac{\partial f}{\partial x}(t,y,u)\| \le \omega(r, |x-y|)$$

for every $u \in U, t \in [0,T]$ and all $x, y \in B_r(0)$. Then $V(t,\cdot)$ is semi-concave for all $t \in [0,T]$.

In order to be able to use the result above we will work with hypotheses (3.2) and:

(3.9)
$$\begin{cases} i) & g : X \to \mathbf{R} \text{ is Fréchet Differentiable on } X \text{ and semiconcave.} \\ ii) & f \text{ is Fréchet differentiable with respect to } x \\ & \text{and there exists a function } \omega \text{ satisfying (2.6) such that:} \\ & \|\frac{\partial f}{\partial x}(t,x,u) - \frac{\partial f}{\partial x}(t,y,u)\| \le \omega(r, |x-y|) \\ & \text{for every } u \in U, t \in [0,T] \text{ and all } x,y \in B_r(0). \end{cases}$$

Theorem 3.3 implies that, under assumptions (3.2), (3.9), the value function is semiconcave with respect to x.

The main result is the following:

THEOREM 3.4. *Assume (3.2), (3.9) and suppose that:*

(3.10)
$$\begin{cases} D(A^*) \subset D(A) \quad \textit{and:} \\ \left|\frac{A^* - A}{2} x\right| \le C \left|(-A^*)^{\frac{1}{2}} x\right| \quad \forall x \in D(A^*) \\ \textit{for some constant } C > 0. \end{cases}$$

Let $(t_0, x_0) \in [0,T[\times X$ *be such that there exists the Gateaux-derivative* $\nabla_x V(t_0, x_0)$*, and let* $\overline{x}$ *be an optimal trajectory for the problem (3.5) such that* $\overline{x}(t_0) = x_0$*.*

Then V *is Fréchet differentiable with respect to* x *at* $(t, \overline{x}(t))$ *for every* $t \in [t_0, T[$*.*

REMARK 3.5. Using the results of [8] it is possible to prove that at the points mentioned above the value function is (t,x) - Fréchet differentiable.

REMARK 3.6. Hypothesis (3.10) is a technical one, due to technical difficulties encountered in the proofs of the Appendix. However it is satisfied for a large class of operators. For parabolic second order operators this assumption is satisfied if the coefficients of the leading part are smooth.

4. Lipschitz regularity of the value function.

As we said in the previous section (see Proposition 3.2), the Lipschitz continuity of of the function $V(t,\cdot)$ for all $t \in [t_0, T]$ (that is with respect the space-like variable x) is a well known result (see [2]), even when the semigroup e^{tA} is just strongly continuous. Indeed, when e^{tA} is analytic, a stronger Lipschitz property holds true for V, as it is proved in [8].

We recall the result of [8] together with its proof for the reader's convenience. We assume that

(4.1)

a) $A : D(A) \subset X \to X$ is the infinitesimal generator of an analytic semigroup of negative type;

b) $f : [0,T] \times X \times U \to \mathbf{R}$ is continuous in both variables and:

$$\begin{aligned} |f(t,x,u)| &\leq C_0(1+|x|) \quad \forall x \in X \\ |f(t,x,u) - f(t,y,u)| &\leq C_0|x-y| \quad \forall x,y \in X \end{aligned}$$

for some constant $C_0 > 0$.

c) g is Lipschitz continuous on all bounded subsets of X.

THEOREM 4.1. *Assume (4.1) a), b), c) hold true and let $R > 0$, $\alpha \in [0,1[$. Then there exists a constant $C = C(\alpha, R, T)$ such that:*

$$|V(t,x_1) - V(t,x_0)| \leq C|(-A)^{-\alpha}(x_1 - x_0)| \tag{4.2}$$

for all $t \in [0, T - \frac{1}{R}]$ and all $x_1, x_0 \in X$ satisfying $|x_1|, |x_0| \leq R$.

PROOF.

Suppose $|(-A)^{-\alpha}(x_1 - x_0)| > 0$ (otherwise there is nothing to prove). Let $u_0 : [t,T] \to U$ be such that:

$$V(t,x_0) + \left|(-A)^{-\alpha}(x_1 - x_0)\right| > g(x(T;t,x_0,u_0))$$

and set $\overline{x}_1(\cdot) = x(\cdot;t,x_1,u_0)$, $\overline{x}_0(\cdot) = x(\cdot;t,x_0,u_0)$. Then recalling c), we have:

$$V(t,x_1) - V(t,x_0) \leq C_R|\overline{x}_1(T) - \overline{x}_0(T)| \tag{4.3}$$

for some constant $C_R > 0$. On the other hand, using (2.2) we obtain:

$$\begin{aligned} |\overline{x}_1(s) - \overline{x}_0(s)| &\leq \left|(-A)^{\alpha} e^{(s-t)A}(-A)^{-\alpha}(x_1 - x_0)\right| \\ &\quad + \left|\int_t^s e^{(t-\sigma)A}\Big[f(\sigma,\overline{x}_1(\sigma),u_0(\sigma)) - f(\sigma,\overline{x}_0(\sigma),u_0(\sigma))\Big]d\sigma\right| \\ &\leq \frac{M_\alpha}{(s-t)^\alpha}\left|(-A)^{-\alpha}(x_1 - x_0)\right| + C_0 M_0 \int_t^s |\overline{x}_1(\sigma) - \overline{x}_0(\sigma)|d\sigma \end{aligned}$$

for all $s \in]t,T]$. Hence, applying Lemma 2.3 we obtain:

$$|\overline{x}_1(s) - \overline{x}_0(s)| \leq C_{\alpha,T}\left[1 + \frac{1}{(s-t)^\alpha}\right]\left|(-A)^{-\alpha}(x_1 - x_0)\right|$$

Taking $s = T$ the conclusion (4.2) follows from the above estimate and (4.3), since the argument is symmetric with respect to x_1, x_0.

Q.E.D.

REMARK 4.2. In view of remark 2.4 we do not expect the previous result hold true in the case $\alpha = 1$.

A very useful consequence of Theorem 4.1 is the following (see [8]):

COROLLARY 4.3. *Assume a), b), c) hold true, and let* $(t_0, x_0) \in [0, T[\times X$. *Then, for all* $\alpha \in [0, 1[$, *we have:*

$$D_x^+ V(t_0, x_0) \subset D((-A^*)^\alpha) \tag{4.4}$$

Moreover, for every $R \in]0, \frac{1}{T}[$, *there exists* $C(R, \alpha, T) > 0$ *such that*

$$|(-A^*)^\alpha p| \le C(R, \alpha, T), \qquad \forall p \in D_x^+ V(t_0, x_0) \quad \forall (t_0, x_0) \in [0, T - \frac{1}{R}] \times B_R(0) \tag{4.5}$$

Furthermore, if A^{-1} *is a compact operator on* X, *then* $D_x^+ V(t_0, x_0)$ *is contained in a fixed compact set of* X, *independent of* (t_0, x_0) *for all* $(t_0, x_0) \in [0, T - \frac{1}{R}] \times B_R(0)$.

PROOF. The proof of (4.4) and (4.5), based on standard functional analysis argument, is given in [8]. As for the last statement of corollary 4.3, we note that, since A^{-1} is compact, then $(-A)^{-\alpha}$ and $(-A^*)^{-\alpha}$ are compact operators. Then $D((-A^*)^\alpha)$ is compactly embedded in X and the conclusion follows from (4.5).

Q.E.D.

If we suppose that (3.2), (3.9) hold true then $V(t_0, \cdot)$ is semiconcave and we obtain the following useful result.

COROLLARY 4.4. *Let (3.2), (3.9) hold true and let* $t_0 > 0$. *Then the function* $V(t, \cdot)$ *is differentiable in the Gateaux sense at* x_0 *if and only if it is Fréchet Differentiable at* x_0.

PROOF. We recall first that, as $V(t_0, \cdot)$ is Gateaux differentiable at x_0, $D_x^+ V(t_0, x_0)$ is a singleton (see Proposition 2.1, iii).

At this point it is enough to apply the second part of Corollary 4.3 and Proposition 2.1, iv) to prove that $V(t_0, \cdot)$ is Fréchet Differentiable at x_0.

Q.E.D.

5. Proof of the main result

We divide the proof into different steps. The first two steps are contained in [7]. We just give an outline of them for completeness.

STEP 1.

Let $(\bar{x}, \bar{u})$ be a trajectory-control pair for problem (3.5). We denote by $G(s, t)$ the solution operator of the linear problem:

$$\begin{cases} \dfrac{\partial G}{\partial s} = \Big(A + \dfrac{\partial f}{\partial x}(s, \bar{x}(s), \bar{u}(s))\Big) G(s, t) & t \le s \le T \\ G(t, t) = \mathbf{I} \end{cases}$$

which is, in fact, the evolution operator of the linearization of (1.1) along $(\bar{x}, \bar{u})$. In the following we denote by $G^*(s,t)$ the adjoint of $G(s,t)$.

If $(\bar{x}, \bar{u})$ is an optimal trajectory-control pair for problem (3.5), then, for every $p \in D^+g(\bar{x}(T))$, the function $\bar{p}(t) = -G^*(T,t)p$ is called the co-state function corresponding to the optimal state $\bar{x}(t)$.

A result is proved in [7] (Theorem 3.1) states that, if (3.2), (3.9) hold true and $(\bar{x}, \bar{u})$ is an optimal trajectory-control pair for problem (3.5), then, for every $p \in D^+g(\bar{x}(T))$, the function $\bar{p}(t) = -G^*(T,t)p$ satisfies the Maximum Principle

$$\langle \bar{p}(t), f(t, \bar{x}(t), \bar{u}(t)) \rangle = \max_{u \in U} \langle \bar{p}(t), f(t, \bar{x}(t), u) \rangle \tag{5.1}$$

for a.e. $t \in [t_0, T]$ and the co-state inclusion

$$-\bar{p}(t) \in D_x^+ V(t, \bar{x}(t)), \ \forall t \in [t_0, T]. \tag{5.2}$$

Finally, $\bar{p}(\cdot)$ is the mild solution of the backward problem

$$\begin{cases} -\bar{p}'(t) = A^*\bar{p}(t) + \left(\dfrac{\partial f}{\partial x}(t, \bar{x}(t), \bar{u}(t))\right)^* \bar{p}(t) & t \in [t_0, T] \\ -\bar{p}(T) = p \end{cases} \tag{5.3}$$

In (5.2), $D_x^+ V(t,x)$ denotes the superdifferential of the function $V(t, \cdot)$ at x, which is non empty due to the semiconcavity of V with respect to x (see section 2, Proposition 2.1, i)).

STEP 2.

Let (3.2), (3.9) and (3.10) hold true, and let $V(t_0, x_0)$ be Gateaux differentiable at x_0. We recall first that, under our hypotheses (see Corollary 4.4) Gateaux Differentiability of $V(t, \cdot)$ is equivalent to Fréchet Differentiability.

We now recall that (see [7], Theorem 5.1), if $\bar{x}$ is an optimal trajectory for problem (3.5) and $\bar{p}$ is the corresponding co-state given by (5.3) then, for all $t \in [t_0, T]$, we have:

$$\bar{p}(t) + D_x^+ V(t, \bar{x}(t)) \subset (G(t, t_0)X)^\perp \tag{5.4}$$

In particular, if $G(t, t_0)$ has dense range, by (5.4) it follows that:

$$D_x^+ V(t, \bar{x}(t)) = \{-\bar{p}(t)\} \tag{5.5}$$

STEP 3.

We apply the result contained in the Appendix.

Setting:

$$B(t) = \frac{\partial f}{\partial t}(t, \overline{x}(t), \overline{u}(t))$$

we have that G satisfies the following operator equation for every $t \geq s \geq 0$:

$$(5.6) \qquad \begin{cases} \dfrac{\partial G}{\partial t}(t,s) = [A + B(t)]\, G(t,s) & \text{on } [s,T]; \\ G(s,s) = I \end{cases}$$

Now, by a well known density criterion in Banach spaces we obtain:

$$\textit{Range}\ (G(t,s)) \text{ is dense in } X \quad \Longleftrightarrow \quad G^*(t,s) \text{ is one-to-one}$$

The adjoint operator $G^*(t,s)$ satisfies the following operator equation in mild form:

$$(5.7) \qquad \begin{cases} \dfrac{\partial G^*}{\partial s}(t,s) = -\left[A + \dfrac{\partial f}{\partial t}(t, \overline{x}(t), \overline{u}(t))\right]^* G^*(t,s) & \text{on } [0,t]; \\ G(t,t) = I \end{cases}$$

and setting for $x \in X$:

$$(5.8) \qquad v(s) = G^*(t, t-s)x$$

we find that the function v satisfies the following initial value problem:

$$(5.9) \qquad \begin{cases} v'(s) = [A^* + B^*(t-s)]v(s) & \text{on } [0,t]; \\ v(0) = x & x \in X \end{cases}$$

Fix now $t > r > 0$. Then $G^*(t,r)$ is one-to-one if and only if, for a given $x \in H$:

$$G^*(t,r)x = 0 \qquad \Longleftrightarrow \qquad x = 0$$

which, in terms of v, reads as follows:

$$v(t-r) = 0 \Longleftrightarrow v \equiv 0 \quad \text{on } [0, t-r]$$

Setting $T := t - r$, we have that v satisfies the problem:

$$(5.10) \qquad \begin{cases} v'(s) = (A^* + B^*(T-s))v(s) & \text{on } [0,T] \\ v(T) = 0 & T > 0. \end{cases}$$

Now one can easily see that theorem A.1 applies to (5.10) with $B^*(T-s) = K(s)$ and exchanging A with A^*. Hence $v \equiv 0$ on $[0,T]$.

In conclusion, the operator $G^*(t,s)$ is one-to-one, which implies:

$$\{-p(t)\} = D_x^+ V(t, \bar{x}(t))$$

along the optimal trajectory $\bar{x}$.

At this point, the claim of Theorem 3.4 easily follows from the properties of the superdifferential of the function $V(t,\cdot)$. More precisely by Proposition 2.1 it follows that $V(t,\cdot)$ is Gateaux Differentiable at $\bar{x}(t)$ and by Corollary 4.4 that $V(t,\cdot)$ is Fréchet Differentiable at $\bar{x}(t)$.

Q.E.D.

6. An example.

Let Ω be an open bounded subset of $\mathbf{R}^n$ with smooth boundary (class C^2), and let $X := L^2(\Omega)$. Consider the following parabolic control system in Ω:

$$(6.1) \qquad \begin{cases} \dfrac{\partial x}{\partial t}(t,\xi) = \Delta_\xi x(t,\xi) + u(t)x(t,\xi) & t \in [t_0, T] \\ x(t,\xi) = 0 & \xi \in \partial\Omega \\ x(t_0,\xi) = x_0(\xi) \end{cases}$$

where $u : [t_0, T] \to [-M, +M]$ for a given positive constant M. Consider the problem of minimizing the terminal cost:

$$g(x(T)) = \int_\Omega x(t,\xi)d\xi$$

over all measurable $u : [t_0, T] \to [-M, M]$.

Then the control set U is given by the interval $[-M, +M]$. Due to the nonlinearity of (6.1) this problem does not fit into the theory of linear quadratic control (see e.g. [2], [17]). Anyway the existence of optimal controls is standard (see e.g. [7]).

Moreover it easy to check that hypotheses (3.2), (3.9) and (3.10) hold true in this case. This fact allows us to state that, if the starting point x_0 is a differentiability point for $V(t,\cdot)$, then the value function:

$$V(t_0, x_0) \stackrel{def}{=} \inf\{\, g(x(T; t_0, x_0, u)) \mid u : [t_0, T] \to U \quad \text{is measurable}\}$$

is Fréchet differentiable with respect to x along the optimal trajectories.

In this example the Hamiltonian of the problem is given by:

$$\begin{aligned} H(x,p) &= \sup_{u\in U}\langle p, f(x,u)\rangle = \\ &= \sup_{u\in U} u \int_\Omega p(\xi)x(\xi)d\xi = \\ &= M\left|\int_\Omega p(\xi)x(\xi)d\xi\right| = \\ &= M\,|\langle p, x\rangle| \end{aligned}$$

where the maximum is attained at:

$$u^* \stackrel{def}{=} M\frac{\langle p, x\rangle}{|\langle p, x\rangle|}$$

Since for $p \neq 0$ we have:

$$\frac{\partial H}{\partial p}(x,p) = M\mathrm{sgn}(\langle p, x\rangle)x$$

by Theorem 3.4 the closed loop equation:

$$\begin{cases} \frac{\partial x}{\partial t}(t,\xi) = \Delta_\xi x(t,\xi) + M\langle \frac{\partial V}{\partial x}(t, x(t,\xi)), x(t,\xi)\rangle x(t,\xi) & t \in [t_0, T] \\ x(t,\xi) = 0 & \xi \in \partial\Omega \\ x(t_0,\xi) = x_0(\xi) \end{cases}$$

possesses a classical solution, provided that the derivative $\frac{\partial V}{\partial x}(t_0, x_0)$ exists and is different from 0.

Appendix: A Backward Uniqueness Result.

Let X be a real Hilbert space and $A : D(A) \subset X \to X$ a linear operator satisfying (3.2) i) and:

(A.0): $D(A) \subset D(A^*)$and:

$$\left|\frac{A - A^*}{2} x\right| \leq C\left|(-A)^{\frac{1}{2}} x\right| \quad \forall x \in D(A)$$

for some constant $C > 0$.

Let

$$K(t) \in \mathcal{L}(X),\ \forall t \in [0,T]$$

$$|K(t)|_{\mathcal{L}(X)} < K_\infty,\ \forall t \subset [0,T]$$

for some constant $K_\infty > 0$.

Consider then the following terminal value problem:

$$(A.1) \qquad \begin{cases} u'(t) = Au(t) + K(t)u(t) & \text{on } [0,T]; \\ u(T) = 0 & T > 0; \end{cases}$$

This is an ill-posed problem if A does not generate a group. It is well-known that problems of this kind may fail to have a solution, in general. Our goal, here, is to show a uniqueness result. This kind of problem has been treated by several authors. They studied the case of $K = 0$ and e^{tA} analytic (see [16] [1],[13]) and the case where A is a special PDE operator (see [14], [19]). By adapting the technique of [19] we can prove the following result:

THEOREM A.1 *Let $u \in L^2(0,T; D(A)) \cap H^1(0,T; X)$ be a solution of the terminal value problem (A.1).*

Then $u(t) = 0$ for every $t \in [0,T]$.

PROOF.

Set:

$$\frac{A+A^*}{2} =: A_+ \,; \quad \frac{A-A^*}{2} =: A_-$$

Then we have, by hypotheses (3.2) and (A.0):

$$A = A_+ + A_- \,; \quad \text{where } A_+^* = A_+; \; A_-^* = -A_-.$$

$$|A_- x| \leq C|(-A)^{\frac{1}{2}} x| \quad \forall x \in D(A)$$

To make the proof clearer, we divide it into several steps, proving two useful lemmas.

LEMMA A.2. *Let $w \in L^2(0,T;D(A)) \cap H^1(0,T;X)$ be a solution of the boundary value problem:*

$$(A.2) \qquad \begin{cases} w'(t) = Aw(t) + K(t)w(t) + ktw(t) + g(t) & \text{on } [0,T]; \\ w(T) = w(T_0) = 0 & T > T_0 > 0; \end{cases}$$

where $g \in L^2(0,T;X)$, $k > 0$, and T_0 is such that

$$(A.3) \qquad \frac{C}{2}(T - T_0) \leq \frac{1}{4}$$

Then there exists a constant C_1 such that for $k > C_1$ the following inequality holds:

$$(A.4) \qquad \int_{T_0}^{T} |w(t)|^2 dt \leq \frac{4}{k - C_1} \int_{T_0}^{T} |g(t)|^2 dt$$

PROOF OF LEMMA A.2. Taking scalar products in equation (A.2) we have:

$$\begin{aligned} |w'(t)|^2 = \langle Aw(t), w'(t)\rangle + \langle K(t)w(t), w'(t)\rangle + kt\langle w(t), w'(t)\rangle + \langle g(t), w'\rangle \leq \\ \leq \frac{1}{2}\frac{d}{dt}\left[\langle A_+ w(t), w(t)\rangle + kt|w(t)|^2\right] - \frac{k}{2}|w(t)|^2 \\ + \langle A_- w(t), w'(t)\rangle + \langle K(t)w(t), w'(t)\rangle + \langle g(t), w'\rangle \end{aligned}$$

Now using hypothesis (A.0), we obtain:

$$\begin{aligned} |w'(t)|^2 + \frac{k}{2}|w(t)|^2 \leq \\ \leq \frac{1}{2}\frac{d}{dt}\left[\langle A_+ w(t), w(t)\rangle + kt|w(t)|^2\right] + \frac{1}{2}|w'(t)|^2 + \\ + \frac{C}{2}|(-A)^{\frac{1}{2}} w(t)|^2 + \frac{1}{2}|w'(t)|^2 + \frac{1}{2}\left[K_\infty |w(t)|^2 + |g|^2\right] \end{aligned}$$

and, integrating on $[T_0, T]$:

$$\int_{T_0}^{T} |w'(t)|^2 + \frac{k}{2}|w(t)|^2 dt \leq$$

(A.5)
$$\leq \int_{T_0}^{T} |w'(t)|^2 dt + \frac{C}{2}\int_{T_0}^{T} |(-A)^{\frac{1}{2}} w(t)|^2 dt +$$
$$+ \frac{1}{2}\left[K_\infty \int_{T_0}^{T} |w(t)|^2 dt + \int_{T_0}^{T} |g|^2 dt\right]$$

Moreover, still taking scalar products in equation (A.2) and integrating, it follows:

(A.6)
$$-\int_{T_0}^{T} \langle Aw(t), w(t)\rangle dt = k\int_{T_0}^{T} t|w(t)|^2 dt + \int_{T_0}^{T} \langle K(t)w(t), w(t)\rangle dt + \int_{T_0}^{T} \langle g(t), w(t)\rangle dt \leq$$
$$\leq \left[k(T - T_0) + K_\infty + \frac{1}{2}\right]\int_{T_0}^{T} |w(t)|^2 dt + \frac{1}{2}\int_{T_0}^{T} |g(t)|^2 dt$$

By combining (A.5) and (A.6) it follows:

$$\left[\frac{k}{2} - \frac{C}{2}k(T - T_0) - \frac{C}{4} - \frac{C}{2}K_\infty - \frac{K_\infty}{2}\right]\int_{T_0}^{T} |w(t)|^2 dt \leq$$
$$\leq \int_{T_0}^{T} |g(t)|^2 dt$$

Now T_0 has been chosen so that:

$$\frac{C}{2}(T - T_0) \leq \frac{1}{4}$$

So, setting:

$$C_1 = C + 2CK_\infty + 2$$

$$k > C_1$$

the claim follows.

Q.E.D.

Lemma A.3. *Let $v \in L^2(0,T;D(A)) \cap H^1(0,T;X)$ be a solution of the boundary value problem:*

$$(A.7) \qquad \begin{cases} v'(t) = Av(t) + K(t)v(t) + f(t) & \text{on } [T_0, T]; \\ v(T) = v(T_0) = 0 & T > T_0 > 0; \end{cases}$$

where $f \in L^2(0,T;X)$, and T_0 verify (A.3). Then there exists a constant C_1 such that for every $k > C_1$ the following inequality holds:

$$(A.8) \qquad \int_{T_0}^{T} e^{kt^2} |v(t)|^2 dt \le \frac{4}{k - C_1} \int_{T_0}^{T} e^{kt^2} |f(t)|^2 dt$$

Proof of Lemma A.3. Setting:

$$w(t) := e^{\frac{1}{2}kt^2} v(t);$$

it easy to see that the function w defined in this way satisfies the hypotheses of Lemma 2.2 with:

$$g(t) := e^{\frac{1}{2}kt^2} f(t)$$

Then, by lemma 2.2 we obtain that, for every $k > C_1$,

$$\int_{T_0}^{T} |w(t)|^2 dt \le \frac{4}{k - C_1} \int_{T_0}^{T} |g(t)|^2 dt$$

and the claim follows by substituting the values of w and g in the above inequality.

Q.E.D.

We now return to the proof of Theorem A.1. Let $T_0 > 0$ as in lemma A.2 and $\theta \in C^1(\mathbf{R})$be such that:

$$\theta(t) = \begin{cases} 1 & \frac{T+T_0}{2} \le t \le T \\ 0 & 0 \le t \le T_0 \end{cases}$$

$$|\theta'(t)| \le \frac{4}{T - T_0}$$

Set:

$$v(t) = \theta(t)u(t)$$

Then:

$$\begin{cases} v \in L^2(0,T;D(A)) \cap H^1(0,T;X) \\ v'(t) = [A + K(t)]v(t) + \theta'(t)u(t) \\ v(T_0) = v(T) = 0 \end{cases}$$

Therefore, by Lemma A.3., for every $k > C_1$:

$$\int_{T_0}^{T} e^{kt^2} |v(t)|^2 dt \leq \frac{4}{k - C_1} \int_{T_0}^{T} e^{kt^2} |\theta'(t)|^2 |u(t)|^2 dt$$

Now remark that:

$$(A.9) \quad \begin{aligned} \int_{T_0}^{T} e^{kt^2} |v(t)|^2 dt &\geq \int_{\frac{T+T_0}{2}}^{T} e^{kt^2} |u(t)|^2 dt \geq \\ &\geq e^{k(\frac{T+T_0}{2})^2} \int_{\frac{T+T_0}{2}}^{T} |u(t)|^2 dt \end{aligned}$$

On the other hand:

$$(A.10) \quad \begin{aligned} \frac{4}{k - C_1} \int_{T_0}^{T} e^{kt^2} |\theta'(t)|^2 |u(t)|^2 dt &= \frac{4}{k - C_1} \int_{T_0}^{\frac{T+T_0}{2}} e^{kt^2} |\theta'(t)|^2 |u(t)|^2 dt \leq \\ &\leq \frac{4}{k - C_1} \left(\frac{4}{T - T_0}\right)^2 e^{k(\frac{T+T_0}{2})^2} \int_{T_0}^{\frac{T+T_0}{2}} |u(t)|^2 dt \end{aligned}$$

Finally (A.9) and (A.10) give:

$$\int_{\frac{T+T_0}{2}}^{T} |u(t)|^2 dt \leq \frac{4}{k - C_1} \left(\frac{4}{T - T_0}\right)^2 \int_{T_0}^{\frac{T+T_0}{2}} |u(t)|^2 dt$$

for every $k > C_1$. Hence $u(t) = 0, \forall t \in [\frac{T+T_0}{2}, T]$. By iterating the reasoning we easily obtain that $u(t) = 0$ on $]0, T]$ and, by the continuity of u, that $u = 0$ on $[0, T]$.

Q.E.D.

REMARK A.4.

1) By a simple change of variables it is possible to prove the above theorem in the case $A \leq \omega I$ for some $\omega > 0$ (see remark 3.1).
2) If $K = 0$ the theorem is easily proved by using the analyticity of A or, alternatively, a logarytmic convexity argument (see [1], [15], [12]).
3) If $K(t)$ commutes with A for every $t \in [0, T]$, then the uniqueness result easily follows by part 2) of this remark and the fact that, in such a case, we have:

$$u(t) = e^{(t-s)A} U_K(t,s)x \qquad \text{for some } x \in X$$

where $U_K(t,s)$ is the evolution operator associated to the family of continuous operators $K(t)$, $t \in [0,T]$.

4) The solution of the equation (A.1), when it exists, does not satisfy, in general, the regularity hypotheses we assumed in Theorem A.1. However, using the analyticity of A we can prove that (see [17], [9]):

$$u \in L^2(\varepsilon, T; D(A)) \cap H^1(\varepsilon, T; X) \qquad \forall \varepsilon > 0$$

This allows us to apply the argument of theorem A.1. without any regularity hypothesis on the solution u as we have done in the proof of the main theorem (see section 5, step 3).

References

[1] S.Agmon, L.Nirenberg — *Properties of solutions of ordinary differential equations in Banach space*, Comm. Pure Appl. Math. 16 (1963), pp.121-139.

[2] V.Barbu and G.Da Prato — *Hamilton-Jacobi equations in Hilbert spaces*, Pitman,London (1983).

[3] V.Barbu and G.Da Prato — *Hamilton-Jacobi equations in Hilbert spaces; variational and semigroup approach,* Ann. Mat. Pura Appl., (IV), Vol. CXLII, (1985), pp 303-349.

[4] P.Cannarsa — *Regularity properties of solutions to Hamilton-Jacobi equations in infinite dimensions and nonlinear optimal control* Differential and Integral Equation, Vol.2, N.4, pp.479-493, 1989.

[5] P.Cannarsa, H. Frankowska — *Quelques caracterizations des trajectoires optimales en théorie de contrôle* C R Acad. Sci Paris 310, Série I, pp. 179-182.

[6] P.Cannarsa, H. Frankowska — *Some characterization of optimal trajectories in control theory* to appear in SIAM J. on Contr. and Opt.

[7] P.Cannarsa, H. Frankowska — *Value functions and optimality conditions for semilinear control problems* submitted.

[8] P.Cannarsa, H. Frankowska — *On the value function of semilinear optimal control problems of parabolyc type*, in preparation.

[9] P.Cannarsa, H.M.Soner — *On the singularities of the viscosity solutions to Hamilton-Jacobi-Bellmann equations*, Indiana Univ. Math. J., 36 (1987), pp.501-524

[10] P.Cannarsa, V.Vespri — *On maximal L^p regularity for the abstract Cauchy problem*, Boll. Un. Mat. Ital. 5-B, pp. 165-175.

[11] F.Clarke — *Optimization and nonsmooth analysis*, Wiley, New York, 1983.

[12] M.G.CRANDALL and P.L.LIONS — *Hamilton-Jacobi equations in infinite dimensions.*

Part I : Uniqueness of viscosity solutions, J. Func. Anal. 62, 1985, pp 379-396.

Part II : Existence of viscosity solutions, J. Func. Anal. 65, 1986, pp 368-405.

Part III, J. Func. Anal. 68, 1986, pp 214-247.

Part IV : Hamiltonians with unbounded linear terms, J. Func. Anal. 90, 1990, pp 237-283.

[13] H.O.FATTORINI — *The Cauchy problem*, Encyclopedia of Mathematics and its applications, Vol.18, Addison-Wesley, 1983.

[14] D. FRIEDMAN — *Partial Differential Equations of Parabolic type*, Prentice-Hall, NJ, 1964.

[15] D.HENRY — *Geometric theory of semilinear parabolic equations,* Lecture Notes in Math. 840, Springer-Verlag, (1981).

[16] S.G.KREIN, O.I.PROZOROVSKAYA — *Analytic semigroups and incorrect problems for evolutionary equations* Doklady Akad. Nauk. SSSR, N. S. 133 (1960), pp. 277-280. English translation, Soviet Mathematics Amer. Math. soc. 1 (1960) pp. 841-844.

[17] J.L.LIONS — *Optimal control of systems governed by partial differential equations,* Springer, Wiesbaden, 1972.

[18] J.L.LIONS, E.MAGENES — *Problèmes aux limites non homogènes et applications II* Paris, Dunod.

[19] J.L.LIONS, B.MALGRANGE — *Sur l'unicité rétrograde dans les problèmes mixtes paraboliques* Math. Scand. 8 (1960), pp.277-286.

[20] A.PAZY — *Semigroups of linear operators and applications to partial differential equations,* Springer-Verlag, New York-Heidelberg- Berlin, 1983

[21] R.R.PHELPS — *Convex functions, monotone operators and differentiability*, Springer-Verlag, Lecture Notes in Mathematics n.1364, Berlin, 1989.

[22] D. PREISS — *Differentiability of Lipschitz functions on Banach spaces,* J. Funct. Anal. 91, (1990), pp. 312-345.

SOME BOUNDARY CONTROL PROBLEMS AND COMPUTATION FOR THE LINEAR ELASTOSTATIC KIRCHHOFF PLATE ON AN EXTERIOR DOMAIN

Goong CHEN* and Jianxin ZHOU*

ABSTRACT

Boundary control of a distributed parameter system on an exterior unbounded domain arises naturally in applications. The questions of existence and uniqueness of solutions for such exterior boundary value problems are very different from those posed on the bounded domains. Also, computationally the unboundedness of the exterior domain causes numerical difficulty. The boundary integral equation and boundary element methods are well suited to tackle such problems. In this paper, we first formulate a combined simple-double layer solution to the elastostatic Kirchhoff plate equation on an exterior domain, and then use boundary elements to compute numerical solutions of linear-quadratic boundary control problems. Computer graphics results are illustrated and discussed.

---

* Department of Mathematics, Texas A&M University, College Station, TX 77843.
Supported in part by the United States Air Force Office of Scientific Research Grant 87-0334

§1. Introduction

The analysis, design and computation of boundary controls are important topics in the research of distributed parameter systems (DPS). In recent years, dramatic advances have been made in this field. So far most of the articles in DPS literature seem to deal with control systems governed by partial differential equations on a *bounded domain*. Nevertheless, in practice there are also many engineering control systems posed on unbounded domains. One can think of the boundary control problem of an electromagnetic or acoustic radiation system in an infinite space with a scatterer, such as the *stealth problem*. In mining engineering, the domain of interest is so large as compared with the size of accessible boundary that such a problem is often formulated as an exterior boundary value problem in solid mechanics. In seismology, the dimension of the tectonic plate or shell is so large that it is also appropriate to formulate mathematical models as exterior boundary value problems. For example, we have learned that seismologists and geologists have proposed setting off small nuclear explosions along the St. Andreas Fault to release the crack and fracture strain energy as a possible solution to reducing potential earthquake damages on major cities in the seismic tectonic region. Indeed, such a problem constitutes a boundary control system in structural mechanics and elasticity over an unbounded domain.

Control and optimization problems on an exterior domain are more difficult to solve than their bounded domain counterparts. First, it is well known that solutions to exterior boundary value problems are generally nonunique unless some additional radiation conditions or growth conditions are imposed. Second, for the purpose of numerical simulation extra care must be exercised when the traditional finite elements or finite differences are used in order to take into account the unboundedness of the domain. Third, for some PDEs, in paticular the plate equation herein, the admissible solutions can grow unbounded pointwise, causing fast loss of accuracy of the numerical solution.

In this paper, we present results of an ongoing study of a boundary control problem governed by the linear elastostatic Kirchhoff plate equation

$$\Delta^2 w(x) = 0, \qquad x \in \Omega \subset \mathbf{R}^2 \tag{1.1}$$

on an exterior domain Ω (whose complement $\mathbf{R}^2\backslash\Omega$ is bounded), where in (1.1), $w(x)$ denotes the vertical deflection at x, and Δ^2 is the biharmonic operator. Let $n(x) = (n_1(x), n_2(x))$ be the unit normal vector at $x \in \partial\Omega$ pointing *into* Ω. Define two boundary operators

$$
\begin{aligned}
(1.2)\quad B_1 w(x) &= \frac{\partial}{\partial n}\Delta w(x) \\
&+ (1-\nu)\frac{\partial}{\partial s}\left[(n_1^2 - n_2^2)\frac{\partial^2}{\partial x_1 \partial x_2}w(x) - n_1 n_2\left(\frac{\partial^2}{\partial x_1^2}w(x) - \frac{\partial^2}{\partial x_2^2}w(x)\right)\right] \\
(1.3)\quad B_2 w(x) &= \nu\Delta w(x) \\
&+ (1-\nu)\left[n_1^2\frac{\partial^2}{\partial x_1^2}w(x) + n_2^2\frac{\partial^2}{\partial x_2^2}w(x) + 2n_1 n_2\frac{\partial^2}{\partial x_1 \partial x_2}w(x)\right]
\end{aligned}
$$

for $x \in \partial\Omega$, where $B_1 w(x)$ corresponds to the transverse force at x, and $B_2 w(x)$ the bending moment at x, $s = (-n_2, n_1)$ is the counterclockwise unit tangent at x. The constant ν: $0 < \nu < 1/2$ is the Poisson ratio. Let u_1, u_2 in

$$
(1.4)\qquad \left\{\begin{array}{l} B_1 w(x) = u_1(x) \\ B_2 w(x) = u_2(x) \end{array}\right\} \qquad x \in \partial\Omega
$$

be the control forces applied at the boundary. Associated with the system (1.1) + (1.4) there is a cost index $J(w,u)$ depending on the state w and the control $u = (u_1, u_2)$ to be minimized:

$$
(1.5)\qquad \inf_{u_1, u_2} J(w,u).
$$

Thus (1.1) + (1.4) + (1.5) constitutes the control/optimization problem to be studied in this paper.

We first note that there seems to be a lack of literature on the existence and uniqueness of solutions for an exterior boundary value problem of the type (1.1) + (1.4). In §II, we first introduce a ***combined simple-double layers potential approach*** to solve the exterior problem (1.1) + (1.4) for given boundary data u_1 and u_2, resulting in the existence and uniqueness of the solution w. This multilayer potential solution provides an *explicit representation* of the state w in terms of the data (u_1, u_2), subject to certain equality constraints.

In §III, we use a *mixed primal-dual method* to handle the governing integral equation and the equality constraints. The Lagrange multipliers are used only for the scalar equality constraints in the duality part, whereas the other state and control variables are substituted directly into the cost functional in the primal part. Discretization by splines for the simple and double layer densities is done by point collection on $\partial\Omega$, and optimization is then performed. This is a *boundary element* approach, resulting in the approximate optimal control and state numerically in a very efficient way while avoiding the difficulty of calculating quadrature on an unbounded domain.

Numerical solutions are illustrated by computer graphics. Some pertinent questions are also discussed. This research is being continued and a more thorough study, containing analysis of convergence and error estimates of the numerical scheme, will appear elsewhere when the investigation is complete.

§II. Solutions of Plate Exterior Boundary Value Problem by Multilayer Potentials

Throughout this paper, we use $H^r, r \in \mathbf{R}$, as the standard notation for the Sobolev space of order r. The strain energy of a thin plate subject to pure bending without lateral loading is given by ([8])

$$\mathcal{E}(w) = \int_\Omega [|\Delta w(x)|^2 + 2(1-\nu)(w^2_{x_1x_2}(x) - w_{x_1x_1}(x)w_{x_2x_2}(x)]dx. \tag{2.1}$$

Associated with (2.1) we also define a bilinear form

$$a(w,v) = \int_\Omega [\Delta w \cdot \Delta v + (1-\nu)(2w_{x_1x_2}v_{x_1x_2} - w_{x_1x_1}v_{x_2x_2} - w_{x_2x_2}v_{x_1x_1})]dx, \tag{2.2}$$

for w, v such that

$$\frac{\partial^2}{\partial x_i \partial x_j} w, \frac{\partial^2}{\partial x_i \partial x_j} v \in L^2(\Omega), \text{ for all } i,j\text{: } i,j = 1 \text{ or } 2. \tag{2.3}$$

When Ω is a *bounded* domain in $\mathbf{R}^2$, calculus of variation [8, pp. 88-91] gives

(2.4)
$$\begin{aligned}\delta\mathcal{E}(w) &= 2a(w,\delta w)\\ &= \int_{\Omega}(\Delta^2 w)(\delta w)dx - \int_{\partial\Omega}\left[B_1 w\cdot(\delta w) - B_2 w\cdot\left(\frac{\partial}{\partial n}\delta w\right)\right]d\sigma\\ &=0.\end{aligned}$$

Therefore we have four types of variational boundary conditions

(2.5)
$$\begin{cases}\Delta^2 w(x) = 0 & \text{on } \Omega\\ \left.\begin{array}{l} w(x) = g_1(x)\\ \frac{\partial w(x)}{\partial n} = g_2(x)\end{array}\right\} & \text{on } \partial\Omega\end{cases}$$

(2.6)
$$\begin{cases}\Delta^2 w(x) = 0 & \text{on } \Omega\\ \left.\begin{array}{l} w(x) = g_1(x)\\ B_2 w(x) = g_2(x)\end{array}\right\} & \text{on } \partial\Omega\end{cases}$$

(2.7)
$$\begin{cases}\Delta^2 w(x) = 0 & \text{on } \Omega\\ \left.\begin{array}{l} \frac{\partial w(x)}{\partial n} = g_1(x)\\ B_1 w(x) = g_2(x)\end{array}\right\} & \text{on } \partial\Omega\end{cases}$$

(2.8)
$$\begin{cases}\Delta^2 w(x) = 0 & \text{on } \Omega\\ \left.\begin{array}{l} B_1 w(x) = g_1(x)\\ B_2 w(x) = g_2(x)\end{array}\right\} & \text{on } \partial\Omega.\end{cases}$$

When the given boundary data g_1 and g_2 are sufficiently regular, it can be shown (see [6], e.g.) that (2.5) and (2.6) always have a unique (strong or weak) solution, whereas (2.7) and (2.8) have (nonunique) solutions if and only if certain orthogonality (or, compatibility) conditions are satisfied, which are

(2.9) $g_2 \perp 1$ in $L^2(\partial\Omega)$ for (2.7),

(2.10) $\begin{bmatrix} g_1\\ g_2\end{bmatrix} \perp \text{span}\left\{\begin{bmatrix}1\\0\end{bmatrix}, \begin{bmatrix}x_1\\ -n_1(x)\end{bmatrix}, \begin{bmatrix}x_2\\ -n_2(x)\end{bmatrix}\right\}$ in $L^2(\partial\Omega)\times L^2(\partial\Omega)$, for (2.8),

respectively. The solutions of (2.7) are unique up to an arbitrrary constant, i.e., they have *one degree of freedom*, whereas the solutions of (2.8) are unique up to an arbitrary first order polynomial $a_0 + a_1x_1 + a_2x_2$, i.e., they have *three degrees of freedom.*

Since our main interest here is exterior BVPs (boundary value problems), let us examine (2.5)-(2.8) when Ω is an *exterior* domain.

Example 1.

Let Ω be the exterior of the closed unit disk in $\mathbf{R}^2$. Then

$$w(x) = |x|^2 \ln |x| - \ln |x|, \qquad x \in \Omega, \tag{2.11}$$

satisfies

$$\begin{cases} \Delta^2 w(x) = 0 \\ \left. \begin{array}{l} w(x) = 0 \\ \frac{\partial w(x)}{\partial n} = \frac{\partial w(x)}{\partial r} = 2r \ln r + r - \frac{1}{r} = 0 \end{array} \right\} \quad \text{on} \quad \partial\Omega. \end{cases} \tag{2.12}$$

Also, $w(x) \equiv 0$ is a trivial solution satisfying the boundary conditions. Therefore the *exterior* boundary value problem (2.5) *does not have uniqueness* of solutions. □

Example 2. Let Ω be the exterior of the closed disk with radius R. Let

$$w(x) = |x|^2 \ln |x| + a_0 + a_1 x_1 + a_2 x_2, \quad \text{for arbitrary} \;\; a_0, a_1, a_2 \in \mathbf{R}. \tag{2.13}$$

Then w satisfies

$$\begin{cases} \Delta^2 w(x) = 0 \\ \left. \begin{array}{l} B_1 w(x) = \frac{1}{R} \equiv g_1(x) \\ B_2 w(x) = 2(1+\nu) \ln R + 3 + \nu \equiv g_2(x) \end{array} \right\} \quad \text{on} \quad \partial\Omega. \end{cases} \tag{2.14}$$

Then obviously (g_1, g_2) violates the compatibility condition (2.10) for the bounded domain case. Yet the exterior BVP (2.8) has (nonunique) solutions. □

To fix the uniqueness for a problem like Example 1, we need to impose some growth rate restriction on the solution. For a problem like Example 2, we not only need to impose the growth rate condition, but also need some additional normalization condition(s).

In an earlier paper by Hsiao and MacCamy [4], the growth rate condition for solutions of the biharmonic equation $\Delta^2 w(x) = 0$ on an exterior domain was set to be

$$w(x) = (A_1x_1 + A_2x_2)\ln|x| + \mathcal{O}(|x|), \quad |x| \text{ large}, \; A_1, A_2 \in \mathbf{R} \tag{2.15}$$

for a linear elasticity (i.e., plate *stretching*) problem in $\mathbf{R}^2$. Such a growth rate would rule out the pathological solutions (2.11) and (2.13), but is still unsuitable for our plate bending problem because the solution w in (2.15) has *unbounded strain energy* when A_1 or A_2 is nonzero. In this paper, we say that a biharmonic function w satisfies the *linear-logarithmic growth condition* if

(2.16)
$$\begin{cases} \text{(a) } w(x) = a_1x_1 + a_2x_2 + A\ln|x| + a_0 + o(1), \text{ for some } a_0, a_1, a_2 \text{ and } A \in \mathbf{R}, |x| \text{ large}, \\ \text{(b) } B_1(r)w(x) = o(r^{-(1+\delta)}), B_2(r)w(x) = o(r^{-(1+\delta)}), \dfrac{\partial w(x)}{\partial r} = \mathcal{O}(1), \\ \qquad \text{for some } \delta > 0 \text{ at } |x| = r, \; r \text{ large}, \end{cases}$$

where

(2.17)
$$\begin{cases} B_1(r) \equiv -\left\{\dfrac{\partial}{\partial r}\Delta - (1-\nu)\dfrac{1}{r}\dfrac{\partial}{\partial\theta}\left[\dfrac{x_1x_2}{r^2}\left(\dfrac{\partial^2}{\partial x_1^2} - \dfrac{\partial^2}{\partial x_2^2}\right) - \dfrac{1}{r^2}(x_1^2 - x_2^2)\dfrac{\partial^2}{\partial x_1\partial x_2}\right]\right\} \\ B_2(r) \equiv -\nu\Delta - \dfrac{(1-\nu)}{r^2}\left[x_1^2\dfrac{\partial^2}{\partial x_1^2} + x_2^2\dfrac{\partial^2}{\partial x_2^2} + 2x_1x_2\dfrac{\partial^2}{\partial x_1\partial x_2}\right], \end{cases}$$

are, respectively, the transverse force and bending moment operators in the radial direction. It is easy to check that the affine part $a_0 + a_1x_1 + a_2x_2$ has zero strain energy and the logarithmic part $A\ln|x|$ of w has *finite strain energy*.

An immediate consequence is that if $w \in H^{s+2}_{\text{loc}}(\Omega)$ for $s \geq 0$ is a biharmonic function on Ω satisfying (2.16), then w has *finite strain energy:* $\mathcal{E}(w) < \infty$.

Let $E(x, \xi)$ be the fundamental solution of the biharmonic equation

(2.18)
$$\Delta_\xi^2 E(x,\xi) = -\delta(x-\xi), \text{ the Dirac delta function,}$$
$$E(x,\xi) = -\frac{1}{8\pi}|x-\xi|^2\ln|x-\xi|, \quad x,\xi \in \mathbf{R}^2.$$

Then we have

Theorem 1. Let Ω be an exterior domain in $\mathbf{R}^2$ with bounded complement and smooth boundary $\partial\Omega$. Consider the exterior BVP

$$(2.19)\qquad \begin{cases} \Delta^2 w(x) = 0 \quad \text{on} \quad \Omega, \\ w(x) = g_1(x) \in H^{r+1}(\partial\Omega), \qquad r \in \mathbf{R}, \\ \dfrac{\partial w(x)}{\partial n} = g_2(x) \in H^r(\partial\Omega). \end{cases}$$

Then there exists a unique solution $w \in H^{r+\frac{3}{2}}_{\text{loc}}(\Omega)$ satisfying the linear-logarithmic growth condition (2.16). The solution can be represented by a combination of simple and double layer potentials and a linear function

$$(2.20)\qquad w(x) = \int\limits_{\partial\Omega} \left[E(x,\xi)\tilde{f}_1(\xi) + \frac{\partial E(x,\xi)}{\partial n_\xi}\tilde{f}_2(\xi) \right] d\sigma_\xi + \tilde{a}_0 + \tilde{a}_1 x_1 + \tilde{a}_2 x_2,$$

where $(\tilde{a}_0, \tilde{a}_1, \tilde{a}_2, \tilde{f}_1, \tilde{f}_2) \in \mathbf{R}^3 \times H^{r-2}(\partial\Omega) \times H^{r-1}(\partial\Omega)$ is the unique solution to the boundary integral equation

$$(2.21)\qquad \mathcal{L}_0 \begin{bmatrix} \tilde{a}_0 \\ \tilde{a}_1 \\ \tilde{a}_2 \\ \tilde{f}_1 \\ \tilde{f}_2 \end{bmatrix} = \begin{bmatrix} 0 \\ 0 \\ 0 \\ g_1 \\ g_2 \end{bmatrix},$$

where

$$(2.22)\qquad \mathcal{L}_0\colon \mathbf{R}^3 \times H^{r-2}(\partial\Omega) \times H^{r-1}(\partial\Omega) \to \mathbf{R}^3 \times H^{r+1}(\partial\Omega) \times H^r(\partial\Omega)$$

is an isomorphism for any $r \in \mathbf{R}$, defined by

(2.23)

$$\mathcal{L}_0 \begin{bmatrix} a_0 \\ a_1 \\ a_2 \\ f_1 \\ f_2 \end{bmatrix}(x) = \begin{bmatrix} \int_{\partial\Omega} f_1(\xi)d\sigma \\ \int_{\partial\Omega} [f_1(\xi)\xi_1 + f_2(\xi)n_1(\xi)]d\sigma \\ \int_{\partial\Omega} [f_1(\xi)\xi_2 + f_2(\xi)n_2(\xi)]d\sigma \\ \int_{\partial\Omega} \left[E(x,\xi)f_1(\xi) + \frac{\partial E(x,\xi)}{\partial n\xi} f_2(\xi)\right] d\sigma_\xi + a_0 + a_1x_1 + a_2x_2 \\ \int_{\partial\Omega} \left[\frac{\partial E(x,\xi)}{\partial n_x} f_1(\xi) + \frac{\partial^2 E(x,\xi)}{\partial n_x \partial n_\xi} f_2(\xi)\right] d\sigma_\xi + a_1n_1(x) + a_2n_2(x) \end{bmatrix}.$$

Remark 1. The integral of the first team on the RHS of (2.20) is a ***simple layer potential***, whereas that of the second term is a ***double layer potential.*** f_1 and f_2 in (2.20) are called, respectively, the ***simple and double layer densities.*** □

Sketch of proof: From the ansatz (2.20), for large $|x|$ we obtain the asymptotic expansion

$$\begin{aligned}
(2.24)\quad w(x) = &-\frac{1}{8\pi}|x|^2 \ln|x| \left\{ \int_{\partial\Omega} \tilde{f}_1(\xi)d\sigma \right\} \\
&+\frac{1}{4\pi}\ln|x| \left\{ x_1 \int_{\partial\Omega} [\xi_1\tilde{f}_1(\xi) + n_1(\xi)\tilde{f}_2(\xi)]d\sigma + x_2 \int_{\partial\Omega} [\xi_2\tilde{f}_1(\xi) + n_2(\xi)\tilde{f}_2(\xi)]d\sigma \right\} \\
&+\frac{1}{8\pi}\left\{ x_1 \int_{\partial\Omega} [\xi_1\tilde{f}_1(\xi) + n_1(\xi)\tilde{f}_2(\xi)]d\sigma + x_2 \int_{\partial\Omega} [\xi_2\tilde{f}_1(\xi) + n_2(\xi)\tilde{f}_2(\xi)]d\sigma \right\} \\
&+\tilde{a}_1x_1 + \tilde{a}_2x_2 - \frac{1}{8\pi}\ln|x| \cdot \left\{ \int_{\partial\Omega} [|\xi|^2\tilde{f}_1(\xi) + 2(\xi\cdot n(\xi))\tilde{f}_2(\xi)]d\sigma \right\} \\
&+\left\{ \tilde{a}_0 - \frac{1}{16\pi} \int_{\partial\Omega} [|\xi|^2\tilde{f}_1(\xi) + 2(\xi\cdot n(\xi))\tilde{f}(\xi)]d\sigma \right\} \\
&-\frac{1}{8\pi}\frac{1}{|x|^2}\left\{ \int_{\partial\Omega} [(x\cdot\xi^2)\tilde{f}_1(\xi) + 2(x\cdot\xi)(x\cdot n(\xi))\tilde{f}_2(\xi)]d\sigma_\xi \right\} + \mathcal{O}(|x|^{-1}).
\end{aligned}$$

The terms in the first three parentheses above vanish because of (2.21). Therefore w satisfies the linear-logarithmic growth condition (2.16). The regularity of the operator $\mathcal{L}_0$ is due to the elliptic pseudodifferential properties of the integral kernel (2.18) [2]. The invertibility of $\mathcal{L}_0$ is a consequence of Theorem 3.2 in [1]. $\square$

Next, we state and prove

Theorem 2. Let Ω be an exterior domain in $\mathbf{R}^2$ with bounded complement and smooth boundary $\partial\Omega$. Consider the exterior BVP

$$(2.25) \qquad \begin{cases} \Delta^2 w(x) = 0 \quad \text{on} \quad \Omega, \\ B_1 w(x) = g_1(x) \in H^r(\partial\Omega), \qquad r \geq 0, \\ B_2 w(x) = g_2(x) \in H^{r+1}(\partial\Omega), \end{cases}$$

where (g_1, g_2) satisfies the orthogonality condition (2.10). Then there exists a weak solution $w \in H_{\mathrm{loc}}^{r+\frac{7}{2}}(\Omega)$ satisfying the linear-logarithmic growth condition (2.16), and w is unique up to a constant. The general solution can be represented by a combination of simple and double layer potentials plus an arbitrary linear function $c_0 + c_1x_1 + c_2x_2$:

(2.26)

$$w(x) = \int_{\partial\Omega} \left[E(x,\xi)\tilde{f}_1(\xi) + \frac{\partial E(x,\xi)}{\partial n_\xi}\tilde{f}_2(\xi) \right] d\sigma_\xi + c_0 + c_1x_1 + c_2x_2, \qquad x \in \Omega, c_0, c_1, c_2 \in \mathbf{R},$$

where $(\tilde{f}_1, \tilde{f}_2) \in H^r(\partial\Omega) \times H^{r+1}(\partial\Omega)$ corresponds to the unique solution in the boundary integral equation

$$(2.27) \qquad \mathcal{L}_1 \begin{bmatrix} \tilde{a}_0 \\ \tilde{a}_1 \\ \tilde{a}_2 \\ \tilde{f}_1 \\ \tilde{f}_2 \end{bmatrix} = \begin{bmatrix} 0 \\ 0 \\ 0 \\ g_1 \\ g_2 \end{bmatrix},$$

where

$$(2.28) \qquad \mathcal{L}_1 \colon \mathbf{R}^3 \times H^r(\partial\Omega) \times H^{r+1}(\partial\Omega) \to \mathbf{R}^3 \times H^r(\partial\Omega) \times H^{r+1}(\partial\Omega)$$

is an isomorphism for any $r \in \mathbf{R}$, defined by

$$(2.29) \qquad \mathcal{L}_1 \begin{bmatrix} a_0 \\ a_1 \\ a_2 \\ f_1 \\ f_2 \end{bmatrix} (x) =$$

$$\begin{bmatrix} \int\limits_{\partial\Omega} f_1(\xi) d\sigma \\ \int\limits_{\partial\Omega} [f_1(\xi)\xi_1 + f_2(\xi)n_1(\xi)] d\sigma \\ \int\limits_{\partial\Omega} [f_1(\xi)\xi_2 + f_2(\xi)n_2(\xi)] d\sigma \\ -\frac{1}{2} f_1(x) + \int\limits_{\partial\Omega} \left[B_{1x} E(x,\xi) \cdot f_1(\xi) + B_{1x} \frac{\partial}{\partial n_\xi} E(x,\xi) \cdot f_2(\xi) \right] \\ d\sigma_\xi + a_0 + a_1 x_1 + a_2 x_2 \\ \frac{1}{2} f_2(x) + \int\limits_{\partial\Omega} \left[B_{2x} E(x,\xi) \cdot f_1(\xi) + B_{2x} \frac{\partial}{\partial n_\xi} E(x,\xi) \cdot f_2(\xi) \right] \\ d\sigma_\xi - a_1 n_1(x) - a_2 n_2(x) \end{bmatrix}.$$

Furthermore, the solution to (2.25) is unique if we either specify c_0 in (2.26), or if we impose three normalization conditions

$$(2.30) \qquad \langle w, \eta_j \rangle_{H^{r+3}(\partial\Omega) \times H^{-(r+3)}(\partial\Omega)} = \alpha_j, \alpha_j \in \mathbf{R}, \quad j = 1,2,3$$

where α_j are given for some $\eta_j \in H^{-(r+3)}(\partial\Omega)$ satisfying

$$(2.31) \qquad \det \begin{vmatrix} \langle 1, \eta_1 \rangle & \langle x_1, \eta_1 \rangle & \langle x_2, \eta_1 \rangle \\ \langle 1, \eta_2 \rangle & \langle x_1, \eta_2 \rangle & \langle x_2, \eta_2 \rangle \\ \langle 1, \eta_3 \rangle & \langle x_1, \eta_3 \rangle & \langle x_2, \eta_3 \rangle \end{vmatrix} \neq 0.$$

Proof: For ν satisfying $0 < \nu < 1/2$, it can be proved [2] that the reduced operator

$$(2.32)\quad \tilde{\mathcal{L}}_1\begin{bmatrix} f_1 \\ f_2 \end{bmatrix}(x) \equiv \begin{bmatrix} -\frac{1}{2}f_1(x) + \int\limits_{\partial\Omega} [B_{1x}E(x,\xi)\cdot f_1(\xi) + B_{1x}\frac{\partial}{\partial n_\xi}E(x,\xi)\cdot f_2(\xi)]d\sigma_\xi \\ \frac{1}{2}f_2(x) + \int\limits_{\partial\Omega} \left[B_{2x}E(x,\xi)\cdot f_1(\xi) + B_{2x}\frac{\partial}{\partial n_\xi}E(x,\xi)\cdot f_2(\xi)\right] d\sigma_\xi \end{bmatrix}$$

is an elliptic pseudodifferential operator mapping *isomorphically* from $H^r(\partial\Omega)\times H^{r+1}(\partial\Omega)$ into itself. Since $\mathcal{L}_1$ is a *finite dimensional augmentation* of $\tilde{\mathcal{L}}_1, \mathcal{L}_1$ is a *Fredholm operator of index 0* [5, Chapter 5]. Therefore if we can show that $\mathcal{L}_1$ is 1-1, then $\mathcal{L}_1$ will also be onto and be an isomorphism. Let $(a_0, a_1, a_2, f_1, f_2) \in Ker\ \mathcal{L}_1$, the kernel of $\mathcal{L}_1$. Then

$$(2.33)\qquad \int\limits_{\partial\Omega} f_1(\xi)d\sigma = 0,$$

$$(2.34)\qquad \int\limits_{\partial\Omega} [f_1(\xi)\xi_1 + f_2(\xi)n_1(\xi)]d\sigma = 0,$$

$$(2.35)\qquad \int\limits_{\partial\Omega} [f_1(\xi)\xi_2 + f_2(\xi)n_2(\xi)]d\sigma = 0,$$

$$(2.36)\qquad -\frac{1}{2}f_1(x) + \int\limits_{\partial\Omega} \left[B_{1x}E(x,\xi)\cdot f_1(\xi) + B_{1x}\frac{\partial}{\partial n_\xi}E(x,\xi)\cdot f_2(\xi)\right] d\sigma_\xi + a_0 + a_1x_1 + a_2x_2 = 0,$$

$$(2.37)\qquad \frac{1}{2}f_2(x) + \int\limits_{\partial\Omega} \left[B_{2x}E(x,\xi)\cdot f_1(\xi) + B_{2x}\frac{\partial}{\partial n_\xi}E(x,\xi)\cdot f_2(\xi)\right] d\sigma_\xi - a_1n_1(x) - a_2n_2(x) = 0.$$

Multiplying (2.36) by $\phi_1(x)$ and (2.37) by $\phi_2(x)$, adding, integrating over $\partial\Omega$ and interchanging the order of integration, we obtain

$$(2.38)\qquad -\int\limits_{\partial\Omega} [f_1(x)\phi_1(x) - f_2(x)\phi_2(x)]d\sigma$$

$$+\int\limits_{\partial\Omega} \left\{ f_1(\xi)\cdot \int\limits_{\partial\Omega} [B_{1x}E(x,\xi)\cdot\phi_1(x) + B_{2x}E(x,\xi)\cdot\phi_2(x)]d\sigma_x \right.$$

$$+f_2(\xi)\cdot\frac{\partial}{\partial n_\xi}\int\limits_{\partial\Omega}[B_{1x}E(x,\xi)\cdot\phi_1(x)+B_{2x}E(x,\xi)\cdot\phi_2(x)]d\sigma_x\Bigg\}d\sigma_\xi$$

$$+\int\limits_{\partial\Omega}[(a_0+a_1x_1+a_2x_2)\phi_1(x)-(a_1n_1(x)+a_2n_2(x))\phi_2(x)]d\sigma=0.$$

Using

$$(2.39)\qquad \begin{bmatrix}\phi_1(x)\\ \phi_2(x)\end{bmatrix}=\begin{bmatrix}1\\0\end{bmatrix},\begin{bmatrix}x_1\\-n_1(x)\end{bmatrix},\begin{bmatrix}x_2\\-n_2(x)\end{bmatrix}$$

successively in (2.38), and noting that ([1, (2.47)]) for $\xi\in\partial\Omega$,

$$\int\limits_{\partial\Omega}[B_{1x}E(x,\xi)\cdot 1+B_{2x}E(x,\xi)\cdot 0]d\sigma_x=-1/2$$

$$\int\limits_{\partial\Omega}[B_{1x}E(x,\xi)\cdot x_j-B_{2x}E(x,\xi)\cdot n_j(x)]d\sigma_x=-\xi_j/2,\ j=1,2.$$

we obtain, by (2.33)-(2.35),

$$(2.40)\qquad \left\langle a_0\begin{bmatrix}1\\0\end{bmatrix}+a_1\begin{bmatrix}x_1\\-n_1(x)\end{bmatrix}+a_2\begin{bmatrix}x_2\\-n_2(x)\end{bmatrix},\begin{bmatrix}\phi_1\\ \phi_2\end{bmatrix}\right\rangle_{L^2(\partial\Omega)\times L^2(\partial\Omega)}=0.$$

Since the three functions for $\left[\begin{smallmatrix}\phi_1\\ \phi_2\end{smallmatrix}\right]$ in (2.39) are linearly independent in $L^2(\partial\Omega)\times L^2(\partial\Omega)$, we see that $a_0=a_1=a_2=0$ in (2.36) and (2.37). Therefore (f_1,f_2) satisfies

$$\tilde{\mathcal{L}}_1\begin{bmatrix}f_1\\ f_2\end{bmatrix}=\begin{bmatrix}0\\0\end{bmatrix}.$$

But we already show that $\tilde{\mathcal{L}}_1$ is an isomorphism. Therefore $(f_1,f_2)=(0,0)$. Hence $(a_0,a_1,a_2,f_1,f_2)=(0,0,0,0,0)$, and $Ker\ \mathcal{L}_1$ is trivial.

Now let $(\tilde{f}_1,\tilde{f}_2)\in H^r(\partial\Omega)\times H^{r+1}(\partial\Omega)$ along with some $(\tilde{a}_0,\tilde{a}_1,\tilde{a}_2)\in\mathbf{R}^3$ satisfy (2.27), we want to show that

$$(2.41)\quad w(x)=\int\limits_{\partial\Omega}\left[E(x,\xi)\tilde{f}(\xi)+\frac{\partial E(x,\xi)}{\partial n_\xi}\tilde{f}_2(\xi)\right]d\sigma_\xi+c_0+c_1x_1+c_2x_2,\qquad x\in\Omega$$

$c_0,c_1,c_2\in\mathbf{R}$ are arbitrary,

is a solution to (2.25) satisfying the linear-logarithmic growth condition (2.16). We take the boundary values $B_1 w$ and $B_2 w$ from (2.41) and obtain ([1, Lemma 2.6])

$$\begin{aligned} B_1 w(x) &= -\frac{1}{2}\tilde{f}_1(x) \\ &+ \int_{\partial\Omega} \left[B_{1x}E(x,\xi)\cdot\tilde{f}_1(\xi) + B_{1x}\frac{\partial E(x,\xi)}{\partial n_\xi}\cdot\tilde{f}_2(\xi)\right] d\sigma_\xi, \quad x \in \partial\Omega, \end{aligned} \tag{2.42}$$

$$\begin{aligned} B_2 w(x) &= \frac{1}{2}\tilde{f}_2(x) \\ &+ \int_{\partial\Omega} \left[B_{2x}E(x,\xi)\cdot\tilde{f}_1(\xi) + B_{2x}\frac{\partial E(x,\xi)}{\partial n_\xi}\cdot\tilde{f}_2(\xi)\right] d\sigma_\xi, \quad x \in \partial\Omega. \end{aligned} \tag{2.43}$$

But from the last two components of equation (2.27) we see that

$$\begin{aligned} -\frac{1}{2}\tilde{f}_1(x) &+ \int_{\partial\Omega} \left[B_{1x}E(x,\xi)\cdot\tilde{f}_1(\xi) + B_{1x}\frac{\partial E(x,\xi)}{\partial n_\xi}\cdot\tilde{f}_2(\xi)\right] d\sigma_\xi \\ &+ \tilde{a}_0 + \tilde{a}_1 x_1 + \tilde{a}_2 x_2 = g_1(x) \end{aligned} \tag{2.44}$$

$$\begin{aligned} \frac{1}{2}\tilde{f}_2(x) &+ \int_{\partial\Omega} \left[B_{2x}E(x,\xi)\cdot\tilde{f}_1(\xi) + B_{2x}\frac{\partial E(x,\xi)}{\partial n_\xi}\cdot\tilde{f}_2(\xi)\right] d\sigma_\xi \\ &- \tilde{a}_1 n_1(x) - \tilde{a}_2 n_2(x) = g_2(x) \end{aligned} \tag{2.45}$$

Therefore w satisfies the boundary conditions $B_1 w = g_1$ and $B_2 w = g_2$ if and only if $\tilde{a}_0 = \tilde{a}_1 = \tilde{a}_2 = 0$ in (2.44) and (2.45). To show this, we again multiply (2.44) by $\phi_1(x)$ and (2.45) by $\phi_2(x)$, using (ϕ_1, ϕ_2) as in (2.39). Repeating similar procedures as before, noting that (g_1, g_2) satisfies (2.10), we again arise at (2.40), with (a_0, a_1, a_2) being replaced by $(\tilde{a}_0, \tilde{a}_1, \tilde{a}_2)$ therein. Therefore $(\tilde{a}_0, \tilde{a}_1, \tilde{a}_2) = (0,0,0)$, and the boundary conditions in (2.25) are satisfied.

To show that $w(x)$ defined by (2.41) satisfies the linear-logarithmic growth condition (2.16), we use the asymptotic expansion (2.24), with $\tilde{a}_1 = c_1, \tilde{a}_2 = c_2$ and $\tilde{a}_0 = c_0$ therein, and immediately obtain (2.16) because $(\tilde{f}_1, \tilde{f}_2)$ satisfies (2.33)-(2.35) by (2.27).

Now we show that w satisfying (2.25) and (2.16) can differ at most by a function $c_0 + c_1 x_1 + c_2 x_2$. Let w_1 and w_2 both satisfy (2.25) and (2.16). Define $w = w_1 - w_2$. Then w satisfies

$$\begin{cases} \Delta^2 w(x) = 0 \\ \left.\begin{array}{l} B_1 w = 0 \\ B_2 w = 0 \end{array}\right\} \quad \text{on} \quad \partial\Omega, \\ \left.\begin{array}{l} w(x) = A \ln|x| + \mathcal{O}(1), \\ B_1(r)w(x) = o(r^{-(1+\frac{\delta}{2})}), B_2(r)w(x) = o(r^{-(1+\frac{\delta}{2})}) \\ \frac{\partial w}{\partial r} = \mathcal{O}(1). \end{array}\right\} \text{ at } |x| = r \text{ large.} \end{cases}$$

Then on $\Omega_R = \{x \in \Omega \mid |x| \leq R\}$, we have

$$\begin{aligned} &\int_{\Omega_R} [|\Delta w|^2 + 2(1-\nu)(w_{x_1x_2}^2 - w_{x_1x_1}w_{x_2x_2})]dx \\ &= -\int_{\partial\Omega} \left[B_1 w \cdot w - B_2 w \cdot \frac{\partial w}{\partial n}\right] d\sigma - \int_{|x|=R} \left[B_1(R)w \cdot w - B_2(w) \cdot \frac{\partial w}{\partial n}\right] d\sigma \\ &\leq 0 + \int_0^{2\pi} \frac{C}{R^{1+\delta/2}} R d\theta = \frac{2\pi C}{R^{\delta/2}} \to 0 \quad \text{as} \quad R \to \infty. \end{aligned}$$

Since $E(w)$ is a seminorm, and a norm up to a linear polynomial, we must have

$$w(x) = c_0 + c_1 x_1 + c_2 x_2, \qquad x \in \Omega.$$

Here c_1 and c_2 must be equal to 0 because w can grow at most logarithmically. Hence $w = w_1 - w_2 = c_0$, a constant.

To fix these arbitrary constants c_0, c_1 and c_2, we write a general solution as

$$w(x) = w_p(x) + c_0 + c_1 x_1 + c_2 x_2,$$

where w_p is any particular solution to (2.25) satisfying (2.16). The regularity of the solution on $\partial\Omega$ is $w \in H^{r+3}(\partial\Omega)$ and by (2.30),

$$\begin{aligned} \alpha_j &= \langle w, \eta_j \rangle_{H^{r+3}(\partial\Omega) \times H^{-(r+3)}(\partial\Omega)} \\ &= \langle w_p + c_0 + c_1 x_1 + c_2 x_2, \eta_j \rangle \\ &= c_0 \langle 1, \eta \rangle + c_1 \langle x_1, \eta_j \rangle + c_2 \langle x_2, \eta_j \rangle + \langle w_p, \eta_j \rangle. \end{aligned}$$

By (2.31), c_0, c_1 and c_2 are uniquely determinable from $\langle w_p, \eta_j \rangle$, $j = 1, 2$ and 3.

Therefore the solution to (2.25) satisfying (2.16) and (2.30) is unique. □

Remark 2.

As shown in Example 2 earlier, the data (g_1, g_2) to the exterior BVP (2.8) need not satisfy the compatibility condition (2.10). But after imposing the growth condition (2.16), the compatibility condition (2.10) for the data (g_1, g_2) in Theorem 2 naturally follows *as a consequence of (2.16).* □

Example 3. As an application of Theorem 2, we apply it to find the numerical solution of

(2.46)

$$\begin{cases} \Delta^2 w(x) = 0, \qquad x \in \Omega \\ B_1 w(x) = \dfrac{2(1-\nu)\cos\theta}{[R^2+\alpha_0^2-2R\alpha_0\sin\theta]^2}\left\{1 - \dfrac{4\alpha_0}{R}\sin\theta + \dfrac{4\alpha_0(3R\sin\theta + 5\alpha_0\cos^2\theta - 3\alpha_0)}{R^2+\alpha_0^2-2R\alpha_0\sin\theta}\right. \\ \qquad \left. - \dfrac{24R\alpha_0^2\cos^2\theta(R-\alpha_0\sin\theta)}{[R^2+\alpha_0^2-2R\alpha_0\sin\theta]^2}\right\} \\ B_2 w(x) = \dfrac{2(1-\nu)\cos\theta}{[R^2+\alpha_0^2-2R\alpha_0\sin\theta]^2}\left\{2\alpha_0\sin\theta - 3R + \dfrac{4R(R-\alpha_0\sin\theta)^2}{R^2+\alpha_0^2-2R\alpha_0\sin\theta}\right\} \\ R = 1/2, \alpha_0 = 1/4, \nu = 1/8, x_1 = R\cos\theta, x_2 = R\sin\theta, \quad 0 \le \theta \le 2\pi, \end{cases}$$

where Ω is the exterior of the closed disk with radius 1/2. It is straighforward to check that the data in (2.46) satisfies the compatibility condition (2.10).

The exact solution satisfying (2.46) and the growth condition (2.16) is (a harmonic function)

$$w(x) = \frac{x_1}{x_1^2 + \left(x_2 - \frac{1}{4}\right)^2}, \qquad x = (x_1, x_2) \in \Omega. \tag{2.47}$$

We compute w by approximating (2.26) and (2.27), where in (2.26) we have specified $c_j = 0$ for $j = 0, 1$ and 2 using some a priori knowledge of w. The linear system (2.27) is approximated by pointwise collocating the boundary integral equations. A (3,2)-system of quadratic B-splines is used, and there are a total of 24 uniformly spaced mesh points. The collocation points are the midpoints between the mesh points. This results in a matrix equation of size 54×54, where

$$24 \times 2 \text{ (one each for } f_1 \text{ and } f_2) + 3 \text{ (compatibility conditions)}$$
$$+3(\text{coefficients } \tilde{a}_0, \tilde{a}_1, \tilde{a}_2) = 54.$$

Graphics of the exact and numerical solutions have no visible differences from each other, as illustrated in Figs. 1 and 2. Numerical values also show good agreement at the selected collocation points on the boundary ($R = 1/2$), and at points on circles with radius $R = 10$ and 20 in Ω, see Table 1. This computation requires very small memory and computer time. □

Note that Example 3 is selected to have just $o(1)$ growth at ∞. If a solution has linear-logarithmic growth (2.16) with $a_1 \neq 0, a_2 \neq 0$ and $A \neq 0$, then the numerical solution usually is accurate only within a small radius of $\partial\Omega$. When $|x|$ becomes large, *the numerical solution quickly loses accuracy with a linear-logarithmic rate*, unless the coefficients a_1, a_2 and A in (2.16) can be computed exactly.

§III. Application to Plate Boundary Control Problems on an Exterior Domain and Computation

Let us discuss some examples and present some numerical results.

Example 4. Let G be an open simply connected subdomain of Ω. Let $\alpha = (\alpha_1, \alpha_2)$ be a unit vector with base point at $x \in G$. Let $n = (\alpha_2, -\alpha_1)$ be a unit vector perpendicular to α. Write

$$n = n(x) = (\alpha_2, -\alpha_1) = (\cos\theta, \sin\theta),$$
$$\alpha = \alpha(x) = (\alpha_1, \alpha_2) = (\sin\theta, -\cos\theta),$$

where θ is the angle formed between n and the positive x_1-axis. Then the bending moment at x along the direction α is

$$\mathcal{M}(x,\alpha) = -D[\partial_n^2 w(x) + \nu\partial_\alpha^2 w(x)] \qquad ([8, (41), \text{ p. } 40]).$$

The transverse force at x along the direction α is

$$\mathcal{T}(x,\alpha) = D[\partial_n \Delta w(x) - (1-\nu)\partial_\alpha\partial_n\partial_\alpha w(x)] \qquad ([8, \text{ p. } 43, 87, 88]),$$

where D is the flexural rigidity of the plate, and $\partial_\alpha, \partial_n$ are the directional derivatives along α and n. If the moment and transverse forces $\mathcal{M}$ and $\mathcal{T}$ become too large along certain direction α, they may cause structural damage to the plate, making cracks and fracture happen, such as the case of earthquakes. Thus we may wish the moment and transverse forces to remain as close to certain safety values as possible. We may consider the minimax problem

$$\inf_{u\in\mathcal{A}} \sup_{\substack{x\in G \\ |\alpha|=1}} [|\mathcal{M}(x,\alpha) - M_0|^2 + \gamma|\mathcal{T}(x,\alpha)|^2] \tag{3.1}$$

where $\gamma > 0$ is a weight to the specified, and $M_0 \in \mathbf{R}$ is given, subject to

$$\begin{cases} \Delta^2 w(x) = 0 & \text{on } \Omega, \\ B_1 w(x) = u_1(x), & x \in \partial\Omega, \nu = 1/8, \\ B_2 w(x) = u_2(x), & x \in \partial\Omega, \\ w \text{ satisfies the linear-logarithmic growth condition (2.16),} \end{cases} \tag{3.2}$$

where $\mathcal{A}$ is an admissible set of controls $u = (u_1, u_2)$, with constraints imposed.

Instead of (3.1), one may also consider the minimization of an integral quadratic cost

$$\inf_{u\in\mathcal{A}} \int_\Omega \int_{|\alpha|=1} [|\mathcal{M}(x,\alpha) - M_0|^2 + \gamma|\mathcal{T}(x,\alpha)|^2] d\alpha dx.$$

The ideas shown above have potential structural applications, but may require large amount of computation. □

Example 5. We consider the following concrete example. Let

$$\Omega = \{x \in \mathbf{R}^2 \mid |x| > 1/2\}$$

and let the state w and control $u = (u_1, u_2)$ satisfy (3.2). Choose

$$G = \{(x_1, x_2) \in \mathbf{R}^2 \mid x_1 = r\cos\theta, x_2 = r\sin\theta - 2, \quad 0 \le r < 1/2,\ 0 \le \theta \le 2\pi\}.$$

The relative geometry of Ω and G is shown in Fig 3.

On G, we define a target function

$$Z(x_1, x_2) = \sin\left(2\pi\sqrt{x_1^2 + (x_2+2)^2}\right)\cos\left(\tan^{-1}\frac{x_1}{x_2+2}\right). \tag{3.3}$$

The profile of Z is shown in Fig. 4.

We wish to minimize the integral quadratic cost

$$J(u) = \int_G |w(x) - Z(w)|^2 dx + \gamma \int_{\partial\Omega} [u_1^2(x) + u_2^2(x)] d\sigma. \tag{3.4}$$

We also define

$$J_1 = \int_G |w(x) - Z(x)|^2 dx \tag{3.5}$$

to be the difference square integral between w and the target Z over G. It is easy to verify that J is coercive and convex with respect to $(u_1, u_2) \in L^2(\partial\Omega) \times L^2(\partial\Omega)$. Inside $\Omega, w \in C^\infty(G)$, so (3.4) is well defined. By [7], there exists a unique optimal control $(\hat{u}_1, \hat{u}_2) \in L^2(\partial\Omega) \times L^2(\partial\Omega)$ minimizing (3.3). By Theorem 2, the corresponding optimal state $\widehat{w} \in H^{5/2}_{\text{loc}}(\Omega)$ holds.

Instead of analyzing $(\hat{u}_1, \hat{u}_2)$ and $\widehat{w}$ further, let us proceed to describe the scheme for computing the optimal control $(\hat{u}_1, \hat{u}_2)$. The governing equation (3.2) is replaced by the boundary integral equation (2.27), which is now rewritten as

$$\mathcal{L}_1 \begin{bmatrix} a_0 \\ a_1 \\ a_2 \\ f_1 \\ f_2 \end{bmatrix} = \begin{bmatrix} 0 \\ 0 \\ 0 \\ u_1 \\ u_2 \end{bmatrix}. \tag{3.6}$$

Define the projection operator

$$\mathbf{P}\colon \mathbf{R}^3 \times [C^\infty(\partial\Omega)]^2 \longrightarrow [C^\infty(\partial\Omega)]^2$$

by

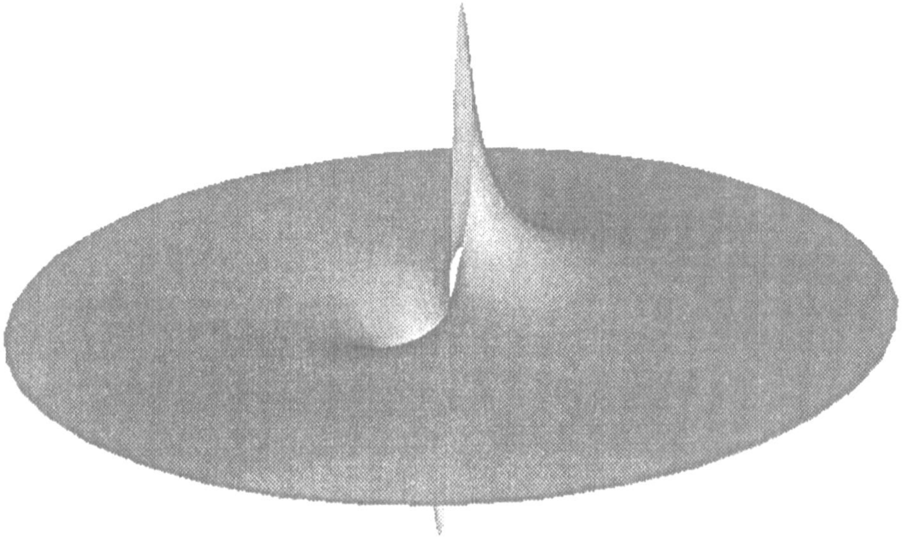

Fig. 1 The exact solution $w(x) = \frac{x_1}{x_1^2+(x-\frac{1}{4})^2}$ as benchmark in Example 3. The graph is plotted on a disk with radius 6.

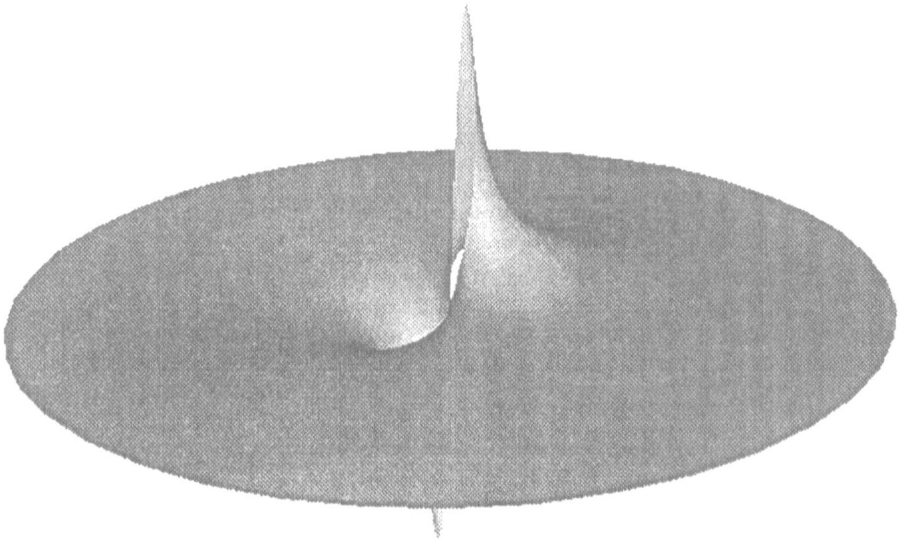

Fig. 2 The graph of the numerical solution for (2.46), obtained by point collocation using quadratic cardinal splines with 24 uniform meshes.

Table 1 Comparison of exact and numerical values for Example 3. The first two columns are the coordinates of $x(=(x_1, x_2) = (x, y))$. Values are tabulated for $R = 0.5$ (the boundary points), and for points on the concentric circles with radii $R = 10$ and 20.

R=0.5				R=20			
X	Y	Numerical	Exact	X	Y	Numerical	Exact
0.4830	0.1294	1.9363	1.9490	19.3185	5.1764	0.0471	0.0486
0.4330	0.2500	2.2981	2.3094	17.3205	10.0000	0.0415	0.0438
0.3536	0.3536	2.5984	2.6050	14.1421	14.1421	0.0334	0.0360
0.2500	0.4330	2.6071	2.6043	10.0000	17.3205	0.0234	0.0255
0.1294	0.4830	1.8194	1.8222	5.1764	19.3185	0.0120	0.0133
0.0000	0.5000	0.0000	0.0000	0.0000	20.0000	0.0000	0.0000
-0.1294	0.4830	-1.8194	-1.8222	-5.1764	19.3185	-0.0120	-0.0133
-0.2500	0.4330	-2.6071	-2.6043	-10.0000	17.3205	-0.0234	-0.0255
-0.3536	0.3536	-2.5984	-2.6050	-14.1421	14.1421	-0.0334	-0.0360
-0.4330	0.2500	-2.2981	-2.3094	-17.3205	10.0000	-0.0415	-0.0438
-0.4830	0.1294	-1.9363	-1.9490	-19.3185	5.1764	-0.0471	-0.0486
-0.5000	0.0000	-1.5875	-1.6000	-20.0000	0.0000	-0.0496	-0.0500
-0.4830	-0.1294	-1.2691	-1.2804	-19.3185	-5.1764	-0.0489	-0.0480
-0.4330	-0.2500	-0.9803	-0.9897	-17.3205	-10.0000	-0.0446	-0.0428
-0.3536	-0.3536	-0.7154	-0.7226	-14.1421	-14.1421	-0.0370	-0.0347
-0.2500	-0.4330	-0.4677	-0.4726	-10.0000	-17.3205	-0.0265	-0.0245
-0.1294	-0.4830	-0.2312	-0.2336	-5.1764	-19.3185	-0.0138	-0.0126
0.0000	-0.5000	0.0000	0.0000	0.0000	-20.0000	0.0000	0.0000
0.1294	-0.4830	0.2312	0.2336	5.1764	-19.3185	0.0138	0.0126
0.2500	-0.4330	0.4677	0.4726	10.0000	-17.3205	0.0265	0.0245
0.3536	-0.3536	0.7154	0.7226	14.1421	-14.1421	0.0370	0.0347
0.4330	-0.2500	0.9803	0.9897	17.3205	-10.0000	0.0446	0.0428
0.4830	-0.1294	1.2691	1.2804	19.3185	-5.1764	0.0489	0.0480
0.5000	0.0000	1.5875	1.6000	20.0000	0.0000	0.0496	0.0500

R=10			
X	Y	Numerical	Exact
9.6593	2.5882	0.0960	0.0978
8.6603	5.0000	0.0862	0.0888
7.0711	7.0711	0.0706	0.0733
5.0000	8.6603	0.0500	0.0522
2.5882	9.6593	0.0260	0.0272
0.0000	10.0000	0.0000	0.0000
-2.5882	9.6593	-0.0260	-0.0272
-5.0000	8.6603	-0.0500	-0.0522
-7.0711	7.0711	-0.0706	-0.0733
-8.6603	5.0000	-0.0862	-0.0888
-9.6593	2.5882	-0.0960	-0.0978
-10.0000	0.0000	-0.0992	-0.0999
-9.6593	-2.5882	-0.0959	-0.0953
-8.6603	-5.0000	-0.0860	-0.0844
-7.0711	-7.0711	-0.0704	-0.0683
-5.0000	-8.6603	-0.0498	-0.0479
-2.5882	-9.6593	-0.0258	-0.0247
0.0000	-10.0000	0.0000	0.0000
2.5882	-9.6593	0.0258	0.0247
5.0000	-8.6603	0.0498	0.0479
7.0711	-7.0711	0.0704	0.0683
8.6603	-5.0000	0.0860	0.0844
9.6593	-2.5882	0.0959	0.0953
10.0000	0.0000	0.0992	0.0999

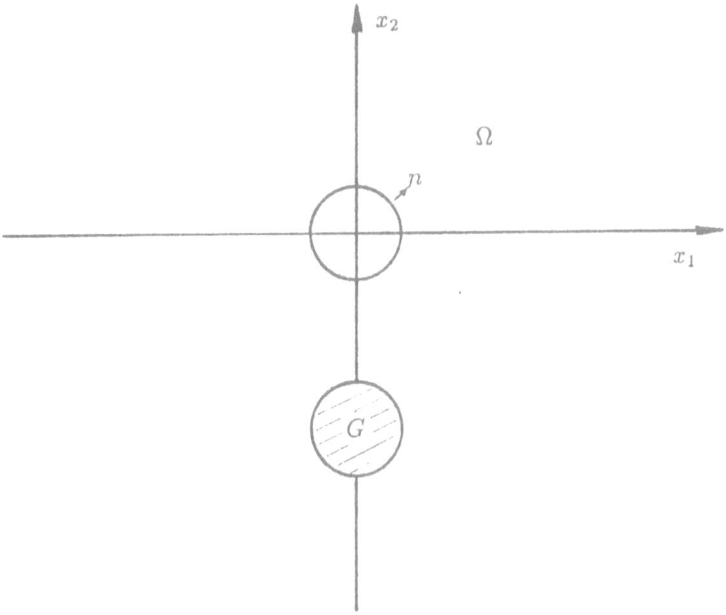

Fig. 3 The exterior domain Ω and a subdomain G.

Fig. 4 The profile of the target function $Z(x)$ in (3.3).

$$\mathbf{P}\begin{bmatrix}\alpha_0\\ \alpha_1\\ \alpha_2\\ \phi_1\\ \phi_2\end{bmatrix} = \begin{bmatrix}\phi_1\\ \phi_2\end{bmatrix}$$

and extend $\mathbf{P}$ by continuity. Then from (3.6) we have

$$\begin{bmatrix}u_1\\ u_2\end{bmatrix} = \mathbf{P}\mathcal{L}_1\begin{bmatrix}a_0\\ a_1\\ a_2\\ f_1\\ f_2\end{bmatrix}. \tag{3.7}$$

Also define

$$\begin{aligned}&V\colon \mathbf{R}\times H^{r-1}(\partial\Omega)\times H^{r}(\partial\Omega)\longrightarrow H_{\mathrm{loc}}^{r+\frac{5}{2}}(\Omega)\\ &V(c_0,f_1,f_2)(x) = \int\limits_{\partial\Omega}\left[E(x,\xi)f_1(\xi) + \frac{\partial E(x,\xi)}{\partial n_\xi}f_2(\xi)\right]d\sigma_\xi + c_0.\end{aligned} \tag{3.8}$$

Introduce three *scalar Lagrange multipliers* λ_0,λ_1 and λ_2 for equality constraints (2.33)-(2.35), but substitute the control and state by (3.6) and (3.7) directly into the cost J, yielding the Lagrangian

$$\begin{aligned}L(\lambda_0,\lambda_1,\lambda_2,c_0,a_0,a_1,a_2,f_1,f_2) &= J(u) + \lambda_0\int\limits_{\partial\Omega} f_1\,d\sigma\\ &+ \lambda_1\int\limits_{\partial\Omega}(f_1\xi_1 + f_2 n_1)d\sigma + \lambda_2\int\limits_{\partial\Omega}(f_1\xi_2 + f_2 n_2)d\sigma\\ &= \int\limits_{G}|V(c_0,f_1,f_2)(x) - Z(x)|^2dx + \gamma\|\mathbf{P}\mathcal{L}_1(a_0,a_1,a_2,f_1,f_2)\|^2_{[L^2(\partial\Omega)]^2}\\ &+ \lambda_0\int\limits_{\partial\Omega} f_1\,d\sigma + \lambda_1\int\limits_{\partial\Omega}(f_1\xi_1 + f_2 n_1)d\sigma + \lambda_2\int\limits_{\partial\Omega}(f_1\xi_2 + f_2 n_2)d\sigma.\end{aligned} \tag{3.9}$$

Thus this is a *mixed primal* (with respect to the variables u_1,u_2 and w) *and dual* (with respect to the equality constraints (2.33)-(2.35)) *approach*. The Lagrangian is *convex* with respect to the variables c_0, a_0, a_1, a_2, f_1 and f_2, but *concave* with respect to the variables λ_0, λ_1 and λ_2. The optimal solution can then be obtained from

$$\text{(3.10)} \quad \begin{cases} \dfrac{\partial L}{\partial \lambda_i} = 0, \dfrac{\partial L}{\partial a_i} = 0, & i = 0,1,2, \\ \dfrac{\partial L}{\partial c_0} = 0, \dfrac{\partial L}{\partial f_i} = 0, & i = 1,2, \end{cases}$$

where the Frechét derivatives are taken. Computationally, the integrals in (3.9) are approximated by certain quadrature rule. The functions f_1 and f_2 are again approximated by a (3,2)-system of quadratic B-splines on $\partial\Omega$, with 24 uniform mesh points chosen on $\partial\Omega$.

We have performed a sequence of computations by varing the weight γ in (3.4). The costs are computed and tabulated in Table 2:

γ	J_1, cf. (3.5)	J, cf. (3.4)
10^{-4}	0.0381	0.0884
10^{-5}	0.0230	0.0739
10^{-6}	0.0211	0.0232

Table 2. Values of cost J and square difference integral J_1 for different weights γ

We have plotted the controlled deformed plate for $\gamma = 10^{-4}, 10^{-5}$ and 10^{-6} successively in Figs. 5,6 and 7, over a circular region of radius 6.5. The corresponding optimal controls u_1 and u_2 are plotted in Figs. 8 an 9, respectively.

First, we remark that Fig. 5,6 and 7 are plotted with different vertical scales, *with roughly a* $1:3:5$ *ratio.*

All these figures have contorted profiles, a little similar to the concave-downward on the left and convex-upward on the right of the target function $Z(x)$ in Fig. 4. In increasing order of figure number, these shapes on the subdomain G approximate Z better and better, as the decreasing magnitude of J_1 in Table 2 shows. We know that a smaller value of the weight γ in (3.4) corresponds to a large *penalty* of the shape (square) difference integral (3.5), resulting in the reduction of J_1.

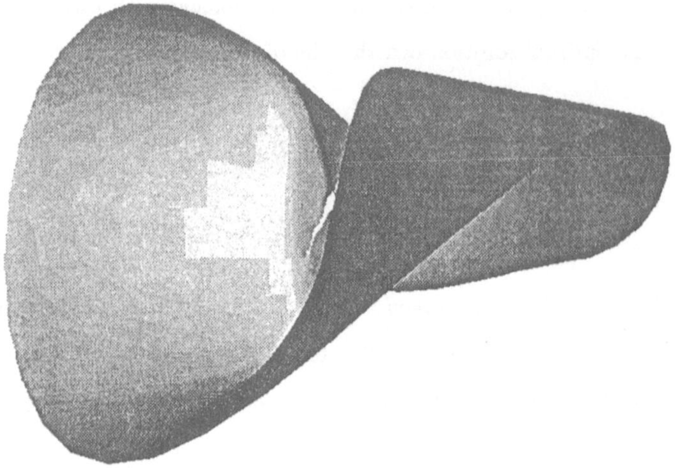

Fig. 5 The deformed state of the plate, using $\gamma = 10^{-4}$ in Example 5.

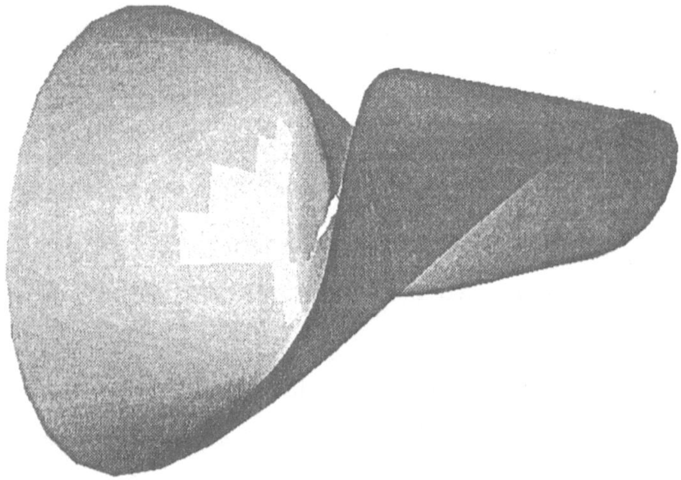

Fig. 6 The deformed state of the plate, using $\gamma = 10^{-5}$ in Example 5.

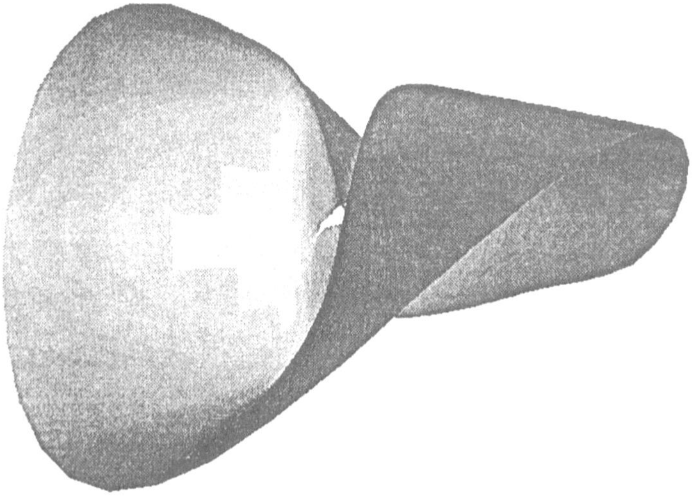

Fig. 7 The deformed state of the plate, using $\gamma = 10^{-6}$ in Example 5.

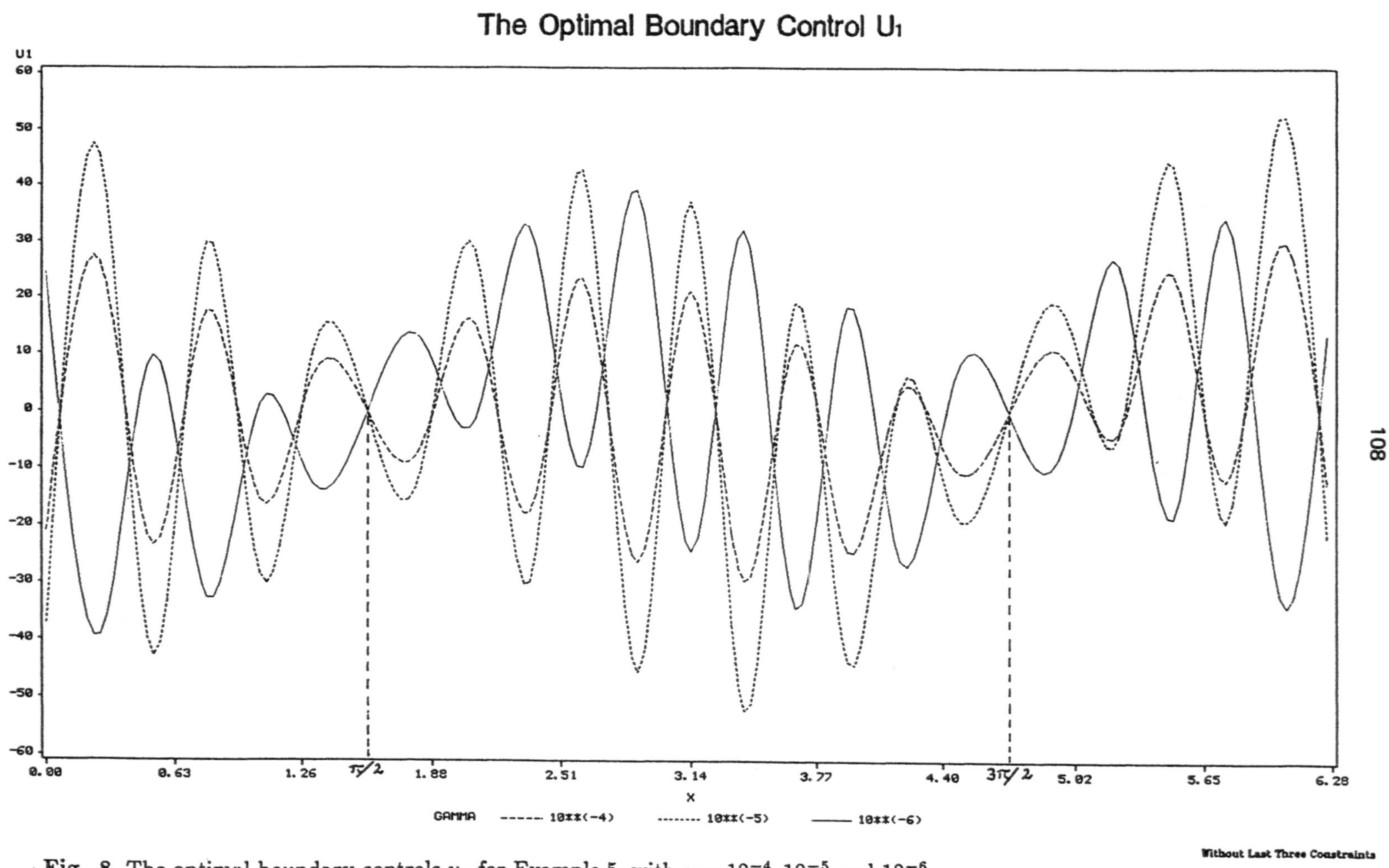

Fig. 8 The optimal boundary controls u_1 for Example 5, with $\gamma = 10^{-4}, 10^{-5}$ and 10^{-6}.

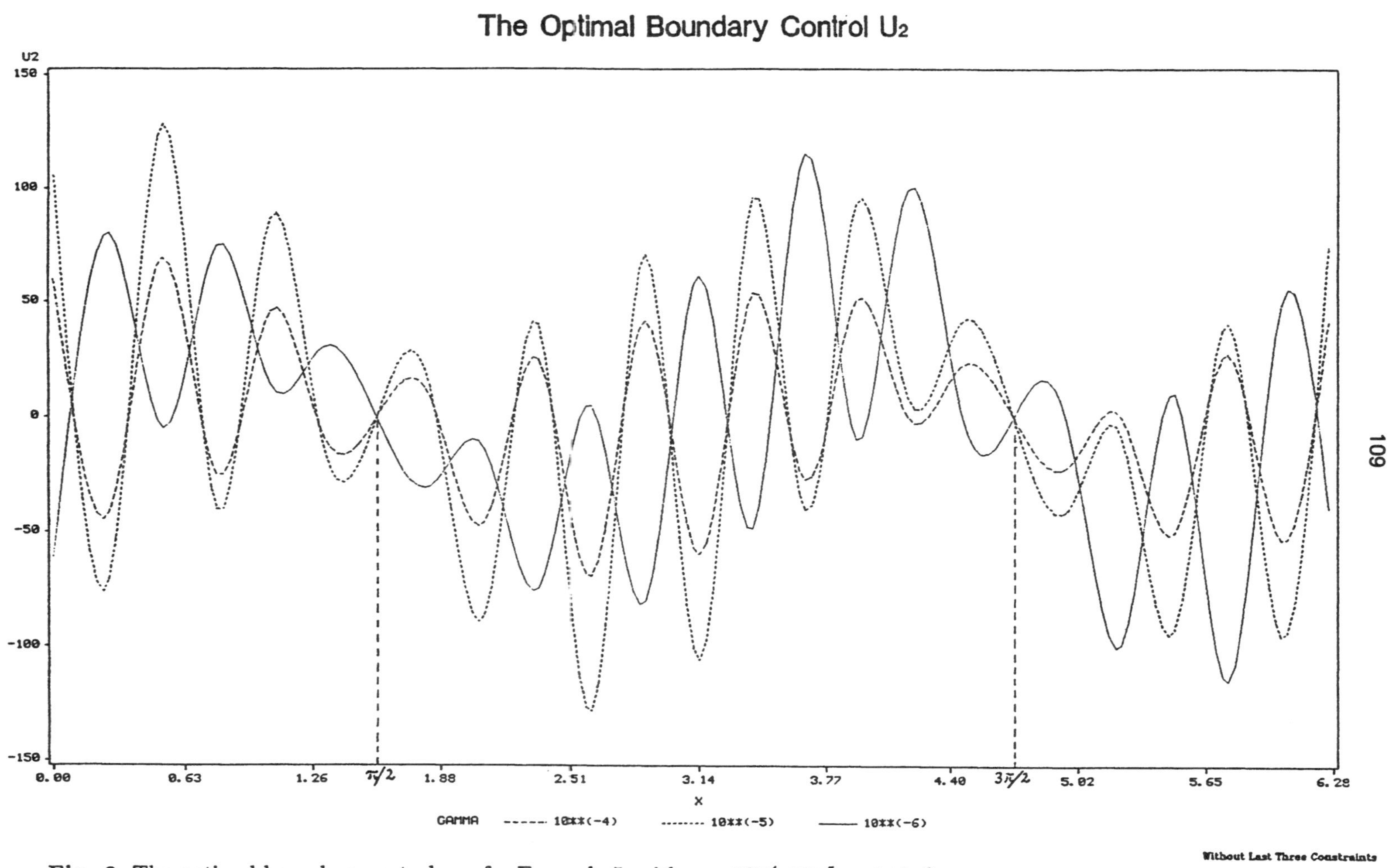

Fig. 9 The optimal boundary controls u_2 for Example 5, with $\gamma = 10^{-4}, 10^{-5}$ and 10^{-6}.

In Fig. 8 and 9, we see that the curves of controls u_1 and u_2 show considerable oscillations (with respect to the spatial variable). These oscillations are commonly related to the *weak convergence* of the controllers. In these figures, we see that all the curves pass through two common points at $\theta = \pi/2$ and $\theta = 3\pi/2$, as indicated therein. Otherwise these curves do not have much else in common behavior. In particular, we do not observe a strong trend of convergences as γ decreases (to zero). □

Example 6. The control problem (3.4) subject to (3.2) in Example 5 has a property that the state $w(x)$ grows with a linear-logarithmic rate (2.16a). This seems to be *physically unreasonable.* We wish to have

$$w(x) = o(1) \qquad \text{for} \qquad |x| \text{ large,} \tag{3.11}$$

namely, *the deflection of the plate at far away points should be negligible.* Obviously, not all controls (u_1, u_2) in (3.2) will make the state w satisfy (3.11). This means that additional constraints must be imposed on the control (u_1, u_2). But how?

This question generally is not easy to deal with. However, by using boundary elements, we can handle it very easily. We need only note from the asymptotic expansion (2.24) that $w(x) = o(1)$ is satisfied if and only if the layer densities (f_1, f_2) also satisfy

$$\begin{cases} \displaystyle\int_{\partial\Omega} [\xi_1^2 f_1(\xi) + 2\xi_1 n_1(\xi) f_2(\xi)] d\sigma = 0, \\ \displaystyle\int_{\partial\Omega} [\xi_2^2 f_1(\xi) + 2\xi_2 n_2(\xi) f_2(\xi)] d\sigma = 0, \\ \displaystyle\int_{\partial\Omega} [\xi_1 \xi_2 f_1(\xi) + (\xi_1 n_2(\xi) + \xi_2 n_1(\xi)) f_2(\xi)] d\sigma = 0, \end{cases} \tag{3.12}$$

in addition to (2.33)-(2.35). Therefore we just introduce three extra Lagrange multipliers λ_3, λ_4 and λ_5, multiply to (3.12) sequentially, and append to (3.9), yielding the new Lagrangian

$$\tilde{\mathcal{L}}(\lambda_0,\lambda_1,\lambda_2,\lambda_3,\lambda_4,\lambda_5,c_0,a_0,a_1,a_2,f_1,f_2)$$
$$\equiv L(\lambda_0,\lambda_1,\lambda_2,c_0,a_0,a_1,a_2,f_1,f_2) \text{ (cf. (3.9))}$$
$$+\lambda_3\int_{\partial\Omega}[\xi_1^2 f_1(\xi)+2\xi_1 n_1(\xi)f_2(\xi)]d\sigma+\lambda_4\int_{\partial\Omega}[\xi_2^2 f_1(\xi)+2\xi_2 n_2(\xi)f_2(\xi)]d\sigma$$
$$+\lambda_5\int_{\partial\Omega}[\xi_1\xi_2 f_1(\xi)+(\xi_1 n_2(\xi)+\xi_2 n_1(\xi))f_2(\xi)]d\sigma.$$

We proceed to carry out the computation just as the rest of Example 5. The control costs are computed and tabulated in Table 3:

γ	J_1, cf. (3.5)	J, cf. (3.4)
10^{-4}	0.0479	0.1206
10^{-5}	0.0398	0.5357
10^{-6}	0.0247	0.6692

Table 3. Values of cost J and square difference integral J_1 for different weights γ

These values show that when γ is decreasing, J_1 is also decreasing, but J becomes *increasing*, in contrast to the values of J in Table 2 of Example 5.

The controlled deformed state $w(x)$ corresponding to $\gamma = 10^{-4}, 10^{-5}$ and 10^{-6} are plotted successively in Figs. 10, 11 and 12, all over a circular region of radius 0.5. The corresponding controls u_1 and u_2 are plotted in Figs. 13 and 14, respectively.

In Figs. 10-12, we see that the state $w(x)$ becomes flat very quickly, namely, satisfying (3.11). Nevertheless, on the subdomain G of interest, we *do not* observe the change of profile of $w(x)$ to match $Z(x)$'s, the target shape. Although all profiles in Figs. 10-12 are similar, they are actually plotted on different verticales of scales, of roughly $1:8:30$ in ratio. (This means that the spike in Fig. 12 is about 30 times as high as that in Fig. 10!)

As for the controls in Figs. 13 and 14, we again see considerable oscillations, and all curves pass two common points at $\theta = \pi/2$ and $3\pi/2$, as indicated. We note that the largest magnitudes of u_1 and u_2 in Figs. 13 and 14 are *considerably larger* than those in Figs. 8 and 9 of Example 5. This, along with the comments in the last paragraph, seems

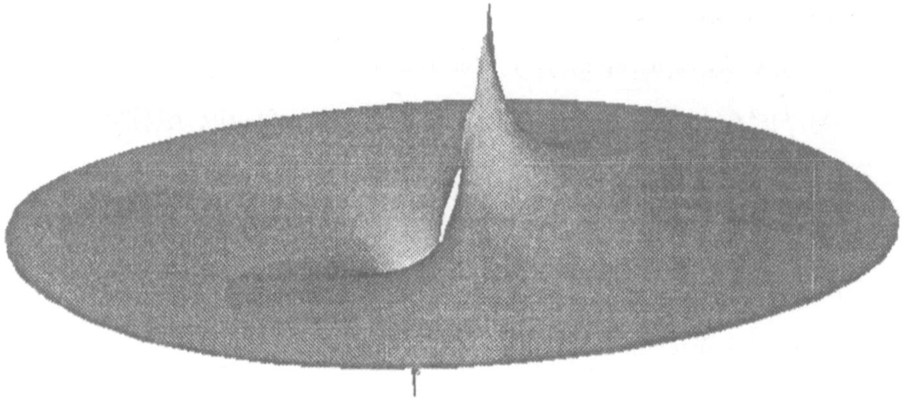

Fig. 10 The deformed state of the plate, using $\gamma = 10^{-4}$ in Example 6.

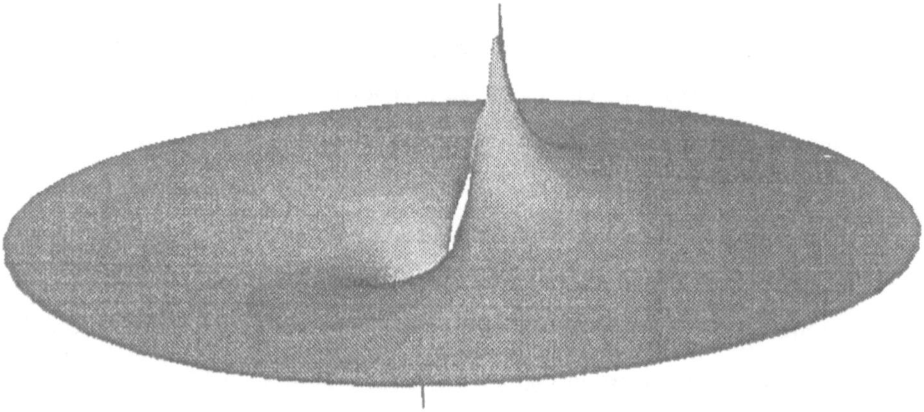

Fig. 11 The deformed state of the plate, using $\gamma = 10^{-5}$ in Example 6.

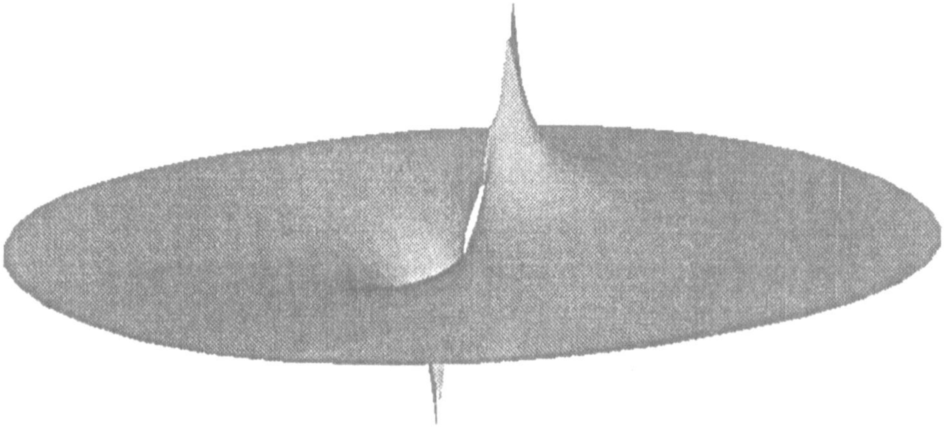

Fig. 12 The deformed state of the plate, using $\gamma = 10^{-6}$ in Example 6.

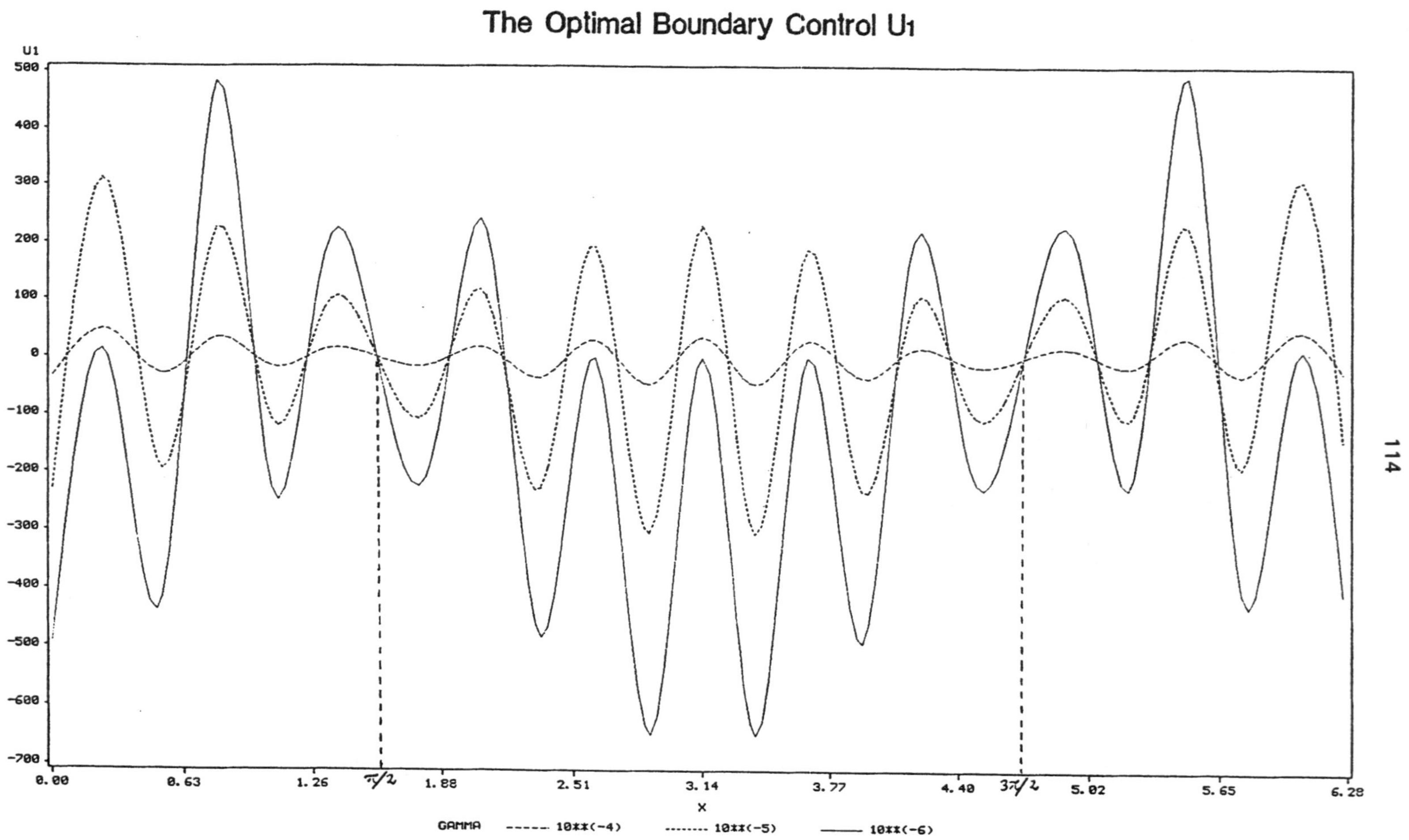

Fig. 13 The optimal boundary controls u_1 for Example 6, with $\gamma = 10^{-4}, 10^{-5}$ and 10^{-6}.

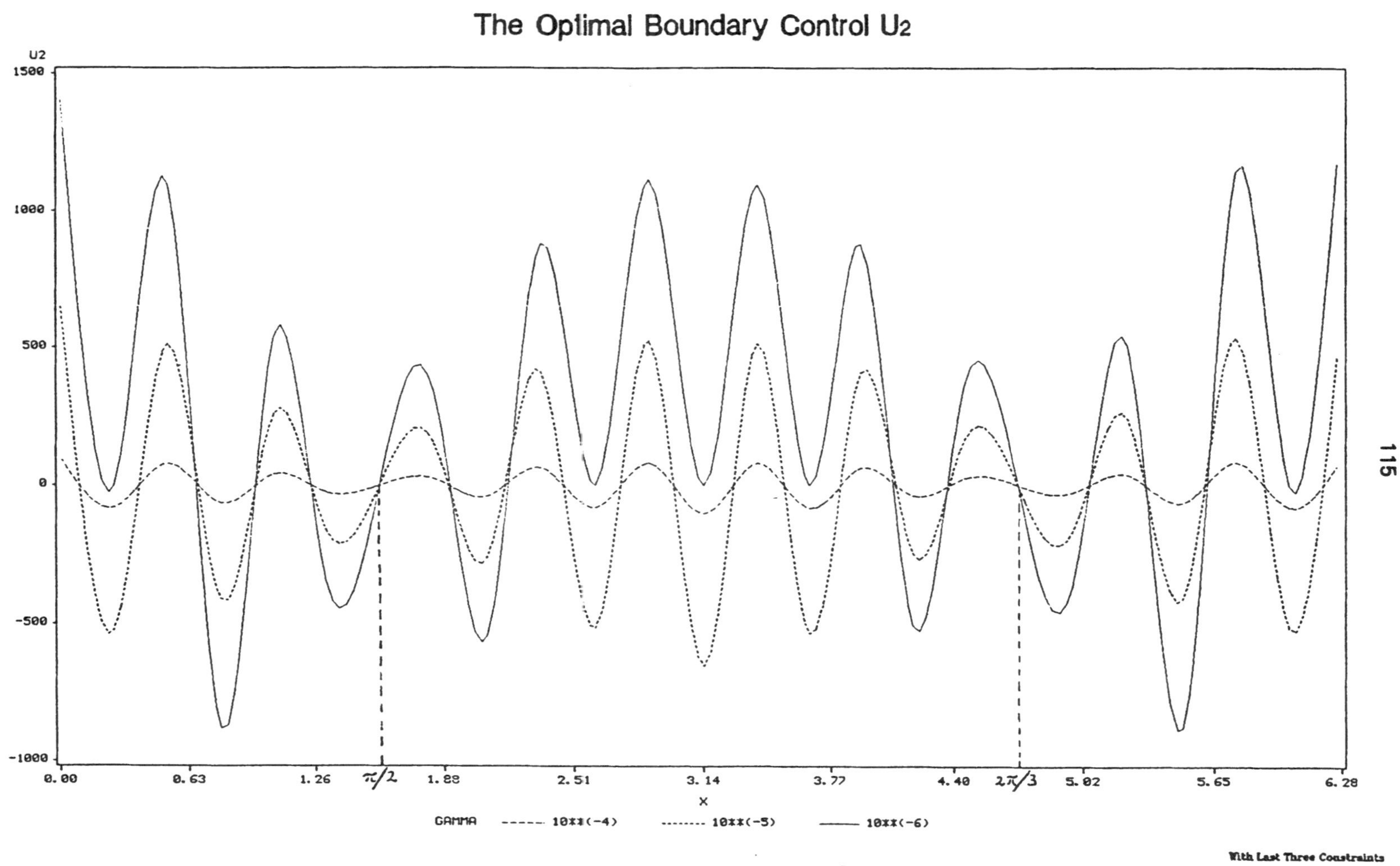

Fig. 14 The optimal boundary controls u_2 for Example 6, with $\gamma = 10^{-4}, 10^{-5}$ and 10^{-6}.
Note that the magnitude becomes rather large, while all the curves "in phrase" with each other.

to indicate that the three constraints (3.12) do impose very severe physical restraints on the state w and controls u_1 and u_2.

It is also interesting to observe in Fig. 14 that all the control u_2 curves are "in phase" with each other, but otherwise we do not see a strong convergence trend such as it has been observed for numerical solutions in [1]. □

We are presently investigating other methods or formulations of the boundary control for an exterior plate, such as a model using bending moment as the only control while holding the boundary displacement fixed.

The questions of convergence and error estimates for the boundary elements adopted and developed here are fundamental. So far no rigorous results are available. They are being studied in [2].

REFERENCES

[1] G. Chen and J. Zhou, Boundary integral equations and boundary element computations for the biharmonic plate equation with applications to boundary shape control, preprint.

[2] G. Chen and J. Zhou, Work in progress.

[3] G. Chen and S. Sun, Augmenting a Fredholm operator of zero index to achieve invertibility for elliptic boundary value problems, to appear in J. Math. Anal. Appl.

[4] G.C. Hsiao and R. McCamy, Solution of boundary value problems by integral equations of the first kind, SIAM Review, 15 (1973), 687-705.

[5] T. Kato, Perturbation Theory for Linear Operators, Springer-Verlag, New York, 1966.

[6] J.E. Lagnese and J.L. Lions, Modelling, Analysis and Control of Thin Platse, Masson, Paris, 1988.

[7] J.L. Lions, Optimal Control of Systems Governed by Partial Differential Equations, Springer-Verlag, New York, 1971.

[8] S.P. Timoshenko and S. Woinowsky-Krieger, Theory of Plates and Shells, 2nd ed., McGraw-Hill, New York, 1959.

APPROXIMATE CONTROLLABILITY FOR THE WAVE EQUATION WITH OSCILLATING COEFFICIENTS

Doina CIORANESCU, Patrizia DONATO and Enrique ZUAZUA

1. INTRODUCTION AND MAIN RESULT.

Let Ω be a bounded open set in $\mathbb{R}^n$, with smooth boundary $\Gamma = \partial\Omega$. Consider the wave equation

$$(1)_\varepsilon \qquad \begin{cases} z_\varepsilon'' + A_\varepsilon z_\varepsilon = 0 & \text{in } \Omega \times (0,T) \\ z_\varepsilon = \varphi & \text{on } \Sigma = \partial\Omega \times (0,T) \\ z_\varepsilon(0) = z^0, \quad z_\varepsilon'(0) = z^1 & \text{in } \Omega \end{cases}$$

where the operator A_ε is defined by

$$(2) \qquad A_\varepsilon = \begin{cases} -\dfrac{\partial}{\partial x_i}(\alpha_{ij}\dfrac{\partial}{\partial x_j}) & \text{in } D_\alpha \\ -\dfrac{\partial}{\partial x_i}(a_{ij}(\dfrac{x}{\varepsilon})\dfrac{\partial}{\partial x_j}) & \text{in } \Omega \setminus \overline{D_\alpha} \end{cases}$$

and D_α denotes the neighbourhood of $\partial\Omega$ of thickness α:

$$D_\alpha = \{x \in \Omega \ / \ d(x, \partial\Omega) \leq \alpha\}.$$

Definition (2) of A_ε signifies that there is a "safety zone" around $\partial\Omega$ where the operator A_ε is independent of ε.

We make the following assumptions on the coefficients of A_ε:

$$(3) \qquad \begin{cases} \text{i. } \alpha_{ij} \text{ are constants with } \alpha_{ij} = \alpha_{ij} \text{ for } i,j = 1,..,n, \\ \quad a_{ij} \in L^\infty(\mathbb{R}^n), \text{ with } a_{ij} = a_{ij} \text{ for } i,j = 1,..,n. \\ \text{ii. There exists a positive number } \lambda \text{ such that} \\ \quad \alpha_{ij}\zeta_i\zeta_j \geq \lambda|\zeta|^2, \quad \forall \zeta \in \mathbb{R}^n, \\ \quad a_{ij}(y)\zeta_i\zeta_j \geq \lambda|\zeta|^2, \quad \forall \zeta \in \mathbb{R}^n, \\ \text{iii. } a_{ij} \text{ is } Y\text{-periodic} \end{cases}$$

where Y denotes the representative cell $Y = [0, l_1[\times ... \times [0, l_n[$.

In $\Omega \setminus \overline{D_\alpha}$ we have an heterogeneous material. By (3)iii, the coefficients $a_{ij}(y)$ are periodic functions in each variable y_k with period l_k. So, the heterogeneities are periodically distributed in $\Omega \setminus \overline{D_\alpha}$ in each direction x_k with the period εl_k, see Figure 1.

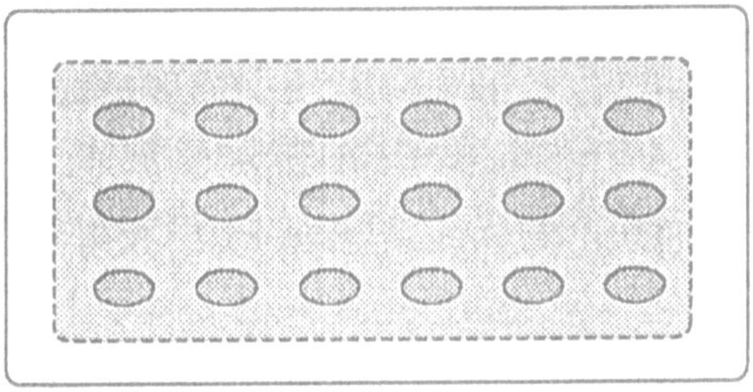

Figure 1

Classical results ensure that for any $\{z^0, z^1\} \in L^2(\Omega) \times H^{-1}(\Omega)$, $\varphi \in L^2(\partial\Omega \times (0,T))$ system $(1)_\varepsilon$ has a unique solution $z_\varepsilon \in C^0(0,T; L^2(\Omega)) \cap C^1(0,T; H^{-1}(\Omega))$.

The problem of exact controllability consists of finding $T > 0$ large enough such that for every $\{z^0, z^1\} \in L^2(\Omega) \times H^{-1}(\Omega)$ there exists a control φ_ε so that the solution of $(1)_\varepsilon$ satisfies

$$z_\varepsilon(T) = z_\varepsilon'(T) = 0.$$

There are many results related to the construction of such exact controls, see for instance J.-L. Lions [9], D. Russell [10] and the references therein. In general, when studying the wave equation with variable coefficients, to obtain an exact control it is necessary to make quite strong assumptions on the coefficients. Actually a $C^\infty(\overline{\Omega})$-regularity is required when applying the method of C. Bardos, G. Lebeau and J. Rauch [2] and multiplier techniques only apply under rather strong assumptions which exclude rapidly oscillating coefficients (see V.Komornik [7]).

We are interested on describing the asymptotic behaviour of z_ε and eventually of the exact controls when the parameter ε tends to zero. To do it, uniform estimates (in ε) of the state z_ε, the control φ_ε and the exact controllability time T are necessary. The exact controllability time depends on the speed of propagation of waves and therefore, under assumption (3), it is uniformly bounded. However, in general, when A_ε has rapidly oscillating coefficients, the controls are not uniformly bounded (see for instance the counter-example of M. Avellaneda, C. Bardos and J. Rauch [1]). It is why we turn our attention to the approximate controllability: we do not construct exact controls for the problems $(1)_\varepsilon$ but give an appropriate boundary value φ in $(1)_\varepsilon$ such that we have exact control of the homogenized equation at the limit, i.e.

$$\begin{cases} z_\varepsilon(T) \to 0 \\ z_\varepsilon'(T) \to 0 \end{cases}$$

(in a sense to be made precise).

Before giving the main result, let us recall a homogenization theorem for the wave equation due to S. Brahim-Otsmane, G. A. Francfort and F. Murat [4].

Proposition 1. *([4]) Let A_ε be given by (2) and suppose that assumption (3) is satisfied. Let u_ε be the solution of*

$$\text{(4)} \qquad \begin{cases} u_\varepsilon'' + A_\varepsilon u_\varepsilon = f_\varepsilon & \text{in } \Omega \times (0,T) \\ u_\varepsilon = 0 \quad \text{in } \partial\Omega \times (0,T) \\ u_\varepsilon(0) = u_\varepsilon^0 \,,\ u_\varepsilon'(0) = u_\varepsilon^1 & \text{in } \Omega \end{cases}$$

with data such that

$$\text{(5)} \qquad \begin{cases} \{u_\varepsilon^0 \,, u_\varepsilon^1\} \rightharpoonup \{u^0 \,, u^1\} & \text{weakly in } H_0^1(\Omega) \times L^2(\Omega) \\ f_\varepsilon \rightharpoonup f \quad \text{weakly in } L^1(0,T; L^2(\Omega)) \end{cases}$$

when $\varepsilon \to 0$. Then

$$\text{(6)} \qquad \begin{cases} u_\varepsilon \rightharpoonup u & \text{weakly} * \text{ in } L^\infty(0,T; H_0^1(\Omega)) \\ u_\varepsilon' \rightharpoonup u' & \text{weakly} * \text{ in } L^\infty(0,T; L^2(\Omega)) \end{cases}$$

and for all $t \in [0,T]$

$$\text{(7)} \qquad \begin{cases} u_\varepsilon(t) \rightharpoonup u(t) & \text{weakly} * \text{ in } H_0^1(\Omega) \\ u_\varepsilon'(t) \rightharpoonup u'(t) & \text{weakly} * \text{ in } L^2(\Omega) \end{cases}$$

where u satisfies the homogenized system

$$\text{(8)} \qquad \begin{cases} u'' + \mathcal{A}u = f & \text{in } \Omega \times (0,T) \\ u = 0 \quad \text{on } \partial\Omega \times (0,T) \\ u(0) = u^0 \,,\ u'(0) = u^1 & \text{in } \Omega \end{cases}$$

with

$$\text{(9)} \qquad \mathcal{A} = \begin{cases} -\alpha_{ij} \dfrac{\partial^2}{\partial x_i \partial x_j} & \text{in } D_\alpha \\ -q_{ij} \dfrac{\partial^2}{\partial x_i \partial x_j} & \text{in } \Omega \setminus \overline{D_\alpha}, \end{cases}$$

$$(10) \qquad q_{ij} = \frac{1}{|Y|} \int_Y (a_{ij} - a_{kj} \frac{\partial \chi^i}{\partial y_k})\, dy,$$

q_{ij} being a positive definite matrix and $\chi^i (i = 1, 2, ..., n)$ the solution of the system

$$(11) \qquad \begin{cases} -\dfrac{\partial}{\partial y_l}(a_{kl}(y)\dfrac{\partial(\chi^i - y_i)}{\partial y_k}) = 0 & in\ Y \\ \chi^i \quad Y-periodic. \end{cases}$$

The homogenized operator is the classical one in homogenization theory, see for details A. Bensoussan, J.-L. Lions and G. Papanicolaou [3]. In this way in $\Omega \setminus \overline{D_\alpha}$, the operator A_ϵ is replaced by $\mathcal{A}$ which has constant coefficients and is an elliptic operator.

Let us remark that due to the definition of A_ϵ which avoids the heterogeneities in D_α, we can make precise the nature of the convergence of $\frac{\partial u_\epsilon}{\partial \nu_{A_\epsilon}}$. From Lemma 4.5 of D. Cioranescu, P. Donato and E. Zuazua [5] we deduce the following result:

PROPOSITION 2. *Let u_ϵ be the solution of system (4). Then, under assumption (5) we have*

$$\frac{\partial u_\epsilon}{\partial \nu_{A_\epsilon}} \rightharpoonup \frac{\partial u}{\partial \nu_{\mathcal{A}}} \quad \text{weakly in } L^2(\partial\Omega \times (O,T)).$$

REMARK 3. From definitions (2) and (9), one has

$$\frac{\partial u_\epsilon}{\partial \nu_{A_\epsilon}}\Big|_\Sigma = \frac{\partial u_\epsilon}{\partial \nu_\alpha}\Big|_\Sigma$$

and

$$\frac{\partial u}{\partial \nu_{\mathcal{A}}}\Big|_\Sigma = \frac{\partial u}{\partial \nu_\alpha}\Big|_\Sigma$$

where

$$\frac{\partial}{\partial \nu_\alpha} = \alpha_{ij}\frac{\partial}{\partial x_j} n_i$$

n being the exterior normal to $\partial\Omega$. □

Of course, we can reformulate the problem of the exact controllability for system (8). Let $\alpha_{ij} = q_{ij}$ and

$$\Gamma^0 = \{x \in \partial\Omega \ / \ (x - x_0) \cdot n(x) > 0\},$$

x_0 being an arbitrary fixed point in $\mathbb{R}^n$. Then one can apply to system (8) the exact controllability result due to V. Komornik [7] which shows that there exists $T_0 = T_0(\mathcal{A}, diam\ \Omega)$

such that when $T > T_0$, for every $\{y^0, y^1\} \in L^2(\Omega) \times H^{-1}(\Omega)$ there exists a control $v \in L^2(\Gamma^0 \times (0,T))$ such that if y is solution of

$$(12)\qquad \begin{cases} y'' + \mathcal{A}y = 0 & \text{in } \Omega \times (0,T) \\ y = v & \text{on } \Gamma^0 \times (0,T) \\ y = 0 & \text{on } (\partial\Omega \setminus \Gamma^0) \times (0,T) \\ y(0) = y^0 \ , \ y'(0) = y^1 & \text{in } \Omega \end{cases}$$

then

$$y(T) = y'(T) = 0.$$

We have the following approximate controllability result:

THEOREM 4. *Suppose that $\alpha_{ij} = q_{ij}$. Let y_ε be the solution of the problem*

$$(13)\qquad \begin{cases} y_\varepsilon'' + A_\varepsilon y_\varepsilon = 0 & \text{in } \Omega \times (0,T) \\ y_\varepsilon = v & \text{on } \Gamma^0 \times (0,T) \\ y_\varepsilon = 0 & \text{on } (\partial\Omega \setminus \Gamma^0) \times (0,T) \\ y_\varepsilon(0) = y^0 \ , \ y_\varepsilon'(0) = y^1 & \text{in } \Omega \end{cases}$$

where A_ε is defined by (2) and (3) and v is the exact control from (12). Then

$$(14)\qquad y_\varepsilon \to y \quad \text{strongly in } L^2(\Omega \times (0,T))$$

and

$$(15)\qquad \begin{cases} \text{i. } y_\varepsilon(T) \to 0 & \text{strongly in } L^2(\Omega) \\ \text{ii. } y_\varepsilon'(T) \to 0 & \text{strongly in } H^{-1}(\Omega) \end{cases}$$

when $\varepsilon \to 0$.

REMARK 4. The same result holds if Γ^0 satisfies the geometric control condition introduced in [2]. □

2. PROOF OF THE MAIN RESULT.

This section is devoted to prove Theorem 4.

The solution y_ε is defined by transposition, i.e. y_ε satisfies

$$(16)\qquad \begin{cases} \displaystyle\int_0^T \int_\Omega y_\varepsilon f_\varepsilon \, dx\, dt = < y_\varepsilon(T), \theta_\varepsilon^1 > - < y^0, \theta_\varepsilon'(0) > - < y_\varepsilon'(T), \theta_\varepsilon^0 > + \\ \qquad\qquad + < y^1, \theta_\varepsilon(0) > + \displaystyle\int_0^T \int_{\partial\Omega} v \frac{\partial \theta_\varepsilon}{\partial \nu_q} \, d\sigma\, dt \end{cases}$$

for all $f_\varepsilon \in L^1(0,T;L^2(\Omega))$ and θ_ε solution of

$$(17)\qquad \begin{cases} \theta_\varepsilon'' + A_\varepsilon\theta_\varepsilon = f_\varepsilon & \text{in } \Omega\times(0,T) \\ \theta_\varepsilon = 0 & \text{on } \partial\Omega\times(0,T) \\ \theta_\varepsilon(T) = \theta_\varepsilon^0\ ,\ \theta_\varepsilon'(T) = \theta_\varepsilon^1 & \text{in } \Omega \end{cases}$$

where $\theta_\varepsilon^0 \in H_0^1(\Omega)$, $\theta_\varepsilon^1 \in L^2(\Omega)$ and $\dfrac{\partial}{\partial\nu_q} = q_{ij}\dfrac{\partial}{\partial x_j}n_i$.

Choose in (16) θ_ε solution of (17) with data satisfying

$$\begin{cases} \{\theta_\varepsilon^0\,,\,\theta_\varepsilon^1\} \rightharpoonup \{\theta^0\,,\,\theta^1\} & \text{weakly in } H_0^1(\Omega)\times L^2(\Omega) \\ f_\varepsilon \rightharpoonup f & \text{weakly in } L^1(0,T;L^2(\Omega)). \end{cases}$$

The above assumptions being exactly hypothesis (5) we can apply to system (17) Propositions 1 and 2. Consequently

$$\theta_\varepsilon \rightharpoonup \theta \quad \text{weakly} * \text{ in } L^\infty(0,T;H_0^1(\Omega))$$

where θ is solution of the homogenized system

$$(18)\qquad \begin{cases} \theta'' + \mathcal{A}\theta = f & \text{in } \Omega\times(0,T) \\ \theta = 0 & \text{on } \partial\Omega\times(0,T) \\ \theta(T) = \theta^0\ ,\ \theta'(T) = \theta^1 & \text{in } \Omega \end{cases}$$

with

$$\begin{cases} \theta_\varepsilon(t) \rightharpoonup \theta(t) & \text{weakly} * \text{ in } H_0^1(\Omega) \\ \theta_\varepsilon'(t) \rightharpoonup \theta'(t) & \text{weakly} * \text{ in } L^2(\Omega) \\ \dfrac{\partial\theta_\varepsilon}{\partial\nu_q} \rightharpoonup \dfrac{\partial\theta}{\partial\nu_q} & \text{weakly in } L^2(\partial\Omega\times(0,T)). \end{cases}$$

With these convergences we can also pass to the limit in identity (16) to obtain

$$(19)\qquad \begin{cases} \lim\limits_{\varepsilon\to 0}\Big[\displaystyle\int_0^T\int_\Omega y_\varepsilon f_\varepsilon\,dx\,dt - < y_\varepsilon(T),\theta_\varepsilon^1 > + < y_\varepsilon'(T),\theta_\varepsilon^0 > \Big] = \\ \qquad\qquad = - < y^0,\theta'(0) > + < y^1,\theta(0) > + \displaystyle\int_0^T\int_{\partial\Omega} v\frac{\partial\theta}{\partial\nu_q}\,d\sigma\,dt. \end{cases}$$

We choose now successively $f_\varepsilon, \theta_\varepsilon^0, \theta_\varepsilon^1$ in order to have in the square brackets only one non vanishing term. First of all let $\theta_\varepsilon^0 = \theta_\varepsilon^1 = 0$. Therefore, the function θ, the corresponding limit of θ_ε with these particular data, satisfies (18) with $\theta^0 = \theta^1 = 0$. Consider now the formulation by transposition of system (12), written with this θ as test function. We have

$$\int_0^T \int_\Omega y\, f\, dx\, dt = < y^0, \theta'(0) > + < y^1, \theta(0) > + \int_0^T \int_{\partial\Omega} v \frac{\partial \theta}{\partial \nu_q}\, d\sigma\, dt$$

which obviously leads to

$$\lim_{\varepsilon \to 0} \int_0^T \int_\Omega y_\varepsilon f_\varepsilon\, dx\, dt = \int_0^T \int_\Omega y\, f\, dx\, dt,$$

hence (14) is proved.

The second choice $f_\varepsilon = \theta^0_\varepsilon = 0$ gives convergence (15)i and finally to prove (15)ii we take $f_\varepsilon = \theta^1_\varepsilon = 0$. □

REMARK 5. Approximated controllability results in perforated domains were obtained by the authors in [6].

REFERENCES.

[1] M. AVELLANEDA, C. BARDOS and J. RAUCH, Contrôlabilité exacte, homogénéisation et localisation d'ondes dans un milieu non-homogène, to appear.

[2] C. BARDOS, G. LEBEAU and J. RAUCH, Contrôle et stabilisation dans les problèmes hyperboliques, Appendix II in J.-L. Lions [8, Tome 1], 492-537.

[3] A. BENSOUSSAN, J.-L. LIONS and G. PAPANICOLAOU, *Asymptotic Analysis for Periodic Structures*, North-Holland, Amsterdam (1978).

[4] S. BRAHIM-OTSMANE, G. A. FRANCFORT and F. MURAT, Correctors for the homogenization of the wave and heat equations, J. Math. Pures et Appl., to appear.

[5] D. CIORANESCU, P. DONATO and E. ZUAZUA, Exact boundary controllability for the wave equation in domains with small holes, J. Math. Pures et Appl., to appear.

[6] D. CIORANESCU, P. DONATO and E. ZUAZUA, Approximate boundary controllability in perforated domains, to appear.

[7] V. KOMORNIK, Exact controllability in short time for the wave equation, Ann. Inst. Henri Poincaré, **6**, 2 (1989), 153-164.

[8] J.-L. LIONS, *Contrôlabilité exacte, perturbations et stabilisation de systèmes distribués. Tome 1. Contrôlabilité exacte.* Masson, RMA8 (1988). *Tome 2. Perturbations.* Masson, RMA9 (1988).

[9] D. L. RUSSELL, Controllability and stabilization theory for linear partial differential equations. Recent progress and open questions, SIAM Review **20** (1978), 639-739.

TRUSS STRUCTURES: FOURIER CONDITIONS AND EIGENVALUE PROBLEM

D. Cioranescu – J. Saint Jean Paulin

We consider here truss-like structures made up by thin bars periodically distributed. The thickness of the material is $\varepsilon\delta$ and is small compared to the period ε. We study the asymptotic behaviour of the solution of an elliptic problem posed in such a structure with Fourier or Dirichlet conditions. The case of Neumann boundary condition is treated in [4]. We also study here the problem of eigenvalues.

To simplify the exposition, we consider here only the bidimensional case but all the results can be extended to higher dimensions and for structures where the material is concentrated along thin layers instead of thin bars.

1. FOURIER BOUNDARY CONDITIONS

Let Ω be an open domain in $\mathbb{R}^2$ with smooth boundary. We cover periodically Ω by square cells $(0,\varepsilon) \times (0,\varepsilon)$, homothetic to a fixed cell $Y = (0,1) \times (0,1)$ which contains a square hole $T_\delta = (\frac{\delta}{2}, 1 - \frac{\delta}{2}) \times (\frac{\delta}{2}, 1 - \frac{\delta}{2})$.

Denote by $Y_\delta = Y \backslash T_\delta$, resp.$\Omega_{\varepsilon\delta}$ the part of Y, resp. of Ω, corresponding to the material. We assume that Ω contains N_ε holes, where N_ε is an integer. This assumption requires a special geometry of Ω and signifies that ε takes for instance, values of the type $C\backslash 2^n$.

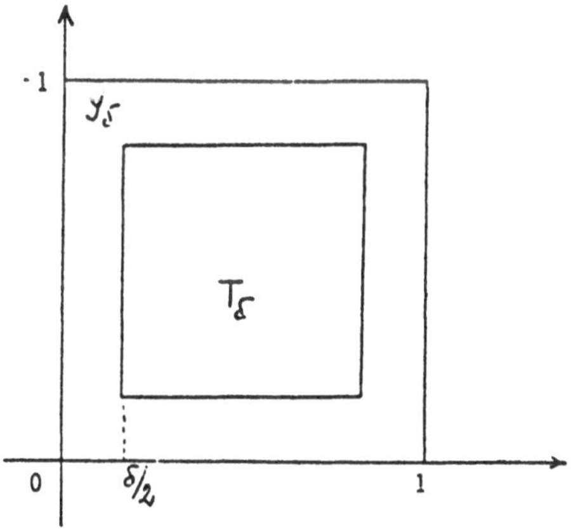

Consider the problem

$$(1.1) \qquad \begin{cases} -\Delta u_{\varepsilon\delta} = f & \text{in } \Omega_{\varepsilon\delta} \\ \dfrac{\partial u_{\varepsilon\delta}}{\partial n} + \varepsilon^\gamma a\, u_{\varepsilon\delta} = g_\varepsilon & \text{on } \partial T_{\varepsilon\delta} \text{ (the boundary of the holes)} \\ u_{\varepsilon\delta} = 0 & \text{on } \partial\Omega \end{cases}$$

where $a \in \mathbb{R}_+$, $\gamma \in \mathbb{R}$, n is the outward normal to $\Omega_{\varepsilon\delta}$ and g_ε is a Y- periodic function defined by

$$g_\varepsilon(x) = g(\frac{x}{\varepsilon}).$$

Let us introduce the following notations:

$$\mathcal{M}_\delta(g) = \int_{\partial T_{\varepsilon\delta}} g(s)\, ds \tag{1.2}$$

$$\mathcal{A}_\delta = q_{ij}^\delta \frac{\partial^2}{\partial x_i \partial x_j}. \tag{1.3}$$

The operator $\mathcal{A}_\delta$ is the homogenized operator corresponding to the Neumann problem , i.e. to the system

$$\begin{cases} -\Delta v_{\varepsilon\delta} = f & \text{in } \Omega_{\varepsilon\delta} \\ v_{\varepsilon\delta} = 0 & \text{on } \partial\Omega \\ \dfrac{\partial v_{\varepsilon\delta}}{\partial n} = 0 & \text{on } \partial T_{\varepsilon\delta}. \end{cases} \tag{1.4}$$

We are interested by the asymptotic behaviour of $u_{\varepsilon\delta}$ when ε and δ tend to zero.Fix δ and let$\varepsilon \to 0$ in (1.4). Cioranescu-Saint Jean Paul [3] proved the following homogenization result:

PROPOSITION 1.1 [3]. *Suppose that $f \in L^2(\Omega)$. Then there exists an extension operator $P_{\varepsilon\delta} \in \mathcal{L}(H^k(\Omega_{\varepsilon\delta}), H^k(\Omega)), k = 0, 1$ such that*

$$P_{\varepsilon\delta} v_{\varepsilon\delta} \rightharpoonup V^* \quad \textit{weakly in } H_0^1(\Omega).$$

where V^* is the solution of the homogenized system

$$\begin{cases} -\mathcal{A}_\delta V^* = \mid Y_\delta \mid f & \text{in } \Omega \\ V^* = 0 & \text{on } \partial\Omega \end{cases} \tag{1.4}$$

and $\mid Y_\delta \mid = 2\delta(1-\delta)$ is the measure of Y_δ.

The coefficients q_{ij}^δ are defined by

$$q_{ij}^\delta = \delta_{ij} \mid Y_\delta \mid - \int_{Y_\delta} \frac{\partial X_\delta^j}{\partial y_i}\, dy, \tag{1.5}$$

where the functions X_δ^j are solutions of the system

$$(1.6)\qquad \begin{cases} -\Delta X_\delta^j = 0 & \text{in } Y_\delta \\ \dfrac{\partial (X_\delta^j - y_j)}{\partial n} = 0 & \text{on } \partial T_\delta \\ X_\delta^j \quad Y-\text{periodic.} \end{cases}$$

□

Obviously, taking $\varepsilon \to 0$ with δ fixed is equivalent to the homogenization of system (1.1). We refer the reader to [1], [6], [7] for general methods of homogenization. As regarding the particular case of problem (1.1), let us recall the following results obtained by Cioranescu-Donato in [2]:

Proposition 1.2 [2]. *Suppose that*

$$(1.6)\qquad f \in L^2(\Omega),\ g \in L^2(\partial T_\delta),$$

$$(1.7)\qquad \mathcal{M}_\delta(g) \neq 0.$$

Then there exists an extension operator $P_{\varepsilon\delta} \in \mathcal{L}(H^k(\Omega_{\varepsilon\delta}), H^k(\Omega)), k = 0,1$ such that

$$\varepsilon\, P_{\varepsilon\delta} u_{\varepsilon\delta} \rightharpoonup U_\delta \qquad \text{weakly in } H_0^1(\Omega)$$

where U_δ is solution of a homogenized system depending on γ.

If $\gamma > 1$, U_δ satisfies

$$(1.8)\qquad \begin{cases} -\mathcal{A}_\delta U_\delta = \mathcal{M}_\delta(g) & \text{in } \Omega \\ U_\delta = 0 & \text{on } \partial\Omega. \end{cases}$$

If $\gamma = 1$, U_δ satisfies

$$(1.9)\qquad \begin{cases} -\mathcal{A}_\delta U_\delta + a \mid \partial T_\delta \mid U_\delta = \mathcal{M}_\delta(g) & \text{in } \Omega \\ U_\delta = 0 & \text{on } \partial\Omega \end{cases}$$

where $\mid \partial T_\delta \mid = 4(1-\delta)$. □

Proposition 1.3 [2]. *Assume (1.6) and suppose that*

$$\mathcal{M}_\delta(g) = 0 \tag{1.10}$$

Then

$$P_{\varepsilon\delta} u_{\varepsilon\delta} \rightharpoonup u_\delta \qquad \text{weakly in } H_0^1(\Omega)$$

where u_δ is solution of a homogenized system depending on γ.

If $\gamma > 1$, u_δ satisfies

$$\begin{cases} -\mathcal{A}_\delta u_\delta = \mid Y_\delta \mid f & \text{in } \Omega \\ u_\delta = 0 & \text{on } \partial\Omega. \end{cases} \tag{1.11}$$

If $\gamma = 1$, u_δ satisfies

$$\begin{cases} -\mathcal{A}_\delta u_\delta + a \mid \partial T_\delta \mid u_\delta = \mid Y_\delta \mid f & \text{in } \Omega \\ u_\delta = 0 & \text{on } \partial\Omega. \end{cases} \tag{1.12}$$

□

In the case $\gamma < 1$, the extension $P_{\varepsilon\delta} u_{\varepsilon\delta}$, eventually rescaled, converges weakly to zero. In [2] the following convergences are established:

- if $-1 \leq \gamma < 1$ and $\mathcal{M}_\delta(g) \neq 0$, then $\varepsilon^{\frac{1+\gamma}{2}} P_{\varepsilon\delta} u_{\varepsilon\delta} \rightharpoonup 0$ weakly in $H_0^1(\Omega)$,
- if $-1 \leq \gamma < 1$ and $\mathcal{M}_\delta(g) = 0$, then $P_{\varepsilon\delta} u_{\varepsilon\delta} \rightharpoonup 0$ weakly in $H_0^1(\Omega)$,
- if $\gamma < -1$, then $P_{\varepsilon\delta} u_{\varepsilon\delta} \to 0$ strongly in $H_0^1(\Omega)$.

Let us remark that the first convergence can be improved by rescaling differently $u_{\varepsilon\delta}$.

Proposition 1.4. *Assume (1.6). Then, if $-1 \leq \gamma < 1$*

$$\varepsilon^\gamma P_{\varepsilon\delta} u_{\varepsilon\delta} \rightharpoonup V_\delta \qquad \text{weakly in } L^2(\Omega)$$

where

$$V_\delta = \frac{1}{a \mid \partial T_\delta \mid} \int_{\partial T_\delta} g(s)\, ds. \tag{1.13}$$

Proof . Recall the a priori estimates obtained in [2] for the solutions of system (1.1)

$$\left\{\begin{array}{l} \| \varepsilon^{\frac{1+\gamma}{2}} u_{\varepsilon\delta} \|_{L^2(\Omega_{\varepsilon\delta})} \leq C \varepsilon^{\frac{1-\gamma}{2}} \\ \| \varepsilon^{\frac{1+\gamma}{2}} \nabla u_{\varepsilon\delta} \|_{L^2(\Omega_{\varepsilon\delta})} \leq C. \end{array}\right. \tag{1.14}$$

From the first inequality it follows

$$\| \varepsilon^{\gamma} u_{\varepsilon\delta} \|_{L^2(\Omega_{\varepsilon\delta})} \leq C,$$

hence

$$\varepsilon^{\gamma} P_{\varepsilon\delta} u_{\varepsilon\delta} \rightharpoonup V_\delta \qquad \text{weakly in } L^2(\Omega). \tag{1.15}$$

The weak formulation of system (1.1) is

$$\int_{\Omega_{\varepsilon\delta}} \nabla u_{\varepsilon\delta} \nabla v \, dx = \int_{\Omega_{\varepsilon\delta}} f v \, dx + \int_{\partial T_{\varepsilon\delta}} u_{\varepsilon\delta} \, g \, v \, ds - a \varepsilon^{\gamma} \int_{\partial T_{\varepsilon\delta}} u_{\varepsilon\delta} \, v \, ds, \tag{1.16}$$
$$\forall v \in H^1(\Omega_{\varepsilon\delta}), v = 0 \text{ on } \partial\Omega.$$

For all Y-periodic function let $\mu_h^{\varepsilon\delta}$ be the measure defined as follows:

$$< \mu_h^{\varepsilon\delta}, \phi > = \varepsilon \int_{\partial T_{\varepsilon\delta}} h(\frac{x}{\varepsilon}) \, \phi \, ds, \qquad \forall \phi \in H_0^1(\Omega).$$

Lemma 3.3 from [2] states that

$$\mu_h^{\varepsilon\delta} \to \mu_h^{\delta} \qquad \text{strongly in } H^{-1}(\Omega),$$

where

$$< \mu_h^{\delta}, \phi > = \mathcal{M}_\delta(h) \int_\Omega \phi \, dx, \qquad \forall \phi \in H_0^1(\Omega). \tag{1.17}$$

Let $\xi_{\varepsilon\delta}$ be the extension by zero of $\nabla u_{\varepsilon\delta}$ in $T_{\varepsilon\delta}$. From (1.14) one has

$$\varepsilon^{\frac{1+\gamma}{2}} \xi_{\varepsilon\delta} \rightharpoonup \xi_\delta^* \qquad \text{weakly in } L^2(\Omega). \tag{1.18}$$

Using the definition of $\mu_h^{\varepsilon\delta}$, (1.16) can be written as follows:

$$\varepsilon^{\frac{1-\gamma}{2}} \int_\Omega \varepsilon^{\frac{1+\gamma}{2}} \xi_{\varepsilon\delta} \nabla \phi = \varepsilon \int_\Omega \chi_{\Omega_{\varepsilon\delta}} f \, \phi \, dx + < \mu_g^{\varepsilon\delta}, \phi > -$$
$$- < \mu_a^{\varepsilon\delta}, (\varepsilon^{\gamma} P_{\varepsilon\delta} u_{\varepsilon\delta}) \, \phi >, \quad \forall \phi \in \mathcal{D}(\Omega)$$

where $\chi_{\Omega_{\varepsilon\delta}}$ is the characteristic function of $\Omega_{\varepsilon\delta}$.

It is now possible to pass to the limit in the last identitity as $\varepsilon \to 0$ in view of convergences (1.15), (1.17) and (1.18). Taking into account that $-1 \leq \gamma < 1$, one has

$$0 = \mathcal{M}_\delta(g) \int_\Omega \phi \, dx - \mathcal{M}_\delta(a) \int_\Omega U_\delta \, \phi \, dx,$$

whence (1.13). □

Our aim is to establish the dependence on δ of the solutions of systems (1.8), (1.9), (1.11) or (1.12), the parameter δ being the thickness of the material in the reference cell Y. To do that we shall follow along the lines Cioranescu-Saint Jean Paulin [4] where the asymptotic behaviour of reinforced or honeycomb structures with homogeneous Neumann boundary condition is studied. We prove the following results:

Theorem 1.5. *Let $f \in L^2(\Omega)$ and suppose that g is a Y-periodic function such that $g \in H^{\frac{1}{2}+\alpha}(\partial Y)$ for $\alpha > 0$. Assume further that*

(1.19)
$$\int_{\partial Y} g(s) \, ds \neq 0.$$

Then

(i) the solution U_δ of system (1.8) satisfies

$$\delta U_\delta \rightharpoonup U^* \quad \textit{weakly in } H_0^1(\Omega)$$

with U^* solution of the system

(1.20)
$$\begin{cases} -\Delta U^* = \displaystyle\int_{\partial Y} g(s) \, ds & \textit{in } \Omega \\ U^* = 0 & \textit{on } \partial\Omega, \end{cases}$$

(ii) the solution U_δ of system (1.9) satisfies

$$U_\delta \rightharpoonup U_0^* \quad \textit{weakly in } L^2(\Omega)$$

with

(1.21)
$$U_0^* = \frac{1}{4a} \int_{\partial Y} g(s) \, ds,$$

(iii) the function V_δ defined by (1.13) satisfies

$$V_\delta \rightharpoonup U_0^* \quad \text{weakly in } L^2(\Omega)$$

with U_0^* given by (1.21).

Proof. It is clear that if (1.19) holds, for δ small enough (1.7) also holds and moreover

$$\mathcal{M}_\delta(g) \to \int_{\partial Y} g(s).$$

On the other hand, as a consequence of the a priori estimates which can be obtained from system (1.5) (see for details [2]), one gets the following convergence for the coefficients q_{ij}^δ defining the operator $\mathcal{A}_\delta$ in formulae (1.8), (1.9), (1.11) or (1.12):

$$\delta^{-1} q_{ij}^\delta \to q_{ij}^* = \delta_{ij}, \tag{1.22}$$

where δ_{ij} is the Kronecker symbol. Then (1.20) follows immediately.

From system (1.9) we deduce the estimate

$$\| U_\delta \|_{L^2(\Omega)} \leq C.$$

Take in (1.9) $\psi \in \mathcal{D}(\Omega)$ as test function and integrate twice par parts. It comes

$$\int_\Omega q_{ij}^\delta U_\delta \nabla\psi \, dx + 4(1-\delta)\, a \int_\Omega U_\delta \psi \, dx = \Big(\int_{\partial T_\delta} g(s)\, ds\Big)\Big(\int_\Omega \psi \, dx\Big).$$

By (1.22) the first integral tends to zero and therefore (1.21) is also proved. □

Theorem 1.6. *Let $f \in L^2(\Omega)$ and suppose that g is a Y-periodic function such that $g \in C^1(\mathbb{R}^2)$ and*

$$\int_{\partial Y} g(s)\, ds = 0. \tag{1.23}$$

Assume further that there exists $\delta_0 > 0$ such that

$$\mathcal{M}_\delta(g) \neq 0, \quad \forall \delta, 0 < \delta < \delta_0. \tag{1.24}$$

Then

(i) the solution U_δ of system (1.8) satisfies

$$U_\delta \rightharpoonup U \quad \text{weakly in } H_0^1(\Omega)$$

with

$$(1.25) \qquad \begin{cases} -\Delta U = -4\,g(0,0) & \text{in } \Omega \\ U = 0 & \text{on } \partial\Omega, \end{cases}$$

(ii) the solution U_δ of system (1.9) satisfies

$$\delta^{-1} U_\delta \rightharpoonup U_0 \quad \text{weakly in } L^2(\Omega)$$

with

$$(1.26) \qquad U_0 = -\frac{1}{a}\,g(0,0),$$

(iii) the function V_δ defined by (1.13) satisfies

$$\delta^{-1} V_\delta \rightharpoonup U_0 \quad \text{weakly in } L^2(\Omega)$$

with U_0 given by (1.26).

Proof. Remark first that in the proof of Theorem 1.5 the only point where we use hypothesis (1.19) is when stating that $\mathcal{M}_\delta(g) \neq 0$ for δ sufficiently small. Now, this property is assumption (1.24). Consequently system (1.20) and relation (1.21) still hold and moreover, as $\int_{\partial Y} g(s)\,ds = 0$, it follows that $U^* = 0$ and $U_0^* = 0$. But we can evaluate U_δ at a higher order by developing $\mathcal{M}_\delta(g)$ in terms of δ. Recall that by hypothesis $g \in C^1(\mathbb{R}^2)$. We have

$$\begin{aligned} \mathcal{M}_\delta(g) = \int_{\partial T_\delta} g(s)\,ds = \int_{\partial Y} g(s)\,ds - \delta\,[g(0,0) + g(1,0) + g(0,1) + g(1,1)] + \\ + \frac{\delta}{2}\Big[\int_0^1 \frac{\partial g}{\partial y_2}(y_1,0)\,dy_1 - \int_0^1 \frac{\partial g}{\partial y_2}(y_1,1)\,dy_1 + \\ + \int_0^1 \frac{\partial g}{\partial y_1}(0,y_2)\,dy_2 - \int_0^1 \frac{\partial g}{\partial y_1}(1,y_2)\,dy_1\Big] + \delta^2\,o(1). \end{aligned}$$

Making use of hypothesis (1.23) and the periodicity of g in Y it comes that

$$(1.27) \qquad \delta^{-1}\mathcal{M}_\delta(g) \to -4\,g(0,0).$$

We pass to the limit in (1.8), (1.9) and (1.13) after rescaling eventually by δ^{-1} and using (1.27) we obtain (1.25) and (1.26). $\square$

THEOREM 1.7. *Let $f \in L^2(\Omega)$ and suppose that g is a Y-periodic function such that $g \in H^{\frac{1}{2}+\alpha}(\partial Y)$ for $\alpha > 0$ and*

$$\int_{\partial Y} g(s)\, ds = 0. \tag{1.28}$$

Assume further that there exists $\delta_0 > 0$ such that

$$\mathcal{M}_\delta(g) = 0 \quad \forall \delta, 0 < \delta < \delta_0. \tag{1.29}$$

Then

(i) the solution u_δ of system (1.11) satisfies

$$u_\delta \rightharpoonup u^* \quad \text{weakly in } H_0^1(\Omega)$$

with

$$\begin{cases} -\Delta u^* = 2f & \text{in } \Omega \\ u^* = 0 & \text{on } \partial\Omega, \end{cases} \tag{1.30}$$

(ii) the solution u_δ of system (1.12) satisfies

$$\delta^{-1} u_\delta \rightharpoonup u_0^* \quad \text{weakly in } L^2(\Omega)$$

with

$$u_0^* = \frac{1}{2a} f. \tag{1.31}$$

Proof. To begin with, remark that (1.11) is exactly the homogenized system given in Proposition 1.1 when treating homogeneous Neumann boundary conditions. The passage to the limit in this system with $\delta \to 0$ was already done in [4]. It remains to prove (1.31).

One has easily from system (1.12) the a priori estimate

$$\| u_\delta \|_{L^2(\Omega)} \leq C\delta$$

and reasoning as for the proof of (1.22) one gets identity (1.31). □

Conclusion. To resume the results obtained below when passing to the limit in system (1.1) successively with $\varepsilon \to 0$ and $\delta \to 0$ we have the following situations:

$$\int_{\partial Y} g(s)\,ds \neq 0$$

$\gamma > 1$ $\qquad \varepsilon\,\delta\, P_{\varepsilon\delta} u_{\varepsilon\delta} \rightharpoonup U^*$

$$\begin{cases} -\Delta U^* = \int_{\partial Y} g(s)\,ds & \text{in } \Omega \\ U^* = 0 & \text{on } \partial\Omega. \end{cases}$$

$\gamma = 1$ $\qquad \varepsilon\, P_{\varepsilon\delta} u_{\varepsilon\delta} \rightharpoonup U_0^* = \frac{1}{4a} \int_{\partial Y} g(s)\,ds.$

$-1 \le \gamma < 1$ $\qquad \varepsilon^{\gamma}\, P_{\varepsilon\delta} u_{\varepsilon\delta} \rightharpoonup U_0^* = \frac{1}{4a} \int_{\partial Y} g(s)\,ds.$

$$\int_{\partial Y} g(s)\,ds = 0$$

$\forall \delta : 0 < \delta < \delta_0 < 0$	$\mathcal{M}_\delta(g) \neq 0$	$\mathcal{M}_\delta(g) = 0$
$\gamma > 1$	$\varepsilon\delta^{-1} P_{\varepsilon\delta} u_{\varepsilon\delta} \rightharpoonup U$	$P_{\varepsilon\delta} u_{\varepsilon\delta} \rightharpoonup u^*$
	$-\Delta U = -4g(0,0)$ in Ω	$-\Delta u^* = 2f$ in Ω
	$U = 0$ on $\partial\Omega$	$u^* = 0$ on $\partial\Omega$
$\gamma = 1$	$\varepsilon\delta^{-1} P_{\varepsilon\delta} u_{\varepsilon\delta} \rightharpoonup U_0$	$\delta^{-1} P_{\varepsilon\delta} u_{\varepsilon\delta} \rightharpoonup u_0^*$
	$U_0 = -\frac{1}{a} g(0,0)$	$u_0^* = \frac{1}{2a} f$

$-1 \le \gamma < 1$ $\qquad \varepsilon^{\gamma}\,\delta^{-1}\, P_{\varepsilon\delta} u_{\varepsilon\delta} \rightharpoonup U_0^* = -\frac{1}{a} g(0,0).$

2. DIRICHLET BOUNDARY CONDITIONS

In the same geometrical framework as in Section 1, let us consider the Dirichlet problem

$$\text{(2.1)} \qquad \begin{cases} -\varepsilon^2 \Delta u_{\varepsilon\delta} = f & \text{in } \Omega_{\varepsilon\delta} \\ u_{\varepsilon\delta} = 0 & \text{on } \partial\Omega_{\varepsilon\delta} \end{cases}$$

Introduce the function w_δ defined on the reference cell Y as solution of the system

$$\text{(2.2)} \qquad \begin{cases} -\Delta w_\delta = 1 & \text{in } Y_\delta \\ w_\delta = 0 & \text{on } \partial T_\delta \\ w_\delta \quad Y - \text{periodic} & \end{cases}$$

In the sequel, for any function v defined for instance, in $L^2(\Omega_{\varepsilon\delta})$, $\tilde{v}$ will denote its extension by zero to the whole of Ω. The asymptotic behaviour of $u_{\varepsilon\delta}$ as $\varepsilon \to 0$ is studied in Lions [6]. Let us recall the result herein obtained:

PROPOSITION 2.1 [6]. *Suppose that* $f \in H_0^1(\Omega)$. *Then there exists a positive constant* $C > 0$ *such that*

$$\| u_{\varepsilon\delta} - w_\delta^\varepsilon f \|_{L^2(\Omega_{\varepsilon\delta})} \leq C\varepsilon$$

where

$$w_\delta^\varepsilon(x) = w_\delta(\frac{x}{\varepsilon}).$$

Moreover,

$$\widetilde{u_{\varepsilon\delta}} \to \Big(\int_{Y_\delta^*} w_\delta(y)\, dy\Big) f \quad \textit{weakly in } L^2(\Omega).$$

□

We let now $\delta \to 0$. The limit behaviour of $u_{\varepsilon\delta}$ is given by the following result:

THEOREM 2.2. *Let* $f \in H_0^1(\Omega)$. *Then*

$$\lim_{\delta\to 0} \lim_{\varepsilon\to 0} (\delta^{-3}\, \widetilde{u_{\varepsilon\delta}}) = \frac{1}{6} f \quad \textit{weakly in } L^2(\Omega).$$

Proof. From system (2.2) defining w_δ and taking into account that the Poincare constant for the domain Y_δ is of order δ, one gets

$$\text{(2.3)} \qquad \begin{cases} \| \nabla w_\delta \|_{[L^2(Y_\delta)]^2} \leq C\, \delta^{\frac{3}{2}} \\ \| w_\delta \|_{L^2(Y_\delta)} \leq C\, \delta^{\frac{5}{2}} \end{cases}$$

We now use the same techniques as in [4]. The structure being periodic in all the directions of $\mathbb{R}^2$, first we translate Y_δ into $\mathcal{Y}_\delta$ with

$$\mathcal{Y}_\delta = H_\delta \cup V_\delta$$

where H_δ and V_δ are respectively the sets

$$H_\delta = \{y \in \mathbb{R}^2 \mid -\frac{1}{2} \leq y_1 \leq +\frac{1}{2},\ -\frac{\delta}{2} \leq y_2 \leq +\frac{\delta}{2}\}$$
$$V_\delta = \{y \in \mathbb{R}^2 \mid -\frac{\delta}{2} \leq y_1 \leq +\frac{\delta}{2},\ -\frac{1}{2} \leq y_2 \leq +\frac{1}{2}\}.$$

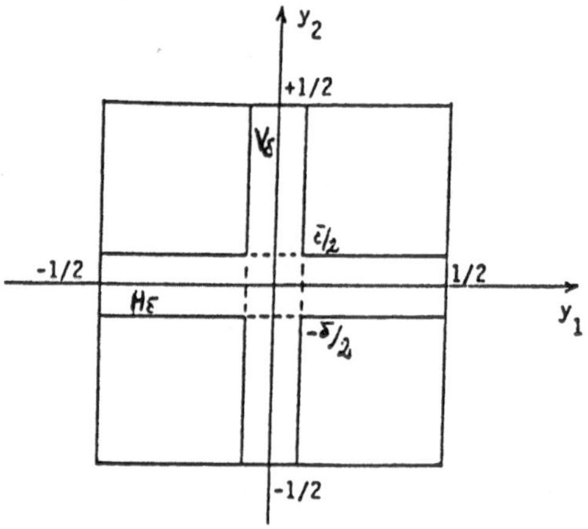

Now, via affinities, H_δ and V_δ are transformed into $\mathcal{Y} = (-\frac{1}{2}, +\frac{1}{2})$. Thus every function φ defined on $\mathcal{Y}_\delta$, is transformed by the affinity in the direction $0y_1$ with ratio δ^{-1} into φ_H, defined by

$$\varphi_H(z_1, z_2) = \varphi(z_1, \delta z_2)$$

where

$$z_1 = y_1, \quad z_2 = \frac{y_2}{\delta}.$$

The notation φ_V has a similar meaning when dealing with the affinity in the direction $0y_2 : z_1' = \frac{y_1}{\delta}, z_2' = y_2$.

From a priori estimates (2.3), one has the following weak convergences in $L^2(\mathcal{Y})$:

$$\text{(2.4)} \qquad \begin{cases} \delta^{-1} \dfrac{\partial w_{\delta,H}}{\partial z_1} \rightharpoonup h_1 & \delta^{-2} \dfrac{\partial w_{\delta,V}}{\partial z_1'} \rightharpoonup v_1 \\ \delta^{-2} \dfrac{\partial w_{\delta,H}}{\partial z_2} \rightharpoonup h_2 & \delta^{-1} \dfrac{\partial w_{\delta,V}}{\partial z_2'} \rightharpoonup v_2 \\ \delta^{-2}\, w_{\delta,H} \rightharpoonup H & \delta^{-2}\, w_{\delta,V} \rightharpoonup V. \end{cases}$$

Moreover, by the peridicity of w_δ

$$\int_{\mathcal{Y}} h_1\, dy = \int_{\mathcal{Y}} v_2\, dy = 0.$$

Obviously, by identifying the L^2-limits we have

$$h_2 = \frac{\partial H}{\partial z_2}, \quad v_1 = \frac{\partial V}{\partial z_1'}. \tag{2.5}$$

Turning back to $u_{\varepsilon\delta}$, passing to the limit in ε first and in δ after, one has

$$\begin{aligned}\lim_{\delta\to 0}\lim_{\varepsilon\to 0}(\delta^{-3}\,\widetilde{u_{\varepsilon\delta}}) &= \lim_{\delta\to 0}\delta^{-3}\Big[\int_{H_\delta} w_\delta(y)\,dy + \int_{V_\delta} w_\delta(y)\,dy\Big]\,f = \\ &= \lim_{\delta\to 0}\delta^{-3}\Big[\int_{\mathcal{Y}} \delta\, w_{\delta,H}(z)\,dz + \int_{\mathcal{Y}} \delta\, w_{\delta,V}(z')\,dz'\Big]\,f = \\ &= \Big[\int_{\mathcal{Y}} (H + V)\,dy\Big]\,f.\end{aligned} \tag{2.6}$$

To compute $\int_{\mathcal{Y}} H\,dy$, multiply (2.2) by a smooth periodic function φ depending only on y_2. Using the affinities defined below as well as convergences (2.4), we obtain at the limit

$$\int_{\mathcal{Y}} h_2\,\frac{\partial\varphi}{\partial z_2}\,dz = \int_{\mathcal{Y}} \varphi\,dz$$

and then from (2.5), we get

$$-\frac{\partial^2 H}{\partial z_2^2} = 1 \quad \text{for} -\frac{1}{2} \le z_2 \le +\frac{1}{2}$$

therefore H is of the form

$$H(z_1, z_2) = -\frac{1}{2}z_2^2 + A_H(z_1)z_2 + B_H(z_1).$$

In fact we will show that

$$H(z_1, z_2) = -\frac{1}{2}\Big(z_2 - \frac{1}{2}\Big)\Big(z_2 + \frac{1}{2}\Big). \tag{2.7}$$

To do that, let $\psi \in H^1(\mathcal{Y})$. From definition (2.4) of $\dfrac{\partial H}{\partial z_2}$, we have by integration by parts in z_2

$$\begin{aligned}\int_{\mathcal{Y}} \frac{\partial H}{\partial z_2}\,\psi\,dz &= \lim_{\delta\to 0}\delta^{-2}\int_{\mathcal{Y}} \frac{\partial w_{\delta,H}}{\partial z_2}\,\psi\,dz = \\ &= \lim_{\delta\to 0}\delta^{-2}\Big[\int_{\{|z_1|\le\frac{1}{2};\,|z_2|=\frac{1}{2}\}} w_{\delta,H}\,\psi\,n_2\,dz_1 - \int_{\mathcal{Y}} w_{\delta,H}\,\frac{\partial\psi}{\partial z_2}\,dz\Big]\end{aligned}$$

where the normal is directed outwards $\mathcal{Y}$. Pass to the limit and reintegrate by parts, it comes

$$\int_{\mathcal{Y}} \frac{\partial H}{\partial z_2} \psi \, dz = \lim_{\delta \to 0} \delta^{-2} \Big[\int_{\{|z_1| \le \frac{1}{2};\, |z_2| = \frac{1}{2}\}} w_{\delta,H} \, \psi \, n_2 \, dz_1 \Big] + \\ + \int_{\mathcal{Y}} \frac{\partial H}{\partial z_2} \psi \, dz - \int_{\{|z_1| \le \frac{1}{2};\, |z_2| = \frac{1}{2}\}} H \, \psi \, n_2 \, dz_1 .$$

It follows that

$$\int_{\{|z_1| \le \frac{1}{2};\, |z_2| = \frac{1}{2}\}} H \, \psi \, n_2 \, dz_1, = \lim_{\delta \to 0} \delta^{-2} \Big[\int_{\{|z_1| \le \frac{1}{2};\, |z_2| = \frac{1}{2}\}} w_{\delta,H} \, \psi \, n_2 \, dz_1 \Big].$$

The function $w_\delta = 0$ on ∂T_δ, therefore $w_{\delta,H} = 0$ on the set $\{\frac{\delta}{2} <| z_1 |\le \frac{1}{2};\, | z_2 |= \frac{1}{2}\}$, thus

$$\int_{\{|z_1| \le \frac{1}{2};\, |z_2| = \frac{1}{2}\}} H \, \psi \, n_2 \, dz_1, = \lim_{\delta \to 0} \delta^{-2} \Big[\int_{\{|z_1| \le \frac{\delta}{2};\, |z_2| = \frac{1}{2}\}} w_{\delta,H} \, \psi \, n_2 \, dz_1 \Big].$$

If ψ is chosen to have compact support in the set $((-\frac{1}{2}, -\frac{\delta}{2}) \cup (+\frac{\delta}{2}, +\frac{1}{2})) \times (-\frac{1}{2}, +\frac{1}{2})$, for δ sufficiently small, the right hand side term vanishes, hence

$$H = 0 \quad \text{a. e. on } \{z \,||\, z_1 | \le \frac{1}{2};\, | z_2 |= \frac{1}{2}\},$$

which proves formula (2.7). Analogously one has

$$V(z_1, z_2) = -\frac{1}{2}\Big(z_1 - \frac{1}{2}\Big)\Big(z_1 + \frac{1}{2}\Big).$$

These expressions used in (2.6) complete the proof of the theorem. □

3. SOME REMARKS ON THE EIGENVALUE PROBLEM

In the same geometrical framework as in Section 1, let now consider the eigenvalue problem

$$\text{(3.1)} \qquad \begin{cases} -\Delta u_{\varepsilon\delta} = \lambda_{\varepsilon\delta} & \text{in } \Omega_{\varepsilon\delta} \\ \dfrac{\partial u_{\varepsilon\delta}}{\partial n} = 0 & \text{on } \partial T_{\varepsilon\delta} \\ u_{\varepsilon\delta} = 0 & \text{on } \partial\Omega \end{cases}$$

$$\text{(3.2)} \qquad \| u_{\varepsilon\delta} \|_{L^2(\Omega_{\varepsilon\delta})} = 1$$

It is known that the problem (3.1)-(3.2) has a family of solutions: the eigenfunctions $u^r_{\varepsilon\delta}$ and their set is an orthogonal basis for the space $L^2(\Omega_{\varepsilon\delta})$. The corresponding eigenvalues are such that

$$0 < \lambda^1_{\varepsilon\delta} \leq \lambda^2_{\varepsilon\delta} \leq \leq \lambda^r_{\varepsilon\delta} \leq, \quad \lim_{r\to\infty} \lambda^r_{\varepsilon\delta} = +\infty.$$

The homogenization (i.e. passing to the limit as $\varepsilon \to 0$ with δ fixed), is done by Vaninathan [9], [10]. The homogenized system is

$$\begin{cases} \mathcal{A}_\delta u_\delta = \lambda_\delta u_\delta & \text{in } \Omega \\ u_\delta = 0 & \text{on } \partial\Omega \end{cases} \tag{3.3}$$

where the operator $\mathcal{A}_\delta$ is defined by (1.3)-(1.5). Let us recall the homogenization result:

Proposition 3.1 [9]. *Let $\varepsilon \to 0$. Then*

(i) for every r fixed

$$\lim_{\varepsilon\to 0} \lambda^r_{\varepsilon\delta} = \lambda^r_\delta$$

where λ^r_δ is the r-th eigenvalue of system (3.3).

(ii) There exists an extension operator $P_{\varepsilon\delta} \in \mathcal{L}(H^1(\Omega_{\varepsilon\delta}, H^1_0(\Omega))$ such that if λ^r_δ is a simple eigenvalue of (3.2) and v^r_δ an associated eigenvector with $\| v^r_\delta \|_{L^2(\Omega)} = 1$, then one can choose an eigenvector $v^r_{\varepsilon\delta}$ of system (3.1) corresponding to the eigenvalue $\lambda^r_{\varepsilon\delta}$ and such that

$$P_{\varepsilon\delta} v^r_{\varepsilon\delta} \rightharpoonup v^r_\delta \quad \text{weakly in } H^1_0(\Omega)$$

for the whole sequence ε. Moreover

$$\| v^r_{\varepsilon\delta} \|_{L^2(\Omega_{\varepsilon\delta})} = | Y_\delta | . \tag{3.4}$$

□

We now let $\delta \to 0$. For passing to the limit in (3.3) we will make use of a theorem due to Kesavan [5]. This result is related to the H-convergence of elliptic operators of the type $A_\eta = \dfrac{\partial}{\partial x_j}(a^\eta_{ij}\dfrac{\partial}{\partial x_i})$ where η is a parameter. The coefficients a^η_{ij} are supposed to be uniformly bounded, symmetric and satisfying a uniform ellipticity condition. One says that the operators A_η H-converge to the operator A if they satisfy the following assertion:

$$\begin{cases} \forall\{U_\eta\} \text{ with } U_\eta \rightharpoonup U \text{ weakly in } H^1(\Omega) \text{ and } A_\eta U_\eta \to f \text{ strongly in } H^{-1}(\Omega) \Longrightarrow \\ a^\eta_{ij}\dfrac{\partial U_\eta}{\partial x_j} \rightharpoonup a_{ij}\dfrac{\partial U}{\partial x_j} \text{ weakly in } L^2(\Omega),\ i = 1, 2. \end{cases}$$

Consider the problem

$$(3.5)\qquad \begin{cases} A_\eta U_\eta = \lambda_\eta\, U_\eta & \text{in } \Omega \\ U_\eta = 0 & \text{on } \partial\Omega \end{cases}$$

One has then

PROPOSITION 3.2 [5]. *Let λ_η^r be the eigenvalues of (3.5) and U_η^r the corresponding family of orthonormed eigenvectors. Assume that A_η H-converges to A when $\eta \to 0$. Then, for all r*

$$\lambda_\eta^r \to \lambda^r \quad \text{and } U_\eta^r \rightharpoonup U^r \text{ weakly in } H_0^1(\Omega)$$

where λ^r is the r-th eigenvalue of the limit system

$$\begin{cases} AU = \lambda U & \text{in } \Omega \\ U = 0 & \text{on } \partial\Omega \end{cases}$$

and U^r is an associated eigenvector. The functions U^r form an orthonormed basis in $L^2(\Omega)$.

Moreover, if λ^r is a simple eigenvalue, then there exists $\eta_0 > 0$ such that for any η such that $0 < \eta < \eta_0$, λ_η^r is an eigenvalue and if further V^r is the eigenvector associated to λ^r with $\| V^r \|_{L^2(\Omega)} = 1$, then there exists V_η^r with $\| V_\eta^r \|_{L^2(\Omega)} = 1$ and the whole sequence V_η^r converges to V^r weakly in $H_0^1(\Omega)$. □

We apply without any difficulty this result in our case as $\dfrac{1}{|Y_\delta|}\dfrac{\partial^2}{\partial x_i \partial x_j}$ H-converges to $-\dfrac{1}{2}\Delta$ (we refer the reader to [4] for details). Combining Propositions 3.1 and 3.2 we have the following result describing the asymptotic behaviour of system (3.1):

THEOREM 3.3. *Let successively $\varepsilon \to 0, \delta \to 0$.*

(i). Let $\lambda_{\varepsilon\delta}^r$ be the r-th eigenvalue of system (3.1), (3.4). Then

$$\lim_{\delta\to 0}\, \lim_{\varepsilon\to 0} \lambda_{\varepsilon\delta}^r = \lambda^r,$$

where λ^r is the r-th eigenvalue of the limit system

$$(3.6)\qquad \begin{cases} -\dfrac{1}{2}\Delta u = \lambda u & \text{in } \Omega \\ u = 0 & \text{on } \partial\Omega. \end{cases}$$

(ii). If λ^r is a simple eigenvalue of (3.6) and u^r is an eigenvector corresponding to λ^r with $\| u^r \|_{L^2(\Omega)} = 1$, then we can choose an eigen vector $v^r_{\varepsilon\delta}$ of system (3.1), (3.4), associated to $\lambda^r_{\varepsilon\delta}$ such that

$$\lim_{\delta\to 0} \lim_{\varepsilon\to 0} P_{\varepsilon\delta} v^r_{\varepsilon\delta} = u^r \text{ weakly in } H^1_0(\Omega)$$

□

REFERENCES.

[1] A. Bensoussan - J. L. Lions - G. Papanicolaou, Asymptotic Analysis for Periodic Structures. North Holland, Amsterdam (1978)

[2] D. Cioranescu - P. Donato, Homogénéisation du problème de Neumann non homogène dans des ouverts perforés ,Asymptotic Analysis 1(1988), 115-138

[3] D.Cioranescu - J.Saint Jean Paulin, Homogenization in open sets with holes, J. Math. Anal. Appl. 71(1978), 590-607

[4] D.Cioranescu - J.Saint Jean Paulin, Reinforced and honeycomb structures, J.Math.Pures et Appl. 65 (1986), 403-422

[5] S. Kesavan, Sur l'approximation des problèmes linéaires et non linéaires de valeurs propres. Thèse d'Etat, Paris (1979)

[6] J. L. Lions, Some Methods in the Mathematical Analysis of Systems and their Control.Science press, Beijing, China, Gordon Breach, New York (1981)

[7] E. Sanchez-Palencia Non-homogeneous Media and Vibration Theory. Lecture Notes in Physics127, Springer Verlag, Berlin (1980)

[8] L. Tartar, Cours Peccot, Collège de France, (1977), partially written in: F. Murat, H-Convergence, Séminaire d'analyse fonctionnelle et numérique de l'Université d'Alger, 1978, duplicated, 34 p.

Doina CIORANESCU
Laboratoire d'Analyse Numérique
Université Pierre et Marie Curie
Tour 55-65, 5ème étage
4 place Jussieu
75252 Paris Cedex 05, France.

Jeannine SAINT JEAN PAULIN
Université de Metz
Département des Mathématiques
Ile de Saulcy
57045 Metz Cedex 01, France

Boundary control and stabilization of the one-dimensional wave equation

Francis Conrad (*), Juliette Leblond (**), Jean-Paul Marmorat (**)

(*) Université de Nancy I, URA CNRS 0750 et Projet NUMATH, INRIA-Lorraine, BP 239, 54506-VANDOEUVRE-lès-NANCY, France.

(**) C.M.A., Ecole des Mines de Paris, Sophia-Antipolis, 06565-VALBONNE, France.

1. Introduction

In this paper we consider in particular the question of exact controllability, by means of boundary control, for the one-dimensional wave equation with *variable coefficients*

$$(1) \quad \begin{cases} b\, y_{tt} - (a\, y_x)_x = 0, \quad 0 < x < 1 \\ y(0, t) = 0 \\ y(1, t) = u(t)\,. \end{cases}$$

Closely related to the problem of exact controllability is the question of uniform stabilization by means of a dissipative boundary feedback. Here we consider the system

$$(2) \quad \begin{cases} b\, y_{tt} - (a\, y_x)_x = 0, \quad 0 < x < 1 \\ y(0, t) = 0 \\ a\, y_x(1, t) = -\, g(y_t(1, t))\,. \end{cases}$$

Proving an exact controllability result for (1) in suitable spaces, or obtaining estimates of the decay for (2) amounts, when general techniques are used, to establish inequalities involving constants, which depend in some way of a and b.

Here we study the following question: *what degree of smoothness is needed on a and b to get exact controllability for (1), or estimates of the decay for (2) ? In other words, on what norms of a and b do the constants of the inequalities depend ?*

Results for the wave equation with variable coefficients are already known, mainly from the viewpoint of exact controllability, to our best knowledge. Let us recall some of them.

Exact controllability for the one-dimensional wave equation with locally distributed control and Dirichlet boundary conditions has been studied in [HO] (see also [LA_1]). The variable coefficients are assumed to be of class C^1. Though this problem, in its nature, is rather far from ours, we mention it since the method of multipliers used there to establish exact controllability by proving the observability of the adjoint system presents some similarity with the technique which will be used here (for more irregular coefficients).

In arbitrary space dimension N, sharp results on the exact controllability for a general wave equation with Dirichlet control in $L^2(\partial\Omega x(0,T))$ and initial data in $L^2(\Omega) \times H^{-1}(\Omega)$ have been given in [KO]. The coefficients $a_{ij}(x)$ of the equation are assumed to be in $W^{1,\infty}(\Omega)$, and satisfy $(2a_{ij} - h.\nabla a_{ij})\, \xi_i\, \xi_j \geq \gamma\, \xi_i\, \xi_j$, with $h(x) = x - x_0$. This condition is needed since the usual multiplier $h.\nabla y$ is used to obtain the basic inequalities required to apply the HUM method of J. L. Lions. An estimate of the exact controllability time T_0 is proportional to $(\min(\gamma, 2N))^{-1}$.
In dimension one, this condition reduces to $2a - (x - x_0)\, a_x > 0$.

In space dimension one, the result has been improved in [ZU_1]. In fact, exact controllability for the wave equation

$$(3) \quad \begin{cases} y_{tt} - (a\, y_x)_x = 0,\ 0 < x < 1 \\ y(0, t) = v(t) \in L^2(0,T) \\ y(1, T) = 0 \\ (y_0, y_1) \in L^2(0,1) \times H^{-1}(0,1) \end{cases}$$

holds for any $a \in W^{1,\infty}(0, 1)$ such that $a \geq a_0 > 0$, with $T_0 \geq \dfrac{2}{\sqrt{a_0}}$. The proof is also based on the HUM method. The key point is the "observability" inequality

$$|\varphi_0|^2_{H^1_0} + |\varphi_1|^2_{L^2} \leq \frac{C}{T - T_0} \int_0^T \varphi_x^2(0, t)\, dt,$$

where φ is a solution of problem (3) with $v \equiv 0$. This inequality is proved in a way very specific of the dimension one and does not use multipliers. An upper bound for C is given by $a(0) \exp \{\frac{|a_x|_{L^\infty}}{a_0}\}$. In [ZU1], it is asked wether this bound is sharp, and already remarked that the result is not optimal since if, for instance, a is nondecreasing, C does not depend on $|a_x|_{L^\infty}$. Also, it is easy to see that the result holds if a is piecewise constant. This suggests that one can expect a result for coefficients more general than $W^{1,\infty}(0, 1)$.

Specifically, we show that a suitable space for the coefficients is BV ([0, 1]), by considering multipliers more general than $h(x) = x - x_0$.

For instance, if a is piecewise constant, the multiplier will also be discontinuous. However, the controllability time T_0 will depend on a in a more complicated way than in [ZU1].

The summary is as follows. First, we construct a suitable "multiplier", well adapted to the case of variable coefficients. Then we use it for the exact controllability problem (1), and also for the stabilization problem (2).

2. Preliminary results

In this paper, we use Riemann-Stieljes integrals as presented for instance in [RU]. Let us recall the definition of such integrals. Other definitions are possible. In view of our applications, the results should be qualitatively the same.

Let f and g be defined on [0, 1] and bounded. If $\sum_{i=1}^{n} f(t_i)\, [g(x_i) - g(x_{i-1})]$ has a limit I as the mesh of the partition $P = \{0 = x_0, x_1, \ldots, x_n = 1\}$ goes to zero, independently of the choice of the points t_i in $[x_{i-1}, x_i]$, then f is said to be integrable with respect to g ($f \in R^*(g)$), and I is denoted by $\int_0^1 f dg$.

If f is continuous on [0, 1] and g is of bounded variation on [0, 1] ($g \in BV([0, 1])$), then $f \in R^*(g)$.

Subsequently, we assume that a and $b \in BV([0, 1])$, $0 < m \le a, b$. Then, if $\varphi \in BV([0, 1])$, it follows that φb and $\frac{\varphi}{a} \in BV([0, 1])$. Thus $f \in R^*(\varphi b)$, and $f \in R^*(\frac{\varphi}{a})$ for any continuous function f.

The following simple result is basic for obtaining exact controllability, or uniform stabilization results.

Lemma 1. Given a and $b \in BV([0, 1])$ such that $a, b \ge m > 0$, there exists $\varphi \in BV([0,1])$, $\varphi \ge 0$, increasing, such that for any $h \in C([0, 1])$, $h \ge 0$:

$$\int_0^1 h \, d\left(\frac{\varphi}{a}\right) \ge \int_0^1 h(x)\,dx, \qquad \int_0^1 h \, d(\varphi b) \ge \int_0^1 h(x)\,dx$$

Proof

We set $a = a(0) + a_+ - a_-$, where a_+ (a_-) is the positive (negative) variation of a [RU]. Similarly, we set $b = b(0) + b_+ - b_-$. The functions a_+, a_-, b_+, b_- are nondecreasing. Let $M = \max\{|a|_{L^\infty}, |b|_{L^\infty}\}$.

We set $H(x) = \max\{M, \frac{1}{m}\}\, x + \frac{M}{m^2} a_+(x) + \frac{1}{m} b_-(x)$, and $\varphi(x) = \exp\{H(x)\}$, $\forall\, x \in [0, 1]$.

Since φ is increasing, $\frac{\varphi}{a}$ and $\varphi b \in BV([0, 1])$ and the integrals in Lemma 1 make sense.

We prove the first inequality.

Let $P = \{0 = x_0, x_1, \ldots, x_n = 1\}$ be a partition of [0, 1], and $t_i \in [x_{i-1}, x_i]$.

We calculate the sum

$$S(P) = \sum_{i=1}^{n} h(t_i)\left[\frac{\varphi}{a}(x_i) - \frac{\varphi}{a}(x_{i-1})\right] = \sum_{i=1}^{n} h(t_i)\frac{\varphi(x_i)\,a(x_{i-1}) - \varphi(x_{i-1})\,a(x_i)}{a(x_i)\,a(x_{i-1})} =$$

$$\sum_{i=1}^{n} h(t_i)\frac{\varphi(x_i) - \varphi(x_{i-1})}{a(x_i)} - \sum_{i=1}^{n} h(t_i)\frac{a(x_i) - a(x_{i-1})}{a(x_i)\,a(x_{i-1})}\,\varphi(x_{i-1}).$$

Now, since $h \ge 0$, $0 < a \le M$, and φ is ≥ 0 and increasing , we have

$$(4) \qquad S(P) \geq \sum_{i=1}^{n} h(t_i) \frac{\varphi(x_i) - \varphi(x_{i-1})}{M} - \sum_{i=1}^{n} h(t_i) \frac{a_+(x_i) - a_+(x_{i-1})}{a(x_i)\, a(x_{i-1})} \varphi(x_{i-1}).$$

But, by convexity of the exponential function

$$(5) \qquad \begin{aligned} \varphi(x_i) - \varphi(x_{i-1}) &= e^{H(x_i)} - e^{H(x_{i-1})} \geq e^{H(x_{i-1})} [H(x_i) - H(x_{i-1})] \\ &= \varphi(x_{i-1}) [H(x_i) - H(x_{i-1})] \end{aligned}$$

Plugging inequality (5) into (4) we get

$$\begin{aligned} S(P) &\geq \sum_{i=1}^{n} h(t_i) \frac{H(x_i) - H(x_{i-1})}{M} \varphi(x_{i-1}) - \sum_{i=1}^{n} h(t_i) \frac{a_+(x_i) - a_+(x_{i-1})}{m^2} \varphi(x_{i-1}) \\ &= \sum_{i=1}^{n} \frac{h(t_i)\, \varphi(x_{i-1})}{M} [H(x_i) - H(x_{i-1}) - M \frac{a_+(x_i) - a_+(x_{i-1})}{m^2}]. \end{aligned}$$

By the definition of H, we easily deduce

$$\begin{aligned} S(P) &\geq \sum_{i=1}^{n} \frac{h(t_i)\, \varphi(x_{i-1})}{M} \max \{M, \frac{1}{m}\} (x_i - x_{i-1}) \\ &\geq \sum_{i=1}^{n} h(t_i)\, \varphi(x_{i-1})\, (x_i - x_{i-1}) \geq \sum_{i=1}^{n} h(t_i)\, (x_i - x_{i-1}). \end{aligned}$$

Taking the limits as the mesh of P goes to zero gives the result.

We prove now the second inequality. In a similar way, and with similar notations

$$\begin{aligned} S(P) &= \sum_{i=1}^{n} h(t_i) [\varphi b(x_i) - \varphi b(x_{i-1})] \\ &= \sum_{i=1}^{n} h(t_i)[\varphi(x_i) - \varphi(x_{i-1})] b(x_i) + \sum_{i=1}^{n} h(t_i)[b(x_i) - b(x_{i-1})] \varphi(x_{i-1}) \\ &\geq \sum_{i=1}^{n} h(t_i)\, \varphi(x_{i-1}) [H(x_i) - H(x_{i-1})]\, b(x_i) \\ &+ \sum_{i=1}^{n} h(t_i)\, \varphi(x_{i-1}) [b(x_i) - b(x_{i-1})] \qquad \text{((5) has been used)} \\ &\geq \sum_{i=1}^{n} h(t_i)\, \varphi(x_{i-1}) [m(H(x_i) - H(x_{i-1})) - (b_-(x_i) - b_-(x_{i-1}))]. \end{aligned}$$

By the definition of H, we easily deduce

$$S(P) \geq \sum_{i=1}^{n} h(t_i)\, \varphi(x_{i-1})\, [m \max \{M, \frac{1}{m}\} (x_i - x_{i-1})]$$

$$\geq \sum_{i=1}^{n} h(t_i)\, \varphi(x_{i-1})\, (x_i - x_{i-1}) \geq \sum_{i=1}^{n} h(t_i)\, (x_i - x_{i-1})$$

which gives the second inequality of Lemma 1 □

Remark 1. If a is nonincreasing and b is nondecreasing, then $a_+ = b_- = 0$, and Lemma 1 is true with $\varphi(x) = e^{\lambda x}$, $\lambda = \max \{M, \frac{1}{m}\}$. In fact, in that case, it is easy to prove directly that Lemma 1 is also true with $\varphi(x) = \mu x$, $\mu = \max \{M, \frac{1}{m}\}$. Thus φ will be the "usual" multiplier and $|\varphi|_{L^\infty} = \mu$ depends only on L^∞ norms.

Remark 2. Assume a and b are piecewiese of class C^1 on [0, 1], and let $0 < x_1 < \ldots < x_n < 1$ be the discontinuities of a and b (we can always suppose that they are common to a and b). Consider the following function, constructed iteratively

- $\varphi(0) = 0 = \varphi(x_0^+)$
- once $\varphi(x_i^+)$ is known, solve $\varphi' \geq \max \{M, \frac{1}{m}\} + \varphi \max \{\frac{a'_+}{a}, \frac{b'_-}{b}\}$ on (x_i^+, x_{i+1}^-)
- chose $\varphi(x_{i+1}^+) \geq \varphi(x_{i+1}^-) \max \{\frac{a(x_{i+1}^+)}{a(x_{i+1}^-)}, \frac{b(x_{i+1}^-)}{b(x_{i+1}^+)}\}$.

It is easy to prove that φ satisfies the inequalities of Lemma 1.

We end this section by recalling two results. First, with our definition of Stieljes integrals, integration by parts takes the following form [RU, Theorem 6.28]

Lemma 2. If $f \in R^*(g)$, then $g \in R^*(f)$, and $\int_0^1 g\, df = [fg]_0^1 - \int_0^1 f\, dg$.

This result is true in particular if f is continuous and $g \in BV([0, 1])$.

Second, if one integrates with respect to an absolutely continuous function, one gets a "usual" (Lebesgue) integral.

Lemma 3. Let f be absolutely continuous and $g \in BV([0,1])$. Then

$$\int_0^1 g\, df = \int_0^1 g(x)\, f'(x)\, dx.$$

Lemma 3 is true when $f \in C^1([0,1])$ [RU p. 108]. The result follows by taking $f_n \to f$ in $W^{1,1}(0,1)$, $f_n \in C^1([0,1])$, and using Lemma 2.

3. An application to exact controllability

Here we consider the same problem as in [ZU1] except that control applies at x =1 instead of x = 0

$$(6) \quad \begin{cases} y_{tt} - (a\, y_x)_x = 0, \quad 0 < x < 1 \\ y(0,t) = 0 \\ y(1,t) = v(t) \\ y(x,0) = y_0\,,\; y_t(x,0) = y_1. \end{cases}$$

We assume $a \geq m > 0$, $a \in BV([0,1])$ (instead of $a \in W^{1,\infty}(0,1)$, as in [ZU1]).

Applying the HUM method as in [ZU1], exact controllability for v in $L^2(0,T)$ and initial data (y_0, y_1) in $L^2(0,1) \times H^{-1}(0,1)$, amounts to get estimates for the uncontrolled problem

$$(7) \quad \begin{cases} z_{tt} - (a\, z_x)_x = 0, \quad 0 < x < 1 \\ z(0,t) = z(1,t) = 0 \\ z(x,0) = z_0 \in H_0^1(0,1)\,,\; z_t(x,0) = z_1 \in L^2(0,1) \end{cases}$$

Proposition 1. There exists $T_0 > 0$, $C > 0$ such that, for any solution z of (7) and any $T > T_0$

$$(8) \quad \|(z_0, z_1)\|^2_{H_0^1 \times L^2} \leq C \int_0^T (a\, z_x)^2 (1,t)\, dt.$$

Proof

For any solution z of (7), with $(z_0, z_1) \in H_0^1(\Omega) \times L^2(\Omega)$, let $E(t) = \frac{1}{2}\int_0^1 [z_t^2 + a z_x^2]\, dx$,

be the "energy" of z. It is easy to show that (7) defines a semi-group on $H_0^1(0,1) \times L^2(0,1)$, equipped with the norm $\|(z_0,z_1)\| = [\int_0^1 a u_x^2 + v^2) dx]^{1/2}$, which is equivalent to the usual norm on $H_0^1(0,1) \times L^2(0,1)$. For regular initial data, the solution is regular. Thus, to establish (8), it is enough to assume $z_t \in H_0^1(0,1)$, and a $z_x \in H^1(0,1)$.

First, we note that the energy is constant, as usual. Then we use the results of Section 2. Let z be a solution of (7) with regular initial data, and φ be defined as in Lemma 1 (with b = 1, which, in fact, is not a restriction).
We use the multiplier φz_x

$$\int_0^T\int_0^1 z_{tt}\, \varphi z_x\, dx\, dt = \int_0^1 [z_t\, \varphi z_x]_0^T\, dx - \int_0^T\int_0^1 \varphi\, z_t\, z_{xt}\, dx\, dt$$

$$= \int_0^1 [z_t\, \varphi z_x]_0^T\, dx - \frac{1}{2}\int_0^T\int_0^1 \varphi\, (z_t^2)_x\, dx\, dt.$$

But $z_t \in H^1(0,1)$ implies $z_t^2 \in H^1(0,1)$, and in particular z_t^2 is absolutely continuous. Applying Lemma 3, since $\varphi \in BV([0,1])$, we get

$$\int_0^T\int_0^1 z_{tt}\, \varphi z_x\, dx\, dt = \int_0^1 [z_t\, \varphi z_x]_0^T\, dx - \frac{1}{2}\int_0^T\int_0^1 \varphi\, d(z_t^2)\, dt$$

(now we have a Stieljes integral).

We apply Lemma 2 ($\varphi \in BV([0,1])$ and z_t^2 is continuous in x)

$$\text{(9)} \quad \int_0^T\int_0^1 z_{tt}\, \varphi z_x\, dx dt = \int_0^1 [z_t\, \varphi z_x]_0^T\, dx - \frac{1}{2}\int_0^T [\varphi\, z_t^2]_0^1\, dt + \frac{1}{2}\int_0^T\int_0^1 z_t^2\, d\varphi dt$$

In a similar way

$$\int_0^T\int_0^1 -(a\,z_x)_x\,\varphi z_x\,dx\,dt = -\frac{1}{2}\int_0^T\int_0^1 \frac{\varphi}{a}[(a\,z_x)^2]_x\,dx\,dt.$$

Since $\frac{\varphi}{a} \in BV([0,1])$, and $(a\,z_x) \in H^1(0,1)$ implies $(a\,z_x)^2 \in W^{1,1}(0,1)$ as before, we can apply Lemma 3

$$\int_0^T\int_0^1 -(a\,z_x)_x\,\varphi z_x\,dx\,dt = -\frac{1}{2}\int_0^T\int_0^1 \frac{\varphi}{a}\,d((a\,z_x)^2)\,dt.$$

Then we apply Lemma 2 since $(a\,z_x)^2$ is continuous

(10) $$\int_0^T\int_0^1 -(a\,z_x)_x\,\varphi z_x\,dxdt = -\frac{1}{2}\int_0^T [\frac{\varphi}{a}(a\,z_x)^2]_0^1\,dt$$

$$+\frac{1}{2}\int_0^T\int_0^1 (a\,z_x)^2\,d(\frac{\varphi}{a})\,dt.$$

Adding (9) and (10) we get, since z satisfies (7)

$$\frac{1}{2}\int_0^T[\int_0^1 z_t^2\,d\varphi + (a\,z_x)^2\,d(\frac{\varphi}{a})]\,dt = \frac{1}{2}\int_0^T[\frac{\varphi}{a}(a\,z_x)^2]_0^1\,dt - \int_0^1 [z_t\varphi z_x]_0^T\,dx.$$

Now we apply Lemma 1

$$\frac{1}{2}\int_0^T\int_0^1 [z_t^2 + (a\,z_x)^2]\,dxdt \le \frac{1}{2}\int_0^T (\frac{\varphi}{a}(a\,z_x)^2)(1)\,dt - \int_0^1 [z_t\,\varphi z_x]_0^T\,dx.$$

The last term can be bounded by Young's inequality

$$\left|\int_0^1 [z_t\,\varphi z_x](t)\,dx\right| \le \frac{1}{2}|\varphi|_{L^\infty}\int_0^1 (z_t^2 + z_x^2)\,dx.$$

Taking into account that $0 < m \le a \le |a|_{L^\infty}$ we get

$$\min(1, m) \int_0^T E(t)\,dt \le \frac{1}{2}\,\frac{\varphi}{a}(1) \int_0^T (a\,z_x)^2\,(1, t)\,dt$$

$$+ |\varphi|_{L^\infty} \max\left(1, \frac{1}{m}\right) [E(0) + E(T)].$$

But, since the energy is constant, the inequality reduces to

$$(11) \qquad \left[\min(1, m)\,T - 2|\varphi|_{L^\infty} \max\left(1, \frac{1}{m}\right)\right] E(0) \le \frac{1}{2}\frac{\varphi}{a}(1) \int_0^T (a\,z_x)^2\,(1, t)\,dt.$$

This proves (8) with $T_0 = 2|\varphi|_{L^\infty} \dfrac{\max\left(1, \frac{1}{m}\right)}{\min(1, m)}$, and the constant involved in the right hand side. of (8) behaves like $\dfrac{\varphi(1)\,(T - T_0)^{-1}}{m \min(1, m)}$, where

$$\varphi(1) = |\varphi|_{L^\infty} = \exp\left(\max\left(M, \frac{1}{m}\right) + \frac{M}{m^2}\,a_+(1)\right) \qquad \square$$

Remark 1. In [ZU1], with $a \in W^{1,\infty}(0, 1)$, $a \ge m > 0$, the estimate of the controllability time is $T_0 = \dfrac{2}{\sqrt{m}}$, and the constant involved in the right hand side of (8) behaves like $\exp\left(\dfrac{|a_x|_{L^\infty}}{m}\right)(T - T_0)^{-1}$.

Remark 2. When $a \in L^\infty(0, 1)$ is nonincreasing, $\varphi = \max(M, 1)\,x$ satisfies Lemma 1, and then T_0 and $\dfrac{\varphi(1)}{m \min(1, m)}$ depend only on m and M. This result is qualitatively similar to the one reported in [ZU1], when a is nondecreasing (but control is applied at $x = 0$).

Proposition 1 implies that exact controllability holds for system (6) with $v \in L^2(0,T)$, $T > T_0$, and for any initial data in $L^2(0,1) \times H^1(0,1)$. In order to prove that the controllability space is not larger than $L^2(0, 1) \times H^{-1}(0, 1)$, one has to derive the converse of inequality (8)

$$(12) \qquad \int_0^T (a\,z_x)^2(1, t)\,dt \le C\,\|(z_0, z_1)\|^2_{H_0^1 \times L^2}.$$

Proposition 2. Let $a \in BV([0,1])$, $a \geq m > 0$. Assume there exists $\varphi \in BV([0,1])$, $\varphi(0) \leq 0$, $\varphi(1) > 0$, such that $\forall\, h \in \mathcal{C}([0,1)]$, $h \geq 0$

$$\int_0^1 h\, d\varphi \leq \int_0^1 h\, dx, \text{ and } \int_0^1 h\, d\left(\frac{\varphi}{a}\right) \leq \int_0^1 h\, dx.$$

Then (12) holds.

Proof

Again, adding (9) and (10), we get

$$\frac{1}{2}\frac{\varphi}{a}(1) \int_0^T (a\, z_x)^2(1,t)\, dt = \frac{1}{2}\int_0^T \Big[\int_0^1 z_t^2\, d\varphi + (a\, z_x)^2\, d\left(\frac{\varphi}{a}\right)\Big]\, dt$$

$$+ \int_0^1 [z_t\, \varphi z_x]_0^T\, dx + \frac{1}{2}\frac{\varphi}{a}(0) \int_0^T (a\, z_x)^2\,(0,t)\, dt$$

$$\leq \max(1, M) \int_0^T E(t)\, dt + \max\left(1, \frac{1}{m}\right) |\varphi|_{L^\infty} [E(0) + E(T)].$$

Since the energy is constant, we get

$$\int_0^T (a\, z_x)^2(1,t)\, dt \leq 2\frac{a}{\varphi}(1)\left[\max(1,M)\, T + 2\ |\varphi|_{L^\infty} \max\left(1, \frac{1}{m}\right)\right] E(0)$$

which is just (12). □

The assumption of Proposition 2 is satisfied if $a \in W^{1,\infty}(0,1)$, in an obvious way (consider $\varphi(x) = \varepsilon x$, with $\varepsilon > 0$ small enough). It is also satisfied if $a \in L^\infty(0,1)$ is nondecreasing (the "bad" case, in view of Proposition 1). Then $\varphi(x) = \eta x$, with $\eta \leq m$, works.

We do not know if the assumption of Proposition 2 holds for any a in $BV([0,1])$, such that $a \geq m > 0$. The strange feature is that in the case of constant coefficients, (12) is considered as the "easy" part of the estimates useful in "HUM", and (8) is the "hard" part. This does not seem to be the case here !

4. An application to uniform stabilization

We consider problem (2) presented in Section 1

$$(13)\quad \begin{cases} b\, y_{tt} - (a\, y_x)_x = 0, \ \ 0 < x < 1 \\ y(0, t) = 0 \\ a\, y_x(1, t) = -\, g(y_t(1, t)) \\ y(x, 0) = y_0(x);\ y_t(x, 0) = y_1(x)\,. \end{cases}$$

where a, b ∈ BV ([0, 1]), a, b ≥ m > 0 ; $g : \mathbb{R} \to \mathbb{R}$ is nondecreasing, continuous and satisfies $g(0) = 0$, $g(\xi) \neq 0$ if $\xi \neq 0$.

The stabilization of the wave equation with constant coefficients has been studied in space dimension $n \le 3$ in [ZU2]. Estimates for the decay of the energy are obtained when g behaves superlinearly near the origin.

Here we prove, in the specific one dimensional case, that the estimates of [ZU2] are still valid when variable coefficients occur. Moreover, we also obtain estimates for the sublinear case. See also [LA2] for similar estimates in a sublinear framework, for a problem of elasticity involving constant coefficients.

We set $H = L^2(0, 1)$, equipped with the norm $\left(\int_0^1 b\, v^2\, dx\right)^{1/2}$,

$V = \{u \in H^1(0, 1) ;\ u(0) = 0\}$, equipped with the norm $\left(\int_0^1 a\, u_x^2\, dx\right)^{1/2}$,

and V x H is equipped with the product norm.

Problem (13) is well-posed on V x H in the sense that it defines a nonlinear semi-group of contractions on V x H, associated with a maximal monotone operator A on V x H, where the domain of A
$\mathrm{dom}(A) = \{(u, v) \in V \times H ;\ v \in V ;\ a\, u_x \in H^1(0, 1) ;\ (a\, u_x)(1) = -g(v(1))\}$,
is dense in V x H. These facts are easily established, using abstract theorems as in [LA3], or the methodology presented in [CO-PI] for a beam equation, and easily adaptable here.

Define the energy of a solution (y, y_t) of (13) by

$$E(t) = \frac{1}{2} \|(y, y_t)\|^2_{V\times H} = \frac{1}{2}\int_0^1 (b\, y_t^2 + a\, y_x^2)\, dx.$$

Then, for initial data in dom (A)

(14) $$\frac{dE}{dt} = - g(y_t(1, t))\, y_t(1, t).$$

The energy in nonincreasing and, using the invariance principle of Lasalle, one establishes that E(t) decays to 0 as t goes to infinity (See also [CO-PI]).

All these results hold for any a, b $\in L^\infty(0, 1)$ such that a, b $\geq m > 0$.

Our purpose here is to give estimates on the decay of the energy. We have to assume a, b $\in$ BV ([0, 1]) in our proof, and, of course, we have to specify the behaviour of g.

The hypotheses on g are (H_1) or (H_2)

(H_1) superlinear case : $\exists\, C_1 > 0, C_2 > 0, p \geq 1$ such that
$C_1 \min\{|\xi|, |\xi|^p\} \leq |g(\xi)| \leq C_2 |\xi|, \forall \xi \in \mathbb{R}$

(H_2) sublinear case : $\exists\, C_1 > 0, C_2 > 0, p \leq 1$ such that
$C_1 |\xi| \leq |g(\xi)| \leq C_2 \max\{|\xi|, |\xi|^p\}, \forall \xi \in \mathbb{R}$.

Remark 1 . If p = 1, (H_1) or (H_2) reduce to: $C_1 |\xi| \leq |g(\xi)| \leq C_2 |\xi|$

Remark 2 . The model case corresponds to
$g(\xi) = |\xi|^{p-1} \xi$ if $|\xi| < 1$, $g(\xi) = \xi$ if $|\xi| \geq 1$.

Theorem 3. Let y(t) be a solution of (13) and E(t) be the associated energy.
(i) If g satisfies (H_1) or (H_2) with p = 1, then
$E(t) \leq M e^{-\mu t} E(0)$, where $M > 0, \mu > 0$ are constants.
(ii) If g satisfies (H_1) with p > 1, then
$E(t) \leq M \left(\frac{1}{1 + \mu t}\right)^{\frac{2}{p-1}} E(0)$, where M > 0 is constant, and $\mu > 0$ depends on E(0).

(iii) If g satisfies (H_2) with $p < 1$, then

$$E(t) \le M \left(\frac{1}{1+\mu t}\right)^{\frac{2p}{1-p}} E(0),$$ where $M > 0$ is constant, and $\mu > 0$ depends on $E(0)$.

Proof

As usual, it is enough to establish the estimates for initial data in dom (A), with constants depending continuously on E(0). Therefore, let y(t) be a regular solution of (13) i.e. $(y, y_t)(t) \in$ dom (A) $\forall\, t > 0$, and $t \to (y, y_t)(t) \in V \times H$ is a.e. differentiable.

We proceed as in Section 3, multiplying the equation in (13) by $\varphi\, y_x$, where φ satisfies Lemma 1, and integrating over $[0, 1] \times [0, T]$. But now, we have to take into account the integral $\int_0^T [\varphi y_t^2]_0^1\, dt$, which is not zero.

Then we get

$$\frac{1}{2}\int_0^T \left(\int_0^1 y_t^2\, d(b\varphi) + (a\, y_x)^2\, d\left(\frac{\varphi}{a}\right)\right) dt = \frac{1}{2}\int_0^T [b\, \varphi y_t^2]_0^1\, dt$$

$$+ \frac{1}{2}\int_0^T \left[\frac{\varphi}{a}(a\, y_x)^2\right]_0^1 dt - \int_0^1 [b\, y_t\, \varphi\, y_x]_0^T\, dx.$$

Now we apply Lemma 1, and the positivity of φ

$$\frac{1}{2}\int_0^T \int_0^1 (y_t^2 + (a\, y_x)^2)\, dx dt \le \frac{1}{2} b\varphi(1) \int_0^T y_t^2(1, t)\, dt +$$

$$\frac{1}{2}\frac{\varphi}{a}(1) \int_0^T (a\, y_x)^2(1, t)\, dt \quad - \int_0^1 [b\, \varphi y_t\, y_x]_0^T\, dx.$$

Using Young's inequality in the last term, the boundary conditions, and the fact that $E(t) \le E(0)$, we get

$$\min(1, m) \int_0^T E(t)\, dt \le \frac{1}{2} b\varphi(1) \int_0^T y_t^2(1, t)\, dt + \frac{1}{2}\frac{\varphi}{a}(1) \int_0^T g(y_t^2(1, t))dt$$

$$+ 2|b\varphi|_{L^\infty} \max(1, \frac{1}{m}) E(0).$$

In other words

$$(15) \qquad \int_0^T E(t)\,dt \le C \int_0^T y_t^2(1,t)\,dt + C \int_0^T g^2(y_t(1,t))\,dt + C\,E(0),$$

where the constants C are of the form

$$C \le \frac{\max(2M, \frac{1}{m}, \frac{2M}{m})}{\min(1, \frac{1}{m})} |\varphi|_{L^\infty}.$$

We recall that $|\varphi|_{L^\infty} = \varphi(1) = \exp[\max(M, \frac{1}{m}) + \frac{M}{m^2}\, a_+(1) + \frac{b_-(1)}{m}]$.

The obtention of the estimates from (15) is more or less standard, at least in the superlinear case. We just sketch the proof, which can also be found in [CO-LE-MA], in the sublinear case.

First, from (14) we deduce

$$E(0) - E(T) = \int_0^T g(y_t(1,t))\, y_t(1,t)\,dt,$$

and thus

$$(16) \qquad \int_0^T g(y_t(1,t))\, y_t(1,t)\,dt \le E(0).$$

Now we consider separately the different cases.

(i) Case $p = 1$. (H_1) or (H_2) imply

$$\xi^2 \le \frac{1}{C_1} g(\xi).\xi, \quad |g(\xi)|^2 \le C_2\, g(\xi).\xi$$

so that

$$\int_0^T y_t^2(1,t)\,dt \le \frac{1}{C_1} \int_0^T g(y_t(1,t))\, y_t(1,t)\,dt \le \frac{1}{C_1} E(0) \text{ by (16)},$$

and also

$$\int_0^T g^2(y_t(1,t))\,dt \le C_2\,E(0).$$

Plugging these estimates into (15) yields

$$\int_0^T E(t)\,dt \le \left[\frac{C}{C_1} + CC_2 + C\right] E(0),$$

and since E is nonincreasing,

$$E(T) \le C\,\frac{E(0)}{T} \tag{17}$$

where C is a constant which does not depend on E(0), and is $\mathcal{O}(|\varphi|_{L^\infty})$.

(ii) Case $p > 1$. From (H_1) and (16), one first deduces that

$$\int_0^T g^2(y_t(1,t))\,dt \le C_2\,E(0)$$

so that, from (15) one obtains

$$\int_0^T E(t)\,dt \le C\int_0^T y_t^2(1,t)\,dt + [CC_2 + C]\,E(0). \tag{18}$$

On the other hand, also from (H_1) and (16), one has

$$\int_{(0,T)\cap\{|y_t|>1\}} y_t^2(1,t)\,dt \le \frac{1}{C_1}\int_{(0,T)\cap\{|y_t|>1\}} g(y_t(1,t))\,y_t(1,t)\,dt \le \frac{1}{C_1}E(0). \tag{19}$$

For the remaining part of the integral, we apply Hölder's inequality with $\alpha = \frac{p+1}{2} > 1$, and $\beta = \frac{p+1}{p-1} > 1$

(20) $$\int_{(0,T)\cap\{|y_t|\le 1\}} y_t^2(1,t)\,dt \le T^{\frac{p-1}{p+1}} \Big(\int_{(0,T)\cap\{|y_t|\le 1\}} y_t^{p+1}(1,t)\,dt\Big)^{\frac{2}{p+1}}$$

$$\le T^{\frac{p-1}{p+1}}\Big(\int_0^T \frac{1}{C_1} g(y_t(1,t))y_t(1,t)\,dt\Big)^{\frac{2}{p+1}}, \text{ by } (H_1)$$

$$\le \Big(\frac{1}{C_1}\Big)^{\frac{2}{p+1}} T^{\frac{p-1}{p+1}} E(0)^{\frac{2}{p+1}} \quad \text{by (16).}$$

Plugging (19) and (20) into (18) yields

$$\int_0^T E(t)\,dt \le \Big[\frac{C}{C_1} + CC_2 + C\Big] E(0) + C.(C_1)^{\frac{-2}{p+1}} T^{\frac{p-1}{p+1}} E(0)^{\frac{2}{p+1}}$$

and, since the energy is nonincreasing

(21) $$E(T) \le C\frac{E(0)}{T} + C\Big(\frac{E(0)}{T}\Big)^{\frac{2}{p+1}}$$

where again C is a constant which is $\mathcal{O}(|\varphi|_{L^\infty})$, and independent of E(0).

(iii) Case $p < 1$. From (H_2) and (16), one first deduces that

$$\int_0^T y_t^2(1,t)\,dt \le \frac{1}{C_1} E(0)$$

so that, from (15) one obtains

(22) $$\int_0^T E(t)\,dt \le C\int_0^T g^2(y_t(1,t))\,dt + \Big[C + \frac{C}{C_1}\Big] E(0).$$

On the other hand, also from (H_2) and (16), one has

(23) $$\int_{(0,T)\cap\{|y_t|>1\}} g^2(y_t(1,t))\,dt \le C_2 \int_{(0,T)\cap\{|y_t|>1\}} g(y_t(1,t))y_t(1,t)\,dt \le C_2\, E(0).$$

For the remaining part of the integral, we apply Hölder's inequality with $\alpha = \frac{p+1}{2p} > 1$, and $\beta = \frac{1+p}{1-p} > 1$

$$\int_{(0,T)\cap\{|y_t|\le 1\}} g^2(y_t(1,t))\,dt \le T^{\frac{1-p}{1+p}} \Big(\int_{(0,T)\cap\{|y_t|\le 1\}} g^{\frac{p+1}{p}}(y_t(1,t))\,dt\Big)^{\frac{2p}{p+1}}$$

$$= T^{\frac{1-p}{1+p}} \Big(\int_{(0,T)\cap\{|y_t|\le 1\}} g(y_t(1,t))\, g^{\frac{1}{p}}(y_t(1,t))\,dt\Big)^{\frac{2p}{p+1}}$$

$$\le T^{\frac{1-p}{1+p}} \Big(\int_{(0,T)\cap\{|y_t|\le 1\}} C_2^{\frac{1}{p}}\, g(y_t(1,t))(y_t(1,t))\,dt\Big)^{\frac{2p}{p+1}} \quad \text{by } (H_2)$$

$$(24) \qquad \le C_2^{\frac{2}{p+1}}\, T^{\frac{1-p}{1+p}}\, E(0)^{\frac{2p}{p+1}} \quad \text{by (15).}$$

Plugging (23) and (24) into (22) yields

$$\int_0^T E(t)\,dt \le \Big[\frac{C}{C_1} + CC_2 + C\Big]\, E(0) + C(C_2)^{\frac{2}{p+1}}\, T^{\frac{1-p}{1+p}}\, E(0)^{\frac{2p}{p+1}}$$

and, since the energy is decreasing,

$$(25) \qquad E(T) \le C\,\frac{E(0)}{T} + C\Big(\frac{E(0)}{T}\Big)^{\frac{2p}{p+1}}$$

where again, C does not depend on E(0), and is $\mathcal{O}(|\varphi|_{L^\infty})$.

From (17), (21) and (25), we derive the estimates of the theorem.

First, from (17), one deduces $E(T_0) \le \frac{1}{2}E(0)$. Uniform exponential decay follows in a standard way:

$E(t) \le Me^{-\mu t}$, where $\mu > 0$ is $\mathcal{O}(\frac{1}{T_0})$, i.e. $\mathcal{O}(\frac{1}{|\varphi|_{L^\infty}})$, and M is constant.

Next, for $p \neq 1$, we remark that (21) and (25) are of the form

$$(26) \qquad E(t) \le C\,\frac{E(0)}{T} + C\Big(\frac{E(0)}{T}\Big)^{\rho},$$

where $\rho = \frac{2}{p+1} < 1$ in case $p > 1$, $\rho = \frac{2p}{p+1} < 1$ in case $p < 1$.

Therefore, the estimates (ii) and (iii) of Theorem 3 are a direct consequence of the following lemma, proved in [CO-LE-MA]

Lemma 4. Let S(t) be a nonlinear semi-group of contractions on a Hilbert space H. Let $E(t) = \frac{1}{2} |S(t)x_0|_H^2$ be the energy. Assume that

$$E(t) \leq C_1 \frac{E(0)}{t} + C_2 \left(\frac{E(0)}{t}\right)^{\rho}, \ 0 < \rho < 1, \text{ for any } t > 0.$$

Then $E(t) \leq M \left(\frac{1}{1+\mu t}\right)^{\frac{\rho}{1-\rho}} E(0)$, where $M > 0$ is constant and $\mu > 0$ depends on E(0).

In the course of the proof of Lemma 4, it appears that the dependence of μ with respect to E(0) and C is the following:

$$\mu = C_1 \left(C_2 C^{\frac{1}{\rho}} [1 + C_3 E(0)^{\rho-1}]\right)^{-1}$$

where C_1, C_2, C_3 are constants independent of E(0) and φ. Thus $\mu \to 0$ as $E(0) \to 0$. For fixed E(0), we see that $\mu = o(C^{\frac{-1}{\rho}}) = o((|\varphi|_{L^\infty})^{\frac{-1}{\rho}})$. We recall that $|\varphi|_{L^\infty} = \varphi(1) = \exp\left(\max\left(M, \frac{1}{m}\right) + \frac{M}{m^2} a_+(1) + \frac{b_-(1)}{m}\right)$, and that M does not depend on E(0) and C. □

Remark 1. Instead of integrating over time and space in order to deduce the estimates in Theorem 3, one can use the modified energy functional (as in [ZU2], for the superlinear case)

$$E_\varepsilon(t) = E(t) + \varepsilon E(t)^{\gamma} \rho(t), \text{ with } \gamma = \frac{p-1}{2} \text{ if } p \geq 1, \text{ or } \gamma = \frac{1-p}{2p} \text{ if } p < 1,$$

where $\rho(t)$ is defined by

$$\rho(t) = 2 \int_0^1 b \varphi y_t y_x \, dx,$$

and where φ is given as in Lemma 1.

One then deduces a differential inequality both in cases $p = 1$, $p > 1$, or $p < 1$, which leads to the same kind of estimates.

Remark 2. The same estimates hold for an Euler-Bernoulli beam controlled by a force, which is a nonlinear function of the velocity at one end [CO]. The mass density and flexural rigidity are assumed to be piecewise regular and satisfy some weak monotonicity conditions. The case of more general BV coefficients is more difficult to handle than in the case of the wave equation and is, to our best knowledge, an open problem.

Aknowledgments. We wish to thank Prof. A. Haraux for fruitful discussions and especially for suggesting the treatment of BV coefficients.

REFERENCES

[CO] F. CONRAD, Stabilization of Euler-Bernoulli beam with nonlinear dissipation on the boundary, Metz Days on Hyperbolic Problems and Exact Controllability, march 1990.

[CO-LE-MA] F. CONRAD, J. LEBLOND, J.P. MARMORAT, Stabilization of second order evolution equations by unbounded nonlinear feedback, IFAC Symp. Control of D.P.S., Perpignan, June 1989.

[CO-PI] F. CONRAD, M. PIERRE, Stabilization of Euler-Bernoulli beam by nonlinear boundary feedback, INRIA Report # 1235 (1990).

[HO] L.F. HO, Exact controllability of the one-dimensional wave equation with locally distributed control, SIAM J. Contr. Opt. 28, 3 (1990), pp. 733-748.

[KO] V. KOMORNIK, Exact controllability in short time for the wave equation, Ann. I.H.P. 6,2 (1989), pp. 153-164.

[LA_1] J. LAGNESE, Control of wave processes with distributed controls supported on a subregion, SIAM J. Contr. Opt. 21,1 (1983), pp. 68-85.

[LA_2] J. LAGNESE, Uniform asymptotic energy estimates for solutions of the equations of dynamic plane elasticity with nonlinear dissipation at the boundary, Nonlinear Anal., T. M. A. 16,1 (1991), pp. 35-54.

[LA_3] I. LASIECKA, Stabilization of wave and plate-like equations with nonlinear dissipation on the boundary, J. Diff. Equs. 79,2 (1989), pp. 340-381.

[RU] W. RUDIN, Principles of mathematical analysis, Mac-Graw Hill (1953).

[ZU_1] E. ZUAZUA, An Introduction to the exact controllability for distributed systems, Textos E Notas # 44, CMAF, Lisboa, 1990.

[ZU_2] E. ZUAZUA, Uniform stabilization of the wave equation by nonlinear boundary feedback, SIAM J. Contr. Opt. 28,2 (1990), pp. 466-477.

Boundary control for inverse free boundary problems

Giuseppe Da Prato
Scuola Normale Superiore di Pisa, Pisa Italy

and

Jean.Paul.Zolésio
I.N.L.N, Université de Nice, Parc Valrose 06034 Nice, France.
& CMA, Ecole des Mines de Paris, Sophia Antipolis, France

1 Introduction and notation

Let D be a bounded open set in $\mathbf{R}^N$ with a sufficiently smooth boundary ∂D, $I = [0,T]$ denotes the time interval. In the cylinder $I \times D$ we consider a given non cylindrical evolution domain defined as follows:

$$Q = \bigcup_{t \in I} \{t\} \times \Omega_t, \tag{1}$$

where Ω_t is an open set in D whose boundary can be written in the form $\partial D \cup \Gamma_t, \partial D \cap \Gamma_t = \emptyset$, Γ_t being a C^1-manifold in D. Let n_t denotes the unitary normal field on Γ_t, outgoing to Ω_t. We shall refer to ∂D as being the outer boundary of Ω_t while Γ_t is the inner one.

Let now consider the outer lateral boundary of Q as

$$S = I \times \partial D$$

and its inner lateral boundary

$$\Sigma = \bigcup_{t \in I} \{t\} \times \Gamma_t, \tag{2}$$

which possesses a unitary normal field ν, $\nu(t,x) \in \mathbf{R}^{N+1}$, outgoing to Q. That normal field can be written in the following form

$$\nu(t,x) = (1 + v^2(t,x))^{-1/2}(v(t,x), n_t(x)). \tag{3}$$

Throghout this paper we assume

$$v \in C^0(\overline{\Sigma}). \tag{4}$$

Given this non cylindrical evolution domain Q with its inner lateral boundary Σ we consider the following inverse free boundary problem.

Given heat equation in Q, determine a forcing term u on the outer lateral boundary S, such that Σ will be the free boundary for the Stefan problem. More precisely let the initial datum w_0 be given in $L^2(\Omega_0)$ and let $w = w(u)$ be the solution of the parabolic problem:

$$\begin{cases} \partial_t w = \Delta w \ \text{ in } \ Q, \\ w(0,\cdot) = w_0 \ \text{ in } \ \Omega_0, \\ w(t,\cdot) = u \ \text{ in } \ S \\ w(t,\cdot) = 0 \ \text{ in } \ \Sigma. \end{cases} \tag{5}$$

The inverse free boundary problem is to determine the forcing term u on S such that the unique solution w of the problem (5) verifies the following extra boundary condition

$$\frac{\partial}{\partial n_t} w(t,\cdot) = 0 \ \text{ on } \ \Sigma. \tag{6}$$

The boundary Σ turns to be the prescribed free boundary (5)-(6) while u is unknown. Our approach is similar to [5], [6], [7]. It consists in regarding u as an optimal control and solving the problem (5)-(6) by least squares approach. We have not existence results for u solving problem (5)-(6) when Σ is given, but we associate the following minimization problem. For each positive number α (P_α) Minimize

$$J(u) = \frac{1}{2} \int_Q (\alpha u^2 + (y - z)^2) dx dt,$$

over all $u \in L^2(S)$ and $(y, z) \in (L^2(Q))^2$

$$\begin{cases} \partial_t y = \Delta y, \ \partial_t z = \Delta z \ \text{ in } \ Q, \\ y(0,\cdot) = z(0,\cdot) = w_0 \ \text{ in } \ \Omega_0, \\ y(t,\cdot) = z(t,\cdot) = u \ \text{ in } \ S \\ y(t,\cdot) = 0, \ \frac{\partial}{\partial n}(t,\cdot) = 0 \ \text{ in } \ \Sigma. \end{cases} \tag{7}$$

The solution of the inhomogeneous moving boundary problem (7), for convenience, will be understood in a suitable *mild* form, that will be precised in section 2. In Section 3 we solve the control problem by using the Dynamic Programming approach. Finally in Section 4 we will study the limit $\alpha \to 0$.

1.1 The inner boundary Σ

In the problem $P_\alpha, \alpha > 0$, the data are the initial condition and the lateral boundary Σ. The are many different ways to give such a boundary. What is essential for the understanding of our approach is to characterize the term v (which appears in the time component of the normal field ν). We assume Σ to be a C^1-manifold in Q, $\overline{\Sigma}$ being connected and

$$\Gamma_t = \overline{\Sigma} \cap \{(t, x); x \in D\},$$

is also connected and $\Lambda_t = \overline{\Omega}^c$ [1]is simply connected in D. It follows that $v \in C^0(\overline{\Sigma})$.

We give now two classical examples for such manifold.

Example 1.1

Level curves. As in [5] and [7] we can assume $\Gamma_t = \sigma^{-1}(t)$ and

$$\Omega_t = \{x \in D; \sigma(x) < t\},$$

for some given function σ in $C^1(\overline{D})$ with $|\nabla\sigma| > 0$ and , for example, $\sigma = -1$ on ∂D. The domain Ω_t is then increasing on time, the horizontal normal field on Γ_t is

$$n_t(x) = |\nabla\sigma(x)|^{-1}\nabla\sigma(x) \text{ if } \sigma(x) = t, \tag{8}$$

while v is given by

$$v(t, x) = |\nabla\sigma(x)|^{-1} \text{ if } \sigma(x) = t. \blacksquare \tag{9}$$

Example 1.2

Coincidence sets. In many situations encountered in Mechanics the open set Ω_t is defined as follows

$$\Omega_t = \{x \in D; \theta(t, x) > 0\}, \tag{10}$$

where θ is some smooth function such that this set is not empty. Then the boundary Γ_t is the coincidence set

$$\Gamma_t = \{x \in D; \theta(t, x) = 0\}. \tag{11}$$

[1] Ω^c denotes the complement of Ω.

The unitary outgoing normal field is given by

$$n_t(x) = |\nabla\theta(x)|^{-1}\nabla\theta(x) \text{ if } \theta(t,x) = 0, \tag{12}$$

while v is given by

$$v(t,x) = -\theta_t(t,x)|\nabla\theta(x)|^{-1} \text{ if } \theta(t,x) = 0. \blacksquare \tag{13}$$

In the sequel we shall need the Change of Variable techniques. We introduce a one parameter family T_t of smooth one- to-one mappings from D onto itself, which maps Γ_0 onto Γ_t, Ω_0 onto Ω_t and ∂D onto itself. The technique introduced in [18] consists in considering the speed vector field

$$V(t,x) = \partial_t T_t \circ T_t^{-1}, \tag{14}$$

from which the tranformation T_t can be recovered as being the flow of V:

$$T_t = x + \int_0^t V(s, T_s(x))ds. \tag{15}$$

From [18] we know that V has just to verifies the two following conditions

$$\begin{cases} (i)\ V(t,\cdot)\cdot n(\cdot) = 0 \text{ on } \partial D, \\ (ii)\ V(t,\cdot)\cdot n_t(\cdot) = v(t,\cdot) \text{ on } \Sigma. \end{cases} \tag{16}$$

More precisely we have the following result

Proposition 1.3 *Let* $V \in C^0([0,T], C^1(\overline{D}; \mathbf{R}^N))$ *verifying* (16). *Then* $T(V) \in C^1([0,T], C^1(\overline{D}; \overline{D}))$, *for any* t, $T_t(V)$ *is a one-to-one mapping from* $\overline{D}$ *onto* $\overline{D}$ *such that* ∂D *is invariant and*

$$T_t(V)(\Omega_0) = \Omega_t,\ T_t(V)(\Gamma_0) = \Gamma_t.$$

The inverse mapping $T_t^{-1}(V)$ *is given by* $T_t(-V_t)$ *(where* $V_t(s)(\cdot) = V(t+s)(\cdot)$*)then is in* $C^1(\overline{D}; \overline{D})$.

$v(\cdot,\cdot)$ is called the *normal speed* of the boundary.

1.2 Change of variables

We introduce now the change of variables.

Let $V \in C^0([0,T], C^1(\overline{D}; \mathbf{R}^N))$ satisfying (16), and let T_t be defined by (15). We set

$$\varphi(t,x) = g(t, T_t(x)), \tag{17}$$

so that, by the same calculations as in [11]

$$\left\{ \begin{array}{l} (\partial_t g - \Delta g) \circ T_t = \partial_t \varphi - \text{ div } (G \nabla \varphi) - < E \cdot \nabla \varphi, V \circ T_t > \\ \\ -J^{-1} < G \nabla J, \nabla \varphi >= 0 \text{ in } Q_0. \end{array} \right. \tag{18}$$

and

$$(\frac{\partial}{\partial n_t}) y \circ T_t = \frac{\partial}{\partial m_t} \varphi, \tag{19}$$

where

$$\begin{array}{l} E(t,x) =^* (DT_t^{-1})(x), \ G(t,x) = E^*(t,x) \cdot E(t,x) \\ \\ J(t,x) = \text{ det } (DT_t(x)), \ m(t,x) = G(t,x) \cdot n_0(x), \end{array} \tag{20}$$

n_0 being the unitary normal field on $\Sigma_0 = I \times \Gamma_0$.

2 Abstract formulation of the problem

We reduce the problem (7) to an evolution equation by proceeding as in [9]. We introduce unbounded operators $\{A(t)\}$ and $\{B(t)\}$ with domain depending on time, but in the fixed space $H = L^2(D)$.

$$\left\{ \begin{array}{l} D(A(t)) = \{ y \in L^2(D); y|_{\Omega_t} \in H^2(\Omega_t), y|_{\Lambda_t} \in H^2(\Lambda_t), \\ y = 0 \text{ on } \partial D, y = 0 \text{ on } \Gamma_t \} \\ A(t)u = \Delta u, \forall\, u \in D(A(t)). \end{array} \right. \tag{21}$$

$$\left\{ \begin{array}{l} D(B(t)) = \{ z \in L^2(D); z|_{\Omega_t} \in H^2(\Omega_t), z|_{\Lambda_t} \in H^2(\Lambda_t), \\ z = 0 \text{ on } \partial D, \frac{\partial z}{\partial n_t} = 0 \text{ on } \Gamma_t \} \\ B(t)u = \Delta u, \forall\, u \in D(B(t)). \end{array} \right. \tag{22}$$

Proposition 2.1 *For the family of linear operators $\{A(t)\}$, $\{B(t)\}$ in $L^2(D)$, there exist evolution operators $U_A(\cdot,\cdot)$, and $U_B(\cdot,\cdot)$ such that:*

(i) $\frac{\partial U_A(t,s)}{\partial t}\omega = A(t) U_A(t,s)\omega, \forall\, t > s, \forall\, \omega \in H.$

(ii) $U_A(s,s)\omega = \omega, \forall\, s \in I.$

(iii) $\frac{\partial U_A(t,s)}{\partial s}\omega = -U_A(t,s)A(s)\omega, \forall\, t > s, \forall\, \omega \in H.$
(iv) $\|\frac{\partial U_A(t,s)}{\partial s}\omega\|_{L^2(D)} \leq \frac{C}{(t-s)^{1-\theta}}\|\omega\|_{H^{2\theta}(D)}, \forall\, \omega \in H^{2\theta}(D), \forall\, \theta \in [0, 1/4[.$

Proof — Setting $g = U_A(t,s)\omega$, g is the solution to the homogeneous parabolic problem

$$\left\{\begin{array}{l} \partial_t g = \Delta g, \text{in } Q_s \\ g(0,\cdot) = \omega \text{ in } \Omega_0, \\ g(t,\cdot) = 0 \text{ in } S_s = [s,T] \times \partial D. \\ g(t,\cdot) = 0, \text{ in } \Sigma_s. \end{array}\right. \tag{23}$$

where $Q_s = \bigcup_{t\in[s,T]}\{t\} \times \Omega_t$ and $\Sigma_s = \bigcup_{t\in[s,T]}\{t\} \times \Gamma_t$ Now by using the change of variable (17) we reduce problem (23) to

$$\left\{\begin{array}{l} \partial_t\varphi - \text{ div } (G\nabla\varphi) - < E\cdot\nabla\varphi, V \circ T_t > \\ -J^{-1} < G\nabla J, \nabla\varphi >= 0 \text{ in } Q_s. \\ \varphi(t,\cdot) = 0 \text{ on } S_s \cup \Sigma_s. \\ \varphi(s,\cdot) = \omega \text{ in } D. \end{array}\right. \tag{24}$$

Now problem (24) is a standard parabolic equation , (in abstract form $\partial_t\varphi = \widehat{A}(t)\varphi$), which has a unique solution $\varphi(t,\cdot) = V_A(\cdot,\cdot)$., where the evolution operator $V_A(\cdot,\cdot)$ fulfils the following conditions:

(i) $\frac{\partial V_A(t,s)}{\partial t}\omega = \widehat{A}(t)V_A(t,s)\omega, \forall\, t > s, \forall\, \omega \in H.$
(ii) $V_A(s,s)\omega = \omega, \forall\, s \in I.$
(iii) $\frac{\partial V_A(t,s)}{\partial s}\omega = -V_A(t,s)\widehat{A}(s)\omega, \forall\, t > s, \forall\, \omega \in H.$
(iv) $\|\frac{\partial V_A(t,s)}{\partial s}\omega\|_{L^2(D)} \leq \frac{C}{(t-s)^{1-\theta}}\|\omega\|_{H^{2\theta}(D)}, \forall\, \omega \in H^{2\theta}(D), \forall\, \theta \in [0, 1/4[.$

see [17]and [16]. Now, the required evolution operator U_A can be obtained by performing the inverse change of variable :

$$U_A(t,s)\omega = (V_A(t,s)\omega) \circ T_{s,t}(V)^{-1},$$

where $T_{s,t}(V) = T_{t-s}(V_s)$ and $V_s(t,x) = V(s+t,x)$. Since V is smooth enough the required estimates follow easily. The statements concerning the family $\{A(t)\}$ are thus proven. We consider now the family $\{B(t)\}$. Setting $h = U_B(t,s)\omega$, h is the solution to the homogeneous parabolic problem

$$\left\{\begin{array}{l} \partial_t h = \Delta h, \text{in } Q_s \\ h(0,\cdot) = \omega \text{ in } \Omega_0, \\ h(t,\cdot) = 0 \text{ in } S_s = [s,T] \times \partial D. \\ \frac{\partial h}{\partial n_t}(t,\cdot) = 0, \frac{\partial}{\partial n}(t,\cdot) = 0 \text{ in } \Sigma_s. \end{array}\right. \tag{25}$$

Now by using the change of variable (17) we reduce problem (24) to

$$\begin{cases} \partial_t\psi - \ \text{div}\,(G\nabla\psi) - < E\cdot\nabla\psi, V\circ T_t > \\ -J^{-1} < G\nabla J, \nabla\psi >= 0 \ \text{in} \ Q_s. \\ \psi(t,\cdot) = 0 \ \text{on} \ S_s. \\ \frac{\partial}{\partial m_t}\psi(t,\cdot) = 0 \ \text{on} \ \Sigma_s.\psi(s,\cdot) = \omega \ \text{in} \ D. \end{cases} \tag{26}$$

Now problem (26) is a parabolic equation with mixed boundary conditions, which is equivalent to an abstract equation $\partial_t\varphi = \widehat{B}(t)\varphi$ with variable domains.). This equation can be solved by using abstract results of [17] or [1]. We have $\varphi(t,\cdot) = V_B(\cdot,\cdot).$, where the evolution operator $V_B(\cdot,\cdot)$ fulfils the following conditions:

(i) $\frac{\partial V_B(t,s)}{\partial t}\omega = \widehat{B}(t)V_B(t,s)\omega, \forall\, t > s, \forall\, \omega \in H.$
(ii) $V_B(s,s)\omega = \omega, \forall\, s \in I.$
(iii) $\frac{\partial V_B(t,s)}{\partial s}\omega = -V_B(t,s)\widehat{B}(s)\omega, \forall\, t > s, \forall\, \omega \in H.$
(iv) $\|\frac{\partial V_B(t,s)}{\partial s}\omega\|_{L^2(D)} \le \frac{C}{(t-s)^{1-\theta}}\|\omega\|_{H^{2\theta}(D)}, \forall\, \omega \in H^{2\theta}(D), \forall\, \theta \in [0,1/4[.:$

For (i)-(iii) see again [17] or [1], while for (iv) see [13]. Finally, the required evolution operator U_B can be obtained by performing as before. The proof is complete. ∎

3 Dynamic Programming

3.1 Introduction

We first recall briefly the Dynamic Programming approach see [8]. Consider the problem

Minimize:

$$J_\alpha(u) = \int_0^T \left(\alpha|u(t)|^2 + < S(s)w(s), w(s) >\right) ds, \tag{27}$$

over all $u \in L^2(0,T;U)$ subject to the differential constraints

$$w'(t) = L(t)w(t) + G(t)u(t);\ w(0) = w_0, \tag{28}$$

where $\{L(t)\}_{t\in[0,T]}$ is a family of linear operators such that an evolution operator $U_L(\cdot,\cdot)$ exists and $\{G(t)\}_{t\in[0,T]}$ are linear operators from a Hilbert space Z in U, which we assume, for the moment, bounded. Then the solution of problem (27) – (28) can be obtained as follows.

(i) We solve the Riccati equation:

$$P'_\alpha + A^\star P_\alpha + P_\alpha A - P_\alpha G G^\star P_\alpha + S = 0;\; P_\alpha(T) = 0, \tag{29}$$

where $P_\alpha(t)$ are symmetric positive operators in Z, and

(ii) the closed loop equation:

$$\frac{d}{dt} w^\star_\alpha = L w^\star_\alpha - \frac{1}{\alpha} G G^\star P_\alpha w^\star_\alpha;\; w^\star_\alpha(0) = w_0. \tag{30}$$

Then the optimal control $u^\star_\alpha$ is given by the feedback formula

$$u^\star_\alpha = -\frac{1}{\alpha} G^\star P_\alpha w^\star_\alpha, t \in [0, T]. \tag{31}$$

In next subsection we shall write problem (7) in the form(27) – (28). We remark that in our case, the operators $G(t)$ are not bounded. Thus, in order to solve Riccati equation we have to write it as a suitable integral equation which can be solved by using the results of [12] and [2].

3.2 The problem

Following a well known device of [4]. To this purpose we introduce the pseudo-differential mappings D and N, as follows.

$$\begin{cases} D(t) : L^2(\partial D) \to L^2(D),\; u \to D(t)u, \\ \Delta D(t)u = 0 \text{ on } \Omega_t; \\ D(t)u = u \text{ on } \partial D;\; D(t)u = 0 \text{ on } \Gamma_t, D(t)u = 0 \text{ on } \Lambda_t. \end{cases}$$

$$\begin{cases} N(t) : L^2(\partial D) \to L^2(D),\; u \to N(t)u, \\ \Delta N(t)u = 0 \text{ on } \Omega_t; \\ N(t)u = u \text{ on } \partial D;\; \frac{\partial}{\partial n_t} N(t)u = 0 \text{ on } \Gamma_t, N(t)u = 0 \text{ on } \Lambda_t. \end{cases}$$

Thus the mild form of problem (7) is the following

$$\begin{cases} y(t) = U_A(t,0)w_0 - \int_0^t \frac{\partial}{\partial s} U_A(t,s) D(s) u(s) ds \\ \\ z(t) = U_B(t,0)w_0 - \int_0^t \frac{\partial}{\partial s} U_B(t,s) N(s) u(s) ds. \end{cases} \tag{32}$$

Now we set

$$H = L^2(D),\; U = L^2(\partial D),\; Z = H \oplus H,\; L(t) A(t) \oplus B(t), \tag{33}$$

$$G(t)u = -\left[\begin{array}{c} D^\star(t)A^\star(t)u \\ N^\star(t)B^\star(t)u \end{array}\right], \ u \in U \tag{34}$$

$$S\left[\begin{array}{c} y \\ z \end{array}\right] = \left[\begin{array}{cc} 1 & -1 \\ -1 & 1 \end{array}\right]\left[\begin{array}{c} y \\ z \end{array}\right] \ y, z \in H. \tag{35}$$

Then problem (7) reduces to problem (27) and the cost function (28). As we said in Section 3, Riccati equation can be solved by proceeding as in [12] and [2]. In conclusion we obtain the synthesis formula

$$\begin{aligned} u^\star &= -D^\star A^\star P_{11} y^\star - N^\star B^\star P_{12} z^\star \\ &-\tfrac{\partial}{\partial n_0}[P_{11} y^\star] - \tfrac{\partial}{\partial n_0}[P_{12} z^\star], \end{aligned} \tag{36}$$

where n_0 is the outgoing normal on ∂D and

$$P(t) = \left[\begin{array}{cc} P_{11}(t) & P_{12}(t) \\ P_{12}^\star(t) & P_{22}(t) \end{array}\right]; \ P_{ij}(t) \in \mathcal{L}(L^2(D)) \tag{37}$$

is the solution to Riccati equation (29).

4 The limit as $\alpha \to 0$

We consider here the setting of Section 3.1, where the operators S, L, G are defined by (33) – (35). We prove the result:

Proposition 4.1 *Let P_α be the mild solution to* (29), *and assume that there exists a constant $C > 0$, independent on α such that*

$$||P_\alpha(t)|| \leq C\alpha, \quad t \in [0,T], \alpha \in]0,1[.$$

Then there exists a control $u \in U$ such that $y = z$, y and z being the solutions of (7).

In other words there exists a control u solving problem (5) – (6) where Σ is given, i.e. u solves the inverse free boundary problem.

Proof —

Set $Q_\alpha = \frac{1}{\alpha}P_\alpha$; then Q_α is the solution of the equation

$$Q'_\alpha = A^\star Q_\alpha + Q_\alpha A - Q_\alpha G G^\star Q_\alpha + \frac{1}{\alpha}S; \ \frac{1}{\alpha}(T) = 0. \tag{38}$$

By well known properties of Riccati equation, Q_α is increasing on α and, due to hypothesis is strongly convergent to a symmetric and positive operator $\overline{Q}$. It follows that there exists the limit as α goes to 0 of the optimal state

$$\lim_{\alpha \to 0} w_\alpha^\star = \overline{w}, \tag{39}$$

in $C([0,T];Z)$. Consequently, there exists also the limit as α goes to 0 of the optimal control

$$\lim_{\alpha \to 0} u_\alpha^\star = \overline{u}. \tag{40}$$

Now, by letting α tend to 0 in the equality

$$< Q_\alpha(T)w_0, w_0 >= \int_0^T \left\{ |u_\alpha^\star(t)|^2 + \frac{1}{\alpha} < Sw_\alpha^\star(s), w_\alpha^\star(s) > \right\} ds,$$

we find $S\overline{w} = 0$, which implies $y = z$ as we required. ■

References

[1] P. ACQUISTAPACE & B. TERRENI (1987) *A Unified Approach to Abstract Linear Nonautonomous Parabolic Equations*, Rend. Sem. Mat. Univ. Padova **78** , 47-107

[2] ACQUISTAPACE P., FLANDOLI F & TERRENI B., (to appear) *Boundary control of non- autonomous parabolic systems*, SIAM J. Control Optimiz.

[3] AGMON S. (1962) *On the eigenfunctions and on the eigenvalues of general elliptic boundary value problems* , Comm. Pure Appl. Math. **15**, 119- 147.

[4] BALAKRISHNAN A.V. (1976), APPLIED FUNCTIONAL ANALYSIS. Springer-Verlag, New- York.

[5] BARBU V. (1990) *The inverse one phase Stefan problem* Differential and Integral Equations, **3**, 209-218.

[6] BARBU V. (to appear) *The approximate solvability of the inverse one phase Stefan problem*

[7] BARBU V. , DA PRATO G.& ZOLÉSIO J.P. (1991) *Feedback controllability of the free boundary of the one phase Stefan problem*, Differential and Integral Equations, Vol.4, 2, 225-239.

[8] BENSOUSSAN A.,DA PRATO G., DELFOUR M. & MITTER S. (to appear) *Representation and Control of Infinite Dimensional Systems*

[9] CANNARSA P., DA PRATO G. & ZOLÉSIO J.P. (1990) *The damped wave equation in a moving domain* J. Differential Equations, **85**,1, 1-16.

[10] DA PRATO G. (1973) *Quelques résultats d'existence unicité et regularité pour un problème de la théorie du contrôle* J. Math. Pures Appl., **52**, 353-375.

[11] DA PRATO G. & ZOLÉSIO J.P. (1988) *An optimal control problem for a parabolic equation in non- cylindrical domains* Systems and Control Letters 11, 73-77.

[12] F.FLANDOLI (1984), *Riccati equation arising in a boundary control problem with distributed parameters.* SIAM J. Control Optimiz. **22**, 76-86

[13] M. FURHMAN (to appear)*Bounded solutions for abstract time periodic parabolic equations with nonconstant domains*, Annali di Matematica Pura e Applicata.

[14] HOFFMAN K.H. & SPREKELS J. (1982) *Real time control of the free boundary in a two phase Stefan problem* Numerical Functional Analysis and Optimization 5, 47-76.

[15] KATO T. & TANABE H. (1962) *On the abstract evolution equation* , Osaka Math. J. , 107- 133.

[16] LUNARDI A. (1989) *Differentiability with respect to (t, s) of the parabolic evolution operator* Israel J. Math. **68**,2,161-184

[17] TANABE H. (1979) EQUATIONS OF EVOLUTION , Pitman, London.

[18] ZOLÉSIO J.P. (1981) *The material derivative, in Optimization of Distributed Parameters Structures* J.Céa, E.J.Haug Eds, Sijthoff & Noordhoff, Alphen aan den Rijn, 1089-1151.

DIFFERENTIABILITY OF MIN MAX AND SADDLE POINTS UNDER RELAXED ASSUMPTIONS*

Michel C. Delfour
Centre de recherches mathématiques et
Département de mathématiques et de statistique,
Université de Montréal, C.P. 6128 A,
Montréal, Québec, Canada, H3C 3J7

Jacqueline Morgan
Dipartimento di Matematica et Applicazioni
Universita di Napoli,
Via Mezzocannone 8
80134 Napoli, Italy

ABSTRACT. The object of this paper is to present new theorems on the differentiability of an Infimum, Supremum, Min Max, or saddle point with respect to a parameter $t \geq 0$ at $t = 0$ under relaxed assumptions. Those technical results have a wide spectrum of applications: Control Theory, Shape Sensitivity Analysis, Game Theory, etc... In non-differentiable situations they provide an interesting description of the non-differentiability.

RÉSUMÉ. On présente de nouveaux théorèmes sur la dérivabilité d'un infimum, supremum, point selle ou Min Max par rapport à un paramètre $t \geq 0$ en $t = 0$ sous des hypothèses relaxées. Ces résultats techniques présentent un large spectre d'applications: théorie du contrôle, analyse de sensitivité de forme, théorie des jeux, etc... Dans les situations non-différentiables, ils donnent une description intéressante de la non-différentiabilité.

1. INTRODUCTION.

The object of this paper is to present new theorems on the differentiability of a Min Max, a saddle point, an infimum and a supremum with respect to a parameter $t \geq 0$ at $t = 0$ under relaxed assumptions. To relax the hypotheses we use concepts of ϵ–solution of an extremum. For saddle points we introduce a concept of ϵ–solution which seems to be new. This type of results has many applications in several interesting areas:
• to compute first and second order derivatives in Shape Optimization problems by *Function Space Parametrization* (cf. DELFOUR–ZOLÉSIO [1–9]), or *Function Space Embedding* (cf. DELFOUR–ZOLÉSIO [10–11]),
• to compute directional derivatives for problems in *Game Theory* (for instance Stackelberg problems), *Multilevel Optimization* and *Control Problems under perturbations* (cf. LIGNOLA–MORGAN [1], J. MORGAN [1]).

* The research of the first author has been supported by a Killam Fellowship from Canada Council, Natural Sciences and Engineering Research Council of Canada operating grant A-8730, and a FCAR grant from the "Ministère de l'éducation du Québec".

In order to make the discussion more precise consider the following typical set–up for a Min Max problem. Given a real number $\tau > 0$, sets X and Y and a functional

$$G : [0, \tau] \times X \times Y \to \mathbf{R}, \tag{1.1}$$

and

$$g(t) = \inf_{x \in X} \quad \sup_{y \in Y} \quad G(t, x, y) \tag{1.2}$$

$$h(t) = \sup_{y \in Y} \quad \inf_{x \in X} \quad G(t, x, y). \tag{1.3}$$

Define the sets

$$X(t) = \{x^t \in X : \sup_{y \in Y} G(t, x^t, y) = g(t)\} \tag{1.4}$$

$$Y(t) = \{y^t \in X : \inf_{x \in X} G(t, x, y^t) = h(t)\} \tag{1.5}$$

and for all x in X

$$Y(t, x) = \{y^t \in Y : G(t, x, y^t) = \sup_{y \in Y} G(t, x, y)\}. \tag{1.6}$$

We are typically interested in the existence and characterization of the limit of the differential quotient of g as $t > 0$ goes to zero

$$dg(0) = \lim_{t \searrow 0} \frac{g(t) - g(0)}{t}. \tag{1.7}$$

DELFOUR–ZOLÉSIO [2] have shown that under certain hypotheses the above limit exists and is characterized by

$$dg(0) = \inf_{x \in X(0)} \sup_{y \in Y(0, x)} \partial_t G(0, x, y). \tag{1.8}$$

If we assume that, for all t in $[0, \tau]$, $g(t) = h(t)$ and the set of saddle points $S(t) = X(t) \times Y(t)$ is not empty. CORREA-SEEGER [1] have shown that under appropriate hypotheses

$$\begin{aligned} dg(0) = &= \inf_{x \in X(0)} \quad \sup_{y \in Y(0)} \quad \partial_t G(0, x, y) \\ &= \sup_{y \in Y(0)} \quad \inf_{x \in X(0)} \quad \partial_t G(0, x, y). \end{aligned} \tag{1.9}$$

The two sets of assumptions leading to the above characterizations are not strictly contained in one another. The saddle point hypothesis is a strong assumption, but in that case the other hypotheses are weaker.

In this paper we present new results and say more about $dg(0)$ under weaker or relaxed hypotheses. Among the interesting outcomes of this work is a sharpening of CORREA-SEEGER [1]'s theorem: there exist points $(x^0, y^0) \in X(0) \times Y(0)$ such that in addition

$$dg(0) = \sup_{y \in Y(0)} \partial_t G(0, x^0, y) = \partial_t G(0, x^0, y^0) = \inf_{x \in X(0)} \partial_t G(0, x, y^0) \tag{1.10}$$

and (x^0, y^0) is a saddle point of $\partial_t G(0, \cdot, \cdot)$ on $X(0) \times Y(0)$. This technical result is quite general and a similar result can also be added to the conclusions of Theorem 3 in DELFOUR-ZOLÉSIO [2]. One of its consequences is an interesting description of the directional derivative for some non-differentiable cost functionals with a state equation constraint. An example is given in DELFOUR–MORGAN [1].

2. A GENERALIZATION OF THE THEOREM OF DELFOUR-ZOLESIO.

The starting point is Theorem 3 in DELOUR-ZOLÉSIO [2, p. 842]. There the existence of a saddle point was not assumed. DELFOUR–MORGAN [1] have shown that under the same assumptions the conclusions of the theorem can be somewhat sharpened.

First introduce the following assumptions.

(H1) There exists $\tau > 0$ such that
  (i) $X(0) \neq \emptyset$, $\forall x_0 \in X(0)$, $\forall t \in [0, \tau]$, $Y(t, x_0) \neq \emptyset$
  (ii) $\forall t \in [0, \tau]$, $X(t) \neq \emptyset$, $\forall x_t \in X(t)$, $Y(0, x_t) \neq \emptyset$.

(H2) For all (x, y) in

$$\bigcup_{x \in X(0)} [\{x\} \times \bigcup_{t \in [0,\tau]} Y(t, x)] \quad and \quad \bigcup_{\substack{x \in X(t) \\ t \in [0,\tau]}} [\{x\} \times Y(0, x)] \tag{2.1}$$

the partial derivative $\partial_t G(t, x, y)$ exists everywhere in $[0, \tau]$

(H3) There exists a topology $\mathcal{T}_Y$ on Y such that for all $x_0 \in X(0)$ and all sequences $\{t_n : 0 < t_n \leq \tau\}$, $t_n \to 0$, $\exists y^0 \in Y(0, x_0)$, $\exists$ a subsequence $\{t_{n_k}\}$ of $\{t_n\}$, and $\exists y^k \in Y(t_{n_k}, x_0)$ such that
  (i) $y^k \to y^0$ in the $\mathcal{T}_Y$-topology
  (ii) $\limsup_{\substack{t \searrow 0 \\ k \to \infty}} \partial_t G(t, x_0, y^k) \leq \partial_t G(0, x_0, y^0)$.

(H4) There exist topologies $\mathcal{T}_X$ on X and $\mathcal{T}_Y$ on Y such that for all sequences $\{t_n : 0 < t_n \leq \tau\}$, $t_n \to 0$, $\exists x^0 \in X(0)$, $\forall y_0 \in Y(0, x^0)$, $\exists$ a subsequence $\{t_{n_k}\}$ of $\{t_n\}$, $\exists x^k \in X(t_{n_k})$, and $\exists y_k \in Y(0, x^k)$ such that
  (i) $y_k \to y_0$ in $\mathcal{T}_Y$, $x^k \to x^0$ in $\mathcal{T}_X$
  (ii) $\liminf_{\substack{t \searrow 0 \\ k \to \infty}} \partial_t G(t, x^k, y_k) \geq \partial_t G(0, x^0, y_0)$.

THEOREM 2.1. *Under assumptions (H1) to (H4)*

$$dg(0) = \min_{x\in X(0)} \sup_{y\in Y(0,x)} \partial_t G(0,x,y) \tag{2.2}$$

and $\exists x^0 \in X(0)$ *and* $\exists y^0 \in Y(0,x^0)$, *such that*

$$dg(0) = \max_{y\in Y(0,x^0)} \partial_t G(0,x^0,y) = \partial_t G(0,x^0,y^0). \quad \square \tag{2.3}$$

To relax the hypotheses of Theorem 2.1 we introduce appropriate concepts of ϵ- and α-solutions. For all $\alpha \geq 0$, define the sets

$$X(t,\alpha) = \{x_\alpha^t \in X : \sup_{y\in Y} G(t,x_\alpha^t,y) \leq g(t) + \alpha t\}. \tag{2.4}$$

In particular $X(t,0) = X(t)$ and $X(0,\alpha) = X(0)$. Similarly for all $\epsilon > 0$ and $x \in X$, define

$$Y(t,x,\epsilon) = \{y_\epsilon^t \in Y : \sup_{y\in Y} G(t,x,y) \leq G(t,x,y_\epsilon^t) + \epsilon t\}. \tag{2.5}$$

Let the following assumptions be verified.

(H1$^\epsilon$) There exists $\tau > 0$ such that
  (i) $X(0) \neq \varnothing$ and $\forall x^0 \in X(0)$, $\forall t \in [0,\tau]$, $\forall \epsilon > 0$, $Y(t,x^0,\epsilon) \neq \varnothing$.
  (ii) $\forall t \in [0,\tau]$, $\forall \alpha > 0$, $X(t,\alpha) \neq \varnothing$ and $\forall x_\alpha^t \in X(t,\alpha)$, $\forall \epsilon > 0$, $Y(t,x_\alpha^t,\epsilon) \neq \varnothing$.

(H2$^\epsilon$) For all (x,y) in

$$\bigcup_{x\in X(0)} \left[\{x\} \times \bigcup_{\substack{\epsilon>0\\ t\in[0,\tau]}} Y(t,x,\epsilon)\right] \quad \text{and} \quad \bigcup_{\substack{x\in X(t,\alpha)\\ \alpha>0\\ t\in[0,\tau]}} \left[\{x\} \times \bigcup_{\epsilon>0} Y(0,x,\epsilon)\right] \tag{2.6}$$

the partial derivative $\partial_s G(s,x,y)$ exists everywhere in $[0,\tau]$.

(H3$^\epsilon$) There exists a topology $\mathcal{T}_Y$ on Y such that for all $x^0 \in X(0)$ and all sequences $\{t_n : 0 \leq t_n \leq \tau\}$, $t_n \to 0$, $\exists y_0 \in Y(0,x^0)$, $\exists$ a subsequence $\{t_{n_k}\}$ of $\{t_n\}$, $\exists$ a sequence $\{\epsilon_k > 0\}$, $\epsilon_k \to 0$, for each $k \geq 1$, $\exists y^k \in Y(t_{n_k}, x^0, \epsilon_k)$ such that
  (i) $y^k \to y_0$ in the $\mathcal{T}_Y$-topology
  (ii) $\limsup_{\substack{t\searrow 0\\ k\to\infty}} \partial_t G(t,x^0,y^k) \leq \partial_t G(0,x^0,y_0)$.

(H4$^\epsilon$) There exist topologies $\mathcal{T}_X$ on X and $\mathcal{T}_Y$ on Y such that for all sequences $\{t_n : 0 \leq t_n \leq \tau\}$, $t_n \to 0$, $\exists x_0 \in X(0)$, $\forall y^0 \in Y(0,x_0)$, $\exists$ a subsequence $\{t_{n_k}\}$ of $\{t_n\}$, $\exists \{\alpha_k > 0\}, \alpha_k \to 0$, $\exists \{\epsilon_k > 0\}$, $\epsilon_k \to 0$, for each $k \geq 1$, $\exists x^k \in X(t_{n_k},\alpha_k)$, $\exists y_k \in Y(0,x^k,\epsilon_k)$ such that
  (i) $y_k \to y^0$ in $\mathcal{T}_Y$, $x^k \to x_0$ in $\mathcal{T}_X$
  (ii) $\liminf_{\substack{t\searrow 0\\ k\to\infty}} \partial_t G(t,x^k,y_k) \geq \partial_t G(0,x_0,y^0)$.

THEOREM 2.1$^\epsilon$. *Under assumptions (H1$^\epsilon$) to (H4$^\epsilon$)*

$$dg(0) = \min_{x \in X(0)} \sup_{y \in Y(0,x)} \partial_t G(0,x,y). \tag{2.7}$$

and $\exists x_0 \in X(0)$ *and* $\exists y_0 \in Y(0, x_0)$, *such that*

$$dg(0) = \max_{y \in Y(0,x_0)} \partial_t G(0,x_0,y) = \partial_t G(0,x_0,y_0). \quad \square \tag{2.8}$$

This theorem is a weaker version of Theorem 3 in DELFOUR–ZOLÉSIO [**2**, p. 842]. Assumption (H1$^\epsilon$) is a relaxion of (H1):
there exists $\tau > 0$ such that

(i) $X(0) \neq \emptyset$, $\forall x_0 \in X(0)$, $\forall t \in [0,\tau]$, $Y(t,x_0) \neq \emptyset$
(ii) $\forall t \in [0,\tau]$, $X(t) \neq \emptyset$, $\forall x_t \in X(t)$, $Y(0,x_t) \neq \emptyset$.

Since $x^0 \notin X(t)$ and $x_t \notin X(0)$, the sets $Y(t,x^0)$ and $Y(0,x_t)$ need not be non empty. But the non emptyness hypothesis on $Y(t,x^0,\epsilon)$, $X(t,\alpha)$ and $Y(t,x^t_\alpha,\epsilon)$ is trivially verified when the corresponding extrema are finite.

3. A GENERALIZATION OF THE THEOREM OF CORREA–SEEGER.

In this section we go over the theorem of CORREA-SEEGER [**1**] for a saddle point and extend its conclusions. Then we specialize to an Infimum.

First introduce the following set of assumptions:

(S1) $S(t) \neq \emptyset$, $0 \leq t \leq \tau$,
(S2) for all (x,y) in $[\cup\{X(t) : 0 \leq t \leq \tau\} \times Y(0)] \cup [X(0) \times \cup\{Y(t) : 0 \leq t \leq \tau\}]$ the partial derivative $\partial_t G(t,x,y)$ exists everywhere in $[0,\tau]$,
(S3) there exists a topology $\mathcal{T}_X$ on X such that for any sequence $\{t_n : 0 < t_n \leq \tau\}$, $t_n \to t_0 = 0$, $\exists x^0 \in X(0)$, $\exists$ a subsequence $\{t_{n_k}\}$ of $\{t_n\}$, and for each $k \geq 1$, $\exists x_k \in X(t_{n_k})$ such that
   (i) $x_k \to x^0$ in the $\mathcal{T}_X$-topology
   (ii) and for all y in $Y(0)$

$$\liminf_{\substack{t \searrow 0 \\ k \to \infty}} \partial_t G(t,x_k,y) \geq \partial_t G(0,x^0,y), \tag{3.1}$$

(S4) there exists a topology $\mathcal{T}_Y$ on Y such that for any sequence $\{t_n : 0 < t_n \leq \tau\}$, $t_n \to t_0 = 0$, $\exists y^0 \in Y(0)$, $\exists$ a subsequence $\{t_{n_k}\}$ of $\{t_n\}$, and for each $k \geq 1$, $\exists y_k \in Y(t_{n_k})$ such that
   (i) $y_k \to y^0$ in the $\mathcal{T}_Y$-topology
   (ii) and for all x in $X(0)$

$$\limsup_{\substack{t \searrow 0 \\ k \to \infty}} \partial_t G(t,x,y_k) \leq \partial_t G(0,x,y^0). \tag{3.2}$$

THEOREM 3.1. *Under assumptions (S1) to (S4)*

$$dg(0) = \inf_{x\in X(0)} \sup_{y\in Y(0)} \partial_t G(0,x,y) = \sup_{y\in Y(0)} \inf_{x\in X(0)} \partial_t G(0,x,y) \tag{3.3}$$

and there exists $(x^0, y^0) \in X(0) \times Y(0)$ *such that*

$$dg(0) = \inf_{x\in X(0)} \partial_t G(0,x,y^0) = \partial_t G(0,x^0,y^0) = \sup_{y\in Y(0)} \partial_t G(0,x^0,y). \tag{3.4}$$

Thus (x^0, y^0) *is a saddle point of* $\partial_t G(0,x,y)$ *on* $X(0) \times Y(0)$. □

REMARK 3.1. In the applications this formulation of the theorem presents some definite technical advantages over its original version.

(i) Equation (3.4) establishes the existence of a saddle point of (3.3).
(ii) Another important feature is the use of subsequences in hypotheses (S3) and (S4). This makes it possible to work with weak topologies in reflexive Banach spaces and use the eventual boundedness of the saddle points.
(iii) Finally hypothesis (S2) and conditions (3.1) and (3.2) in (S3) and (S4) need only be verified on the family of saddle points at $t = 0$. For instance the first part of hypotheses (S3) and (S4) could be verified in $H^1(\Omega) \times H^1(\Omega)$. Yet, if the saddle points are smoother, say in $H^2(\Omega) \times H^2(\Omega)$ this extra smoothness can be used to verify (S2) and (3.1) and (3.2) in (S3) and (S4). □

Theorem 3.1 can be specialized to the derivative of an infimum.

THEOREM 3.2. *Let the real number* $\tau > 0$, *the set* X *and the functional* $G : [0,\tau] \times X \to \mathbf{R}$ *be given and let*

$$g(t) \overset{\text{def}}{=} \inf_{x\in X} G(t,x), \qquad X(t) = \{x^t \in X : G(t,x^t) = \inf_{x\in X} G(t,x)\}. \tag{3.5}$$

Assume that the following hypotheses hold:

(I1) $X(t) \neq \emptyset$, $0 \le t \le \tau$
(I2) *for all* x *in* $X(0)$, $\partial_t G(t,x)$ *exists everywhere in* $[0,\tau]$
(I3) *there exists a topology* $\mathcal{T}_X$ *on* X *such that for any sequence* $\{t_n : 0 < t_n \le \tau\}$, $t_n \to t_0 = 0$, $\exists x_0 \in X(0)$, $\exists$ *a subsequence* $\{t_{n_k}\}$ *of* $\{t_n\}$, *and for each* $k \ge 1$, $\exists x_k \in X(t_{n_k})$ *such that*
  (i) $x_k \to x^0$ *in the* $\mathcal{T}_X$*-topology*
  (ii) *and*
$$\liminf_{\substack{t\searrow 0\\ k\to\infty}} \partial_t G(t,x_k) \ge \partial_t G(0,x^0) \tag{3.6}$$
(I4) *for all* $x \in X(0)$
$$\limsup_{t\searrow 0} \partial_t G(t,x) \le \partial_t G(0,x). \tag{3.7}$$

Then

$$dg(0) = \lim_{t \searrow 0} \frac{g(t) - g(0)}{t} = \inf_{x \in X(0)} \partial_t G(0, x) \tag{3.8}$$

and there exists $x^0 \in X(0)$ *such that*

$$dg(0) = \partial_t G(0, x^0). \quad \square \tag{3.9}$$

REMARK 3.2. This theorem including property (3.9) extends for instance an older result by B. LEMAIRE [1, Th. 2.1, p. 38] where sequential compactness of the set X was assumed. It also completes and extends Theorem 1 in DELFOUR-ZOLÉZIO [6] and J.P. ZOLÉZIO [1]. □

We now relax the hypotheses of Theorems 3.1 and 3.2 by going to ϵ–solutions. The first step is to make sense of an ϵ–saddle point.

DEFINITION 3.1. *Given* $\epsilon > 0$ *and* t, $0 \leq t \leq \tau$, *we say that* $(\hat{x}, \hat{y})$ *is an* ϵ*–saddle point of the functional* $G(t, \cdot, \cdot)$ *if*

$$-\epsilon t + \sup_{y \in Y} G(t, \hat{x}, y) \leq G(t, \hat{x}, \hat{y}) \leq \inf_{x \in X} G(t, x, \hat{y}) + \epsilon t. \quad \square \tag{3.10}$$

We shall also need the following sets

$$X(t, \epsilon) = \{x \in X \ : \ \sup_{y \in Y} G(t, x, y) \leq \ h(t) + \epsilon t\} \tag{3.11}$$

$$Y(t, \epsilon) = \{y \in Y \ : \ \inf_{x \in X} G(t, x, y) \geq \ g(t) - \epsilon t\}, \tag{3.12}$$

where h and g are the sup inf and the inf sup as defined in (1.2)–(1.3). Notice that this definition of $X(t, \epsilon)$ differs from the definition of the set $X(t, \alpha)$ in (2.4) since it uses h instead of g. Moreover we would like to point out that the above definitions are given in the context of a family of functionals indexed by t. In the case of a single functional $F(x, y)$ set $t = 1$ and substitute the functional $F(x, y)$ to $G(1, x, y)$ in the above definitions.

LEMMA 3.1. *Fix* t, $0 \leq t \leq \tau$.

(i) *If* $(\hat{x}, \hat{y})$ *is an* ϵ*–saddle point of the functional* $G(t, \cdot, \cdot)$ *for some* $\epsilon > 0$, *then* $(\hat{x}, \hat{y}) \in X(t, 2\epsilon) \times Y(t, 2\epsilon)$.

(ii) *Conversely if* $(\hat{x}, \hat{y}) \in X(t, \epsilon) \times Y(t, \epsilon)$ *for some* $\epsilon > 0$, *then* $(\hat{x}, \hat{y})$ *is a* 2ϵ*–saddle point of the functional* $G(t, \cdot, \cdot)$.

(iii) *If there exists a sequence* $\{\epsilon_k > 0\}$, $\epsilon_k \searrow 0$, *such that* $X(t, \epsilon_k) \times Y(t, \epsilon_k) \neq \emptyset$, *then*

$$g(t) = h(t).$$

(iv) *Conversely if* $g(t) = h(t)$, *then for all* $\epsilon > 0$, $X(t, \epsilon) \times Y(t, \epsilon) \neq \emptyset$. □

Now the corresponding ϵ–assumptions are:

(S1$^\epsilon$) $S(0) \neq \emptyset$, and $\forall t,\ 0 < t \leq \tau$ and $\forall \epsilon > 0$, $X(t,\epsilon) \times Y(t,\epsilon) \neq \emptyset$.

(S2$^\epsilon$) for all (x, y) in

$$[\bigcup_{\substack{0\leq t\leq \tau\\ \epsilon>0}} \{X(t,\epsilon)\} \times Y(0)] \cup [X(0) \times \bigcup_{\substack{0\leq t\leq \tau\\ \epsilon>0}} \{Y(t,\epsilon)\}]$$

the partial derivative $\partial_t G(t, x, y)$ exists everywhere in $[0, \tau]$,

(S3$^\epsilon$) there exists a topology $\mathcal{T}_X$ on X such that for any sequence $\{t_n : 0 < t_n \leq \tau\}$, $t_n \to t_0 = 0$, $\exists x^0 \in X(0)$, $\exists$ a subsequence $\{t_{n_k}\}$ of $\{t_n\}$ and a sequence $\{\epsilon_k\}$, $\epsilon_k \to 0$, and for each $k \geq 1$, $\exists x_k \in X(t_{n_k}, \epsilon_k)$ such that

(i) $x_k \to x^0$ in the $\mathcal{T}_X$-topology

(ii) and for all y in $Y(0)$

$$\liminf_{\substack{t\searrow 0\\ k\to\infty}} \partial_t G(t, x_k, y) \geq \partial_t G(0, x^0, y), \tag{3.13}$$

(S4$^\epsilon$) there exists a topology $\mathcal{T}_Y$ on Y such that for any sequence $\{t_n : 0 < t_n \leq \tau\}$, $t_n \to t_0 = 0$, $\exists y^0 \in Y(0)$, $\exists$ a subsequence $\{t_{n_k}\}$ of $\{t_n\}$ and a sequence $\{\epsilon_k\}$, $\epsilon_k \to 0$, and for each $k \geq 1$, $\exists y_k \in Y(t_{n_k}, \epsilon_k)$ such that

(i) $y_k \to y^0$ in the $\mathcal{T}_Y$-topology

(ii) and for all x in $X(0)$

$$\limsup_{\substack{t\searrow 0\\ k\to\infty}} \partial_t G(t, x, y_k) \leq \partial_t G(0, x, y^0). \tag{3.14}$$

THEOREM 3.1$^\epsilon$. *Under assumptions (S1$^\epsilon$) to (S4$^\epsilon$)*

$$dg(0) = \inf_{x\in X(0)} \sup_{y\in Y(0)} \partial_t G(0, x, y) = \sup_{y\in Y(0)} \inf_{x\in X(0)} \partial_t G(0, x, y) \tag{3.15}$$

and there exists $(x^0, y^0) \in X(0) \times Y(0)$ *such that*

$$dg(0) = \min_{x\in X(0)} \partial_t G(0, x, y^0) = \partial_t G(0, x^0, y^0) = \max_{y\in Y(0)} \partial_t G(0, x^0, y). \tag{3.16}$$

Thus (x^0, y^0) *is a saddle point of* $\partial_t G(0, \cdot, \cdot)$ *on* $X(0) \times Y(0)$. $\square$

Theorem 3.1$^\epsilon$ can also be specialized to the derivative of an infimum.

THEOREM 3.2$^\epsilon$. *Let the real number $\tau > 0$, the set X and the functional $G : [0,\tau]\times X \to \mathbf{R}$ be given and let*

$$g(t) \stackrel{\text{def}}{=} \inf_{x\in X} G(t,x), \quad X(t) = \{x^t \in X : G(t,x^t) = \inf_{x\in X} G(t,x)\}. \tag{3.17}$$

Assume that the following hypotheses hold:

(I1$^\epsilon$) $X(0) \neq \varnothing$, and $\forall t,\ 0 < t \le \tau$ and $\forall \epsilon > 0\ X(t,\epsilon) \neq \varnothing$

(I2$^\epsilon$) *for all x in*

$$\bigcup_{\substack{0\le t\le\tau \\ \epsilon>0}} X(t,\epsilon),$$

$\partial_t G(t,x)$ *exists everywhere in* $[0,\tau]$

(I3$^\epsilon$) *there exists a topology $\mathcal{T}_X$ on X such that for any sequence $\{t_n : 0 < t_n \le \tau\}$, $t_n \to t_0 = 0$, $\exists x_0 \in X(0)$, $\exists$ a subsequence $\{t_{n_k}\}$ of $\{t_n\}$, a sequence $\{\epsilon_k > 0\}$, $\epsilon_k \to 0$, and for each $k \ge 1$, $\exists x_k \in X(t_{n_k},\epsilon_k)$ such that*

(i) $x_k \to x^0$ in the $\mathcal{T}_X$-topology

(ii) and

$$\liminf_{\substack{t\searrow 0 \\ k\to\infty}} \partial_t G(t,x_k) \ge \partial_t G(0,x^0) \tag{3.18}$$

(I4$^\epsilon$) *for all $x \in X(0)$*

$$\limsup_{t\searrow 0} \partial_t G(t,x) \le \partial_t G(0,x). \tag{3.19}$$

Then

$$dg(0) = \lim_{t\searrow 0} \frac{g(t)-g(0)}{t} = \inf_{x\in X(0)} \partial_t G(0,x) \tag{3.20}$$

and there exists $x^0 \in X(0)$ such that

$$dg(0) = \partial_t G(0,x^0). \quad \square \tag{3.21}$$

[4], *Differentiability of a MinMax and Application to Optimal Control and Design Problems, Part II*, in "Control Problems for Systems Described as Partial Differential Equations and Applications," I. Lasiecka and R. Triggiani, eds., Springer-Verlag, New York, 1987, pp. 220-229.

[5], *Further Developments in Shape Sensitivity Analysis via a Penalization Method*, in "Boundary Control and Boundary Variations," J. P. Zolésio, ed., Springer-Verlag, Berlin, Heidelberg, New York, Tokyo, 1988, pp. 153-191.

[6], *Shape Sensitivity Analysis via a Penalization Method*, Annali di Matematica Pura ed Applicata **CLI** (1988), 179-212.

[7], *Analyse des problèmes de forme par la dérivation des Min Max*, in "Analyse Non Linéaire," H. Attouch, J.P. Aubin, F.H. Clarke and I. Ekeland, eds, Série Analyse Non Linéaire, Annales de l'Institut Henri-Poincaré, Special volume in honor of J.-J. Moreau, Gauthier-Villars, Bordas, Paris, France, 1989, pp. 211-228.

[8], *Anatomy of the shape Hessian*, Annali di Matematica Pura et Applicata **CLVIII** (1989).

[9], *Computation of the shape Hessian by a Lagrangian method*, in "Fifth Symp. on Control of Distributed Parameter Systems," A. El Jai and M. Amouroux, eds., Pergamon Press, 1989, pp. 85–90.

[10], *Shape Hessian by the velocity method: a Lagrangian approach*, in "Stabilization of Flexible Structures," J.P. Zolésio, ed., Springer–Verlag, Berlin, New–York, 1990, pp. 255–279.

[11], *Velocity method and Lagrangian formulation in the computation of the shape Hessian*, SIAM J. on Control and Optimization **29** 6 (1991) (to appear).

B. LEMAIRE [1], "Problèmes Min-Max et applications au contrôle optimal de systèmes gouvernés par des équations aux dérivées partielles linéaires," Thèse de doctorat d'état, Montpellier, France, 1970.

LIGNOLA AND J. MORGAN [1], *Topological existence and stability for MinSup problems*, J. Math. Anal. Appl. (to appear).

J. MORGAN [1], *Constrained well posed two level optimization problems*, in "E. Majorana School (Erice) on Non Smooth Optimization," Plenum Press, 1989.

J. P. ZOLÉSIO [1], *Domain variational formulation for free boundary problems*, in "Optimization of Distributed Parameter Structures, vol II," E.J. Haug and J. Céa, eds., Sijhofff and Nordhoff, Alphen aan den Rijn, The Netherlands, 1981, pp. 1089–1151.

NEW NONITERATIVE APPROXIMATIONS TO THE OLD RICCATI DIFFERENTIAL EQUATION*

M.C. DELFOUR
Centre de recherches mathématiques et
Département de mathématiques et de statistique,
Université de Montréal, C.P. 6128 Succ. A,
Montréal, Québec, Canada, H3C 3J7

A. OUANSAFI
Département de mathématiques,
Faculté des Sciences,
Université Ibn Tofail, Kénitra, Maroc

ABSTRACT. The object of this paper is to present new approximation schemes for the non-linear matrix Riccati differential equation. They are obtained from any one-step or multistep method applied to the original linear quadratic control problem. They lead to the same type of scheme. However only one matrix inversion is required at each discretization node even if the Riccati equation is a non-linear equation. This important computational advantage is obtained without altering the original nodal asymptotic convergence rate. It is proved that this rate is the same as the one of the initially chosen scheme.

1. INTRODUCTION

The object of this paper is the old matrix Riccati differential equation

$$(1)\qquad \begin{cases} \dfrac{d\Pi}{dt} + A^*(t)\Pi + \Pi A(t) - \Pi R(t)\Pi + Q(t) = 0 \ \text{ in } \ [0,T] \\ \Pi(T) = F, \quad R(t) = B(t)N^{-1}(t)B^*(t) \end{cases}$$

over a finite time horizon $[0,T]$, $T > 0$, which naturally occurs in the finite time horizon Linear Quadratic Optimal Control problem

$$(2)\qquad \operatorname{Inf}\ \{J(u,x^0) : u \in L^2(0,T;\mathbf{R}^m)\}$$

for the quadratic cost function

$$(3)\qquad J(u,x^0) = (Fx(T) + 2\ell) \bullet x(T) + (Qx + 2q, x)_{L^2} + (Nu,u)_{L^2},$$

where the state x is the solution of the linear differential equation

$$(4)\qquad \begin{cases} \dfrac{dx}{dt}(t) = A(t)x(t) + B(t)u(t) + f(t) \ \text{ in } \ [0,T] \\ x(0) = x^0, \end{cases}$$

"$\bullet$" is the inner product in $\mathbf{R}^d$, $d \geq 1$ is an integer, $(\cdot,\cdot)_{L^2}$ the inner product in $L^2(0,T;\mathbf{R}^m)$ or $L^2(0,T;\mathbf{R}^d)$, and A, B, F, Q, N are appropriate matrices and operators which are defined in the next section on Notation and section 2.

* The research of the second author has been supported in part by a Killam fellowship from Canada Council, National Sciences and Engineering Research Council of Canada operating grants OGP-8730 and INF-7939 and a FCAR grant from the "Ministère de l'Education du Québec.

The matrix Riccati differential equation can be numerically solved by a wide spectrum of well-established techniques. For instance it can be solved directly by using standard iterative codes for systems of nonlinear differential equations. For most purposes this is sufficient. However when the size of $\Pi(t)$ becomes large the iterative process can become costly and the numerical stability of the method difficult to control. For a highly structured non-linear equation it is wise to take advantage of its structure in the construction of the numerical scheme. To achieve this we first discretize the state equation (4) and the associated cost function (3) by any one step or multistep approximation scheme. We obtain a discrete time control problem for which a discrete time backward matrix Riccati equation can be constructed by *mesh dependent Invariant Embedding* of the continuous and discrete time systems. This last equation can then be solved using only one or two matrix inversions at each step even though the Riccati equation is non-linear. We prove that this important computational advantage is obtained without altering the nodal asymptotic convergence rate of the original scheme. This approach is both very natural and widespread in the Engineering Control community.

The contribution of this paper is twofold. First we give the precise construction which associates with a given one-step (and even multistep) discretization of the state equation (4), the corresponding discrete time matrix Riccati equation and the nodal approximation to the solution of (1). Then we show that the pointwise or nodal rate of convergence to the solution of the matrix Riccati differential equation is equal to the pointwise or nodal asymptotic rate of convergence of the chosen discretization scheme for the state equation. In this paper we shall only cover the family of mesh-dependent (or discontinuous piecewise polynomial) approximation schemes studied by DELFOUR–DUBEAU [1]. They include Euler explicit, HAMMER and HOLLINGWORTH [1] and other high order $(2k+2)$ one-step methods all in the same framework. The whole theory also extends to multistep methods such as Adams–Moulton or Adams–Bashford. For more details and specific constructions the reader is referred to DELFOUR–DUBEAU [1].

Notation 1.1. $\mathbf{R}$ will denote the field of real numbers and E and U the respective Euclidean finite dimensional spaces $\mathbf{R}^d$ and $\mathbf{R}^m$ of respective dimensions $d \geq 1$ and $m \geq 1$. The product space and norms in E or U will be

$$x \bullet y = \sum_{i=1}^{d} x_i y_i, \quad |x| = (x \bullet y)^{1/2} \quad x = (x_1, \ldots, x_d), \quad y = (y_1, \ldots, y_d) \in E$$

$$u \bullet v = \sum_{i=1}^{m} u_i v_i, \quad |u| = (u \bullet u)^{1/2} \quad u = (u_1, \ldots, u_m), \quad v = (v_1, \ldots, v_m) \in U.$$

Given an interval $[a,b]$, the following spaces of functions $f : [a,b] \to E$ will be used: $L^p(a,b;E)$, the space of p-integrable $(1 \leq p < \infty)$ or essentially bounded function $(p = \infty)$, $H^p(a,b;E)$, the Sobolev space of functions with distributional derivatives up to order p in $L^2(a,b;E)$, $C(a,b;E)$, the space of continuous functions on $[a,b]$, $W^{1,1}(a,b;E)$, the space of functions in $L^1(a,b;E)$ with a distributional first derivative in $L^1(a,b;E)$, and $P^k(a,b;E)$, the space of all polynomials of degree less or equal to

k, $k \geq 0$. Similar definitions will hold with U or $\mathbf{R}$ in place of E. Given $T > 0$ and a partition

$$0 = t_0 < t_1 < \cdots < t_N = T$$

of the interval $[0,T]$, denote by I_n the subinterval $[t_{n-1}, t_n]$, $1 \leq n \leq N$. For $f \in L^\infty(I_n; E)$, denote by $||f||_{\infty,n}$ the essential supremum of $|f(t)|$ over t in I_n. The inner product in $L^2(I_n, E)$ will be denoted by $(\cdot,\cdot)_n$ and in $L^2(0,T;E)$ by $(\cdot,\cdot)$. For f in $H^p(I_n; E)$, define

$$||f||_{p,n}^2 = \sum_{i=1}^{p} (f^{(i)}, f^{(i)})_n$$

where $f^{(i)}$ denotes the i–th derivative of f. We also define the norms

$$||\cdot||_p^2 = \sum_{n=1}^{N} ||\cdot||_{p,n}^2, \text{ and } ||\cdot||_\infty = \max\ \{||\cdot||_{\infty,n} : 1 \leq n \leq N\}.$$

When $E = \mathbf{R}$, we shall drop E in the above function space notation.

The topological dual of a real Banach space B will always be written B' and the transpose of a continuous linear transformation $L : B_1 \to B_2$ between two Banach spaces will be written $L^* : B_2' \to B_1'$. The transpose of a matrix M will be written $M^\top$. A symmetrical positive definite (resp. semi definite) matrix M will be denoted $M > 0$ (resp. $M \geq 0$).

In each section the equations are numbered starting from (1). Cross references will be written (3.7) meaning equation (7) in section 3.

2. CONTINUOUS AND DISCRETE PROBLEMS

Let $E = \mathbf{R}^d$ and $U = \mathbf{R}^m$ be the Euclidean spaces of respective dimension $d \geq 1$ and $m \geq 1$ and let $[0,T]$, $T > 0$, be a fixed time interval. Consider the following system of linear differential equations

$$(1) \qquad \begin{cases} \dfrac{dx}{dt}(t) = A(t)x(t) + B(t)u(t) + f(t), & \text{a.e. in } [0,T] \\ x(0) = x^0 \end{cases}$$

where $f \in L^2(0,T;E)$, $x^0 \in E$, $u \in L^2(0,T;U)$ and $A(t)$ and $B(t)$ are $d \times d$ and $d \times m$ matrices of bounded measurable elements on $[0,T]$. Associate with the solution x in $H^1(0,T;E)$ the cost function

$$(2) \qquad J(u, x^0) = [Fx(T) + 2\ell] \bullet x(T) + (Qx + 2q, x) + (Nu, u).$$

The vector $\ell \in E$, the function $q \in L^2(0,T;E)$ and the matrices F, $Q(t)$ and $N(t)$ are fixed. Moreover it is assumed that

$$(3) \qquad F^* = F \geq 0, \quad Q(t)^* = Q(t) \geq 0 \text{ and } N(t)^* = N(t) \quad \text{a.e. in } [0,T]$$

and there exists a constant $\lambda > 0$ such that

$$N(t)u \bullet u \geq \lambda |u|^2. \tag{4}$$

2.1 CONTINUOUS PROBLEM

Given $x^0 \in E$, we want to find the control u^* in $L^2(0,T;U)$ which minimizes the cost function (2)

$$\text{Inf } \{J(u,x^0) : u \in L^2(0,T:U)\}. \tag{5}$$

The minimizing control u^* is completely characterized by the Optimality System

$$\begin{cases} \dfrac{dx}{dt}(t) = A(t)x(t) + B(t)u^*(t) + f(t), & \text{a.e. in } [0,T] \\ x(0) = x^0, \end{cases} \tag{6}$$

$$\begin{cases} \dfrac{dp}{dt}(t) + A^*(t)p(t) + Q(t)x(t) + q(t) = 0, & \text{a.e. in } [0,T] \\ p(T) = Fx(T) + \ell, \end{cases} \tag{7}$$

$$u^*(t) = -N(t)^{-1}B^*(t)p(t) \quad \text{a.e. in } [0,T]. \tag{8}$$

The above system is usually decoupled by introducing a family of symmetrical positive semi definite $d \times d$ matrices $\{\Pi(t) : 0 \leq t \leq T\}$ and d–vectors $\{\rho(t) : 0 \leq t \leq T\}$

$$p(t) = \Pi(t)x(t) + \rho(t) \quad \text{in } [0,T] \tag{9}$$

and the optimal control can be expressed in feedback form

$$u^*(t) = -N(t)^{-1}B^*(t)[\Pi(t)x(t) + \rho(t)] \quad \text{in } [0,T], \tag{10}$$

where

$$K(t) = -N(t)^{-1}B^*(t)\Pi(t) \tag{11}$$

is the matrix of feedback gains.

THEOREM 1.

(i) *The matrix function $\Pi(t)$ is the unique symmetrical positive semi definite solution to the matrix differential Riccati equation*

$$\begin{cases} \dfrac{d\Pi}{dt}(t) + A^*(t)\Pi(t) + \Pi(t)A(t) - \Pi(t)R(t)\Pi(t) + Q(t) & \text{a.e. in } [0,T] \\ \Pi(T) = F, \quad R(t) = B(t)N(t)^{-1}B^*(t). \end{cases} \tag{12}$$

(ii) *The vector $\rho(t)$ is the unique solution in $H^1(0,T;E)$ to the linear equation*

$$\begin{cases} \dfrac{d\rho}{dt}(t) + [A(t) - R(t)\Pi(t)]^*\rho(t) + q(t) + \Pi(t)f(t) = 0, & \text{a.e. in } [0,T] \\ \rho(T) = \ell. \quad \square \end{cases} \tag{13}$$

2.2 DISCRETE PROBLEM

First introduce $N+1$ nodes for a fixed integer $N > 1$

$$0 = t_0 < t_1 < \cdots < t_N = T,$$

which partition the time interval $[0,T]$ into N intervals $I_n = [t_{n-1}, t_n]$, $1 \leq n \leq N$. Define the parameters

$$h_n = t_n - t_{n-1}, \quad 1 \leq n \leq N, \quad \text{and} \quad h = \max_{1 \leq n \leq N} h_n. \tag{15}$$

In this paper we assume that the families of partitions indexed by h are uniform, that is,

$$\exists c > 0, \quad \forall h, \quad ch \leq \min_{1 \leq n \leq N} h_n. \tag{16}$$

This is a standard hypothesis in the asymptotic analysis of the error as h goes to zero.

Numerous approximation schemes are available for the linear system (1). It has been established that most classical schemes can be obtained from a mesh-dependent weak formulation of (1) (cf. DELFOUR–HAGER–TROCHU [1], DELFOUR–DUBEAU [1] and their bibliographies) by using polynomials x_n of degree $k \geq 0$ on each interval I_n, traces X_n at each node t_n, and appropriate quadrature formulae. They include both one-step and multistep methods, and explicit and implicit schemes. This weak formulation is not just a mathematical toy. It turns out to be extremely important in Optimal Control problems where it automatically provides the natural constructions not only for the discretization of the state equation, but also for the associated adjoint state equation and ultimately for the matrix differential Riccati equation. It tells in what sense the computed vectors or matrices will converge. This type of information is critical in the establishment of optimal asymptotic rates of convergence.

For one-step methods the weak formulation of the state equation (1) yields a discrete time system of the form

$$\begin{cases} \underline{\Gamma}_n \underline{x}_n = \underline{C}_n \underline{x}_{n-1} + B_n u + \underline{f}_n, & 1 \leq n \leq N \\ \underline{x}_0 = \underline{x}^0, \quad \underline{x}^0 = (0, \ldots, 0, x^0), & \underline{x}_{N+1} = \underline{C}_{N+1} \underline{x}_N \end{cases} \tag{17}$$

where $\underline{x}_n \in E^{k+2}$ is the state vector at time n, $\underline{x}^0 \in E^{k+2}$ is the initial condition vector, $\underline{\Gamma}_n$ and $\underline{C}_n$ are $(k+2)d \times (k+2)d$ matrices, and $\underline{f}_n \in E^{k+2}$ corresponds to the discretization of the function f. It is important to notice that in this paper we do not discretize the control variable in order to isolate the effect of the discretization of the state equation. The control will naturally disappear in the final discretization scheme and provide a direct approximation to the solution of the matrix Riccati differential equation. So in (17) $u \in L^2(0,T;U)$ and $B_n : L^2(I_n, U) \to E^{k+2}$ are the operators resulting from the discretization of the state equation.

The cost function resulting from the approximation scheme will be of the form

$$J_h(u, \underline{x}^0) = [\underline{F}\, \underline{x}_{N+1} + 2\underline{\ell}] \bullet \underline{x}_{N+1} \sum_{n=1}^{N} [\underline{Q}_n \underline{x}_n + 2\underline{q}_n] \bullet \underline{x}_n + (Nu, u), \tag{18}$$

where $\underline{F}$ and $\underline{Q}_n$ are $(k+2)d \times (k+2)d$–matrices and $\underline{\ell}$ and $\underline{q}_n$ are $(k+2)d$–vectors.

2.2.1 *Optimality system*

The minimization problem

$$\min_{u \in L^2(0,T;U)} J_h(u, \underline{x}^0) \tag{19}$$

has a unique solution u^* which is completely characterized by the optimality system:

$$\begin{cases} \underline{\Gamma}_n \underline{x}_n = \underline{C}_n \underline{x}_{n-1} + B_n \hat{u}_n + \underline{f}_n, & 1 \le n \le N, \\ \underline{x}_0 = \underline{x}^0, \quad \underline{x}_{N+1} = \underline{C}_{N+1} \underline{x}_N \end{cases} \tag{20}$$

$$\begin{cases} \underline{\Gamma}_n^\top \underline{p}_n = \underline{C}_{n+1}^\top \underline{p}_{n+1} + \underline{Q}_n \underline{x}_n + \underline{q}_n, & 1 \le n < N, \\ \underline{p}_{N+1} = \underline{F}\,\underline{x}_{N+1} + \underline{\ell}, \quad \underline{p}_0 = C_1^\top \underline{p}_1, \end{cases} \tag{21}$$

$$\hat{u}_n = \hat{u}|_{I_n} = -N^{-1} B_n^* p_n, \quad \text{in } I_n, \quad 1 \le n \le N, \tag{22}$$

where $\hat{u}|_{I_n}$ is the restriction of $\hat{u}$ to I_n. The substitution of $\hat{u}_n$ in (20) yields

$$B_n \hat{u}_n = -B_n N^{-1} B_n^* \underline{p}_n = -\underline{R}_n \underline{p}_n, \tag{23}$$

where $\underline{R}_n$ is the matrix

$$\underline{R}_n = \int_{I_n} B_n(t) N^{-1}(t) B_n^*(t)\, dt. \tag{24}$$

2.2.2 *Decoupling of the optimality system and discrete Riccati equation*

This optimality system can be decoupled and a discrete time Riccati equation can be obtained. By invariant embedding there exist a sequence of $(k+2)d \times (k+2)d$–matrices $\{\underline{P}_n : 1 \le n \le N+1\}$ and a sequence of $(k+2)d$–vectors $\{\underline{r}_n : 1 \le n \le N+1\}$ such that

$$\underline{p}_n = \underline{P}_n \underline{x}_{n-1} + \underline{r}_n, \quad 1 \le n \le N+1. \tag{25}$$

Moreover

$$(\underline{C}_n^\top \underline{P}_n)^\top = \underline{C}_n^\top P_n \ge 0, \quad 1 \le n \le N+1. \tag{26}$$

The sequence of matrices $\underline{P}_n$ is the unique solution of the system of matrix equations

$$\begin{cases} \underline{P}_n = \left[\underline{\Gamma}_n^\top + (\underline{C}_{n+1}^\top \underline{P}_{n+1} + \underline{Q}_n)\underline{\Gamma}_n^{-1} \underline{R}_n\right]^{-1} (\underline{C}_{n+1}^\top \underline{P}_{n+1} + \underline{Q}_n)\underline{\Gamma}_n^{-1} \underline{C}_n, \\ \qquad\qquad\qquad\qquad\qquad\qquad\qquad\qquad 1 \le n \le N, \\ \underline{P}_{N+1} = \underline{F} \end{cases} \tag{27}$$

and the sequence of vectors $\underline{r}_n$ is the unique solution of the equation

$$(28)\qquad \begin{cases} \left[\underline{\Gamma}_n^{\mathsf{T}} + (\underline{C}_{n+1}^{\mathsf{T}}\underline{P}_{n+1} + \underline{Q}_n)\underline{\Gamma}_n^{-1}\underline{R}_n\right]\underline{r}_n \\ = \left[\underline{C}_{n+1}^{\mathsf{T}}\underline{r}_{n+1} + \underline{q}_{n+1} + (\underline{C}_{n+1}^{\mathsf{T}}\underline{P}_{n+1} + \underline{Q}_n)\underline{\Gamma}_n^{-1}\underline{f}_n\right], \quad 1 \le n \le N, \\ \underline{r}_{N+1} = \underline{\ell}. \end{cases}$$

REMARK 2.1: The matrix

$$(29)\qquad \underline{\tilde{\Pi}}_n = \left[\underline{\Gamma}_n^{\mathsf{T}} + (\underline{C}_{n+1}^{\mathsf{T}}\underline{P}_{n+1} + \underline{Q}_n)\underline{\Gamma}_n^{-1}\underline{R}_n\right]^{-1}(\underline{C}_{n+1}^{\mathsf{T}}\underline{P}_{n+1} + \underline{Q}_n)\underline{\Gamma}_n^{-1}$$

is symmetrical

$$(30)\qquad \underline{\tilde{\Pi}}_n^{\mathsf{T}} = \underline{\tilde{\Pi}}_n$$

and by definition

$$(31)\qquad \underline{P}_n = \underline{\tilde{\Pi}}_n\underline{C}_n.$$

Thus equation (25) can also be written

$$(32)\qquad \begin{cases} \underline{P}_n = (\underline{\Gamma}_n^{-1})^{\mathsf{T}}(\underline{C}_{n+1}^{\mathsf{T}}\underline{P}_{n+1} + \underline{Q}_n)\left[\underline{\Gamma}_n + \underline{R}_n(\underline{\Gamma}_n^{-1})^{\mathsf{T}}(\underline{C}_{n+1}^{\mathsf{T}}\underline{P}_{n+1} + \underline{Q}_n)\right]^{-1}\underline{C}_n, \\ \qquad\qquad 1 \le n \le N, \\ \underline{P}_{N+1} = \underline{F}. \quad \square \end{cases}$$

3. WEAK FORMULATION, CONVERGENCE AND ERROR ESTIMATES

The starting point is a mesh-dependent weak formulation of the state equation (2.1) for the uniform mesh or partition (2.14) to (2.16) introduced in section 2. The natural connection between the continuous and discrete time problems is then provided by a *mesh dependent Invariant Embedding* with respect to an initial time interval, an initial point and an initial function on that interval. This differs from the standard Invariant Embedding with respect to an initial point at time t for all t in $[0, T]$. From this most one-step and multistep methods can be recovered and new ones can be designed.

3.1 MESH-DEPENDENT WEAK FORMULATION.

Define the spaces

$$(1)\qquad V = E \times \prod_{n=1}^{N} H^1(I_n; E) \times E, \ \mathcal{X} = E^{N+1} \times L^2(0, T; E) \times E$$

and the continuous bilinear form

$$(2) \quad \mathcal{B} : \mathcal{X} \times V \to \mathbf{R}$$

$$(3) \quad \begin{cases} \mathcal{B}(\tilde{x}, v) = (x, \sum_{n=1}^{N} [\dot{v}_n + A^*(t) v_n] \chi_n) + \sum_{n=1}^{N-1} X_n \bullet [v_{n+1}(t_n) - v_n(t_n)] \\ \qquad + X_0 \bullet [v_1(t_0) - v_0] + X_N \bullet [v_{N+1} - v_N(t_N)] - x_{N+1} \bullet v_{N+1} \end{cases}$$

for all

$$(4) \quad \tilde{x} = (X_0, X_1, \ldots, X_N, x, x_{N+1}), \quad \text{and} \quad v = (v_0, v_1, \ldots, v_N, v_{N+1}) \in V,$$

where χ_n is the characteristic function of the interval I_n. The variable v_0 (resp. v_{N+1}) is to be interpreted as $v(t_0^-)$ (resp. $v(t_N^+)$) the left (resp. right) hand side value of the piecewise continuous function $v = \sum_{n=0}^{N+1} v_n \chi_n$ at $t = t_0$ (resp. $t = t_N$) with the convention that χ_0 (resp. χ_{N+1}) is the characteristic function of $]-\infty, 0]$ (resp. $[t_N, +\infty[$). Similarly x_{N+1} is to be interpreted as $x(t_N^+)$.

The weak problem

$$(5) \quad \begin{cases} \text{to find} \quad \tilde{x} \in \mathcal{X} \quad \text{such that for all} \quad v \in V \\ \mathcal{B}(\tilde{x}, v) + (Bu + f, v) + x^0 \bullet v_0 = 0 \end{cases}$$

has a unique solution and

$$(6) \quad \tilde{x} = (x(t_0), x(t_1), \ldots, x(t_N), x, x(t_N))$$

where x is the unique solution in $H^1(0, T; E)$ of the state equation (2.1).

Notation 3.1. In (5) we have used the notation v to designate the L^2–function

$$(7) \quad \sum_{n=1}^{N} v_n \chi_n$$

on $[0, T]$. This "convenient notation" is not to be confused with the same notation for the element $v = (v_0, v_1, \ldots, v_N, v_{N+1})$ of V. □

For the weak formulation (5), the associated cost function becomes

$$(8) \quad J(u, x^0) = [F x_{N+1} + 2\ell] \bullet x_{N+1} + (Qx + 2q, x) + (Nu, u)$$

and the minimization problem is

$$(9) \quad \inf_{u \in L^2(0,T;U)} J(u, x^0).$$

3.2 CHOICE OF THE APPROXIMATION SCHEME.

Though the method is quite general, we shall specialize to a family of approximation schemes which are of order h^{2k+2}, $k \geq 0$ (cf. Appendic A). Such schemes can be obtained from the mesh dependent weak formulation used in DELFOUR–DUBEAU [1, case $L = 0$].

Introduce the following finite dimensional subspaces

$$\mathcal{X}^h = \{\tilde{x}^h = (X_0, X_1, \ldots, X_N, x, x_{N+1}) \in \mathcal{X} : x|_{I_n} \in P^k(I_n, E), 1 \leq n \leq N\} \tag{10}$$

$$V^h = \{v^h = (v_0, v_1, \ldots, v_N, v_{N+1}) \in V : v_n \in P^{k+1}(I_n, E), 1 \leq n \leq N\} \tag{11}$$

of $\mathcal{X}$ and V. Notice that

$$\dim \mathcal{X}^h = \dim V^h = [2 + (k+2)N] \dim E. \tag{12}$$

DELFOUR–DUBEAU [1] have established that, for h small enough, there exists a unique solution $\tilde{x}^h$ in $\mathcal{X}^h$ to the equation

$$\mathcal{B}(\tilde{x}^h, v^h) + (Bu + f, v^h) + x^0 \bullet v_0^h = 0, \quad \forall v^h \in V^h. \tag{13}$$

To $\tilde{x}_h^h$ associate the cost function

$$J^h(u, x^0) = \left[F x_{N+1}^h + 2\ell\right] \bullet x_{N+1}^h + (Qx^h + 2q, x^h) + (Nu, u) \tag{14}$$

and the minimization problem

$$\inf_{u \in L^2(0,T;U)} \quad J^h(u, x^0). \tag{15}$$

For the solution $\tilde{x}^h$ of (13) we have the following convergence and asymptotic convergence results.

THEOREM 3.1. *(DELFOUR–DUBEAU [1, Thm. 4.1]).*

Assume that the solution x of (2.1) belongs to $H^{k+1}(0, T; E)$ for some $k \geq 0$, and that the partition of $[0, T]$ is uniform.

Then, as h goes to zero, there exists a constant $c > 0$ (independent of h) such that

$$\|x^h - x\|_{L^2(0,T;E)} \leq ch^{k+1} \|x^{(k+1)}\|_{L^2(0,T;E)} \tag{16}$$

and

$$\left\{ \begin{array}{l} \max \left\{ \max\limits_{n=0,\ldots,N} |X_n^h - x(t_n)|, \ |x_{N+1}^h - x(t_N)| \right\} \\ \leq ch^{2k+2} \|x^{(k+1)}\|_{L^2} \left[1 + c\|x^{(k+1)}\|_{L^2}\right]. \quad \square \end{array} \right. \tag{17}$$

3.3 CONSTRUCTION OF THE DISCRETE TIME CONTROL PROBLEM.

In this section we construct matrices and vectors to express the weak formulation (13) to (15) of the control problem in the form (2.17) to (2.19) of section 2.2.1. This will require the constructions of bijections

$$\text{(18)}\qquad \left\{\begin{array}{ll} i_{\mathcal{X},n} & : P^k(I_n,E)\times E \to E^{k+2} \\ i_{V,n} & : P^{k+1}(I_n,E) \to E^{k+2} \end{array}\right\},\quad 1\le n\le N,$$

between spaces of vectors and spaces of polynomials. The formulation (2.17) to (2.19) will be used for computations, and the weak formulation (13) to (15) over polynomial spaces will be used to prove convergence and asymptotic error estimates. They will also tell in what way the computed vectors and matrices converge to their counterparts in the original continuous problem.

For simplicity, we drop the index h in the remainder of this section. The first step is to decompose the weak equation (13) into $N+2$ weak equations. For $n, 1\le n\le N$, we obtain on the interval I_n the equation

$$\text{(19)}\qquad \left\{\begin{array}{l} \text{to find}\quad (x_n, X_n)\in P^k(I_n,E)\times E,\quad \text{such that}\quad \forall v_n\in P^{k+1}(I_n,E) \\ X_n\bullet v_n(t_{n-1}) - (x_n, \dot{v}_n + A^* v_n) = X_{n-1}\bullet v_n(t_{n-1}) + (Bu+f, v_n). \end{array}\right.$$

and for $n=0$ and $N+1$

$$\text{(20)}\qquad \left\{\begin{array}{l} \text{to find}\quad X_0\in E,\quad \text{such that} \\ \forall v_0\in E,\quad -X_0\bullet v_0 + x^0\bullet v_0 = 0 \end{array}\right\},\quad \Rightarrow X_0 = x^0$$

$$\text{(21)}\qquad \left\{\begin{array}{l} \text{to find}\quad x_{N+1}\in E,\quad \text{such that} \\ \forall v_{N+1}\in E,\quad X_N\bullet v_{N+1} - x_{N+1}\bullet v_{N+1} = 0 \end{array}\right\},\quad \Rightarrow x_{N+1} = X_N.$$

The last two equations yield $X_0 = x^0$ and $x_{N+1} = X_N$.

The second step is to construct the appropriate bases for $P^k(I_n;E)$ and $P^{k+1}(I_n;E)$, $I\le n\le N$, to express (18) in the standard form of a one-step method. It is customary to work on a reference interval $[0,1]$ and map everything on I_n. Choose a numerical $(k+1)$–point quadrature formula on $[0,1]$

$$\text{(22)}\qquad \int_0^1 g(\tau)d\tau = \sum_{\ell=1}^{k+1} a_\ell g(\tau_\ell)$$

where $0\le\tau_1\le\cdots\le\tau_{k+1}\le 1$ are the quadrature points and $\{a_\ell : 1\le\ell\le k+1\}$ are the weights. Assume that the formula is exact for all polynomials of degree less or equal to $2k+1$ (cf. V. I. KRYLOV [1]). Denote by $\{\varphi_\ell : 1\le\ell\le k+1\}$ the Lagrange interpolating polynomials associated with the points $\{\tau_\ell : 1\le\ell\le k+1\}$

$$\text{(23)}\qquad \varphi_\ell(\tau) = \prod_{\substack{i=1\\ i\ne\ell}}^{k+1} \frac{\tau-\tau_i}{\tau_\ell-\tau_i},\quad 1\le\ell\le k+1,\quad \tau\in[0,1].$$

This family of polynomials is a basis of $P^k(0,1)$. For $P^{k+1}(0,1)$ we construct the following basis

$$\left\{\begin{array}{l}\psi_j(\tau)=\frac{1}{a_j}\int_\tau^1\varphi_j(\xi)d\xi,\quad 1\le j\le k+1\\ \psi_{k+2}(\tau)=1,\quad \tau\in[0,1].\end{array}\right.\tag{24}$$

From the φ_ℓ's and the ψ_j's we can now construct bases for $P^k(I_n)$ and $P^{k+1}(I_n)$

$$\left.\begin{array}{ll}\varphi_{n\ell}(t)=\varphi_\ell\left(\dfrac{t-t_{n-1}}{h_n}\right), & 1\le\ell\le k+1,\\ \psi_{n\ell}(t)=\psi_\ell\left(\dfrac{t-t_{n-1}}{h_n}\right), & 1\le\ell\le k+2,\end{array}\right\}\quad t\in I_n,\tag{25}$$

and define the parameters

$$t_{n\ell}=t_{n-1}+h_n\tau_\ell,\quad 1\le\ell\le k+2,\quad 1\le n\le N.\tag{26}$$

The polynomials $\{\varphi_{n\ell}:1\le\ell\le k+1\}$ and $\{\psi_{n\ell}:1\le\ell\le k+2\}$ form a basis of $P^k(I_n)$ and $P^{k+1}(I_n)$, respectively. For each n, $1\le n\le N$, we introduce the following isomorphisms

$$(x,X)\mapsto i_{\mathcal{X},n}(x,X)=\begin{bmatrix}x(t_{n1})\\ x(t_{n2})\\ \vdots\\ x(t_{n,k+1})\\ X\end{bmatrix}:P^k(I_n,E)\times E\to E^{k+2}\tag{27}$$

$$v\mapsto i_{V,n}(v)=\begin{bmatrix}-h_na_1\dot v(t_{n1})\\ -h_na_2\dot v(t_{n2})\\ \vdots\\ -h_na_{k+1}\dot v(t_{n,k+1})\\ v(t_n)\end{bmatrix}:P^{k+1}(I_n,E)\to E^{k+2}.\tag{28}$$

It is easy to check that

$$i_{\mathcal{X},n}^{-1}(\underline{x})=i_{\mathcal{X},n}^{-1}(x_1,\ \dots\ ,x_{k+2})=\left(\sum_{i=1}^{k+1}x_i\varphi_{ni},x_{k+2}\right)\tag{29}$$

$$i_{V,n}^{-1}(\underline{v})=i_{V,n}^{-1}(v_1,\ \dots\ ,v_{k+2})=\sum_{i=1}^{k+2}v_i\psi_{ni},\tag{30}$$

and notice that

$$(31) \qquad \left[i_{V,n}^{-1}(\underline{v})\right](t_{n-1}) = \sum_{i=1}^{k+2} v_i \psi_{ni}(t_{n-1}) = \sum_{i=1}^{k+2} v_i.$$

With the above notation and definitions, introduce the matrices $\underline{\Gamma}_n$ and $\underline{C}_n, 1 \leq n \leq N$,

$$(32) \qquad \begin{cases} \underline{\Gamma}_n \underline{x} \bullet \underline{v} = X \bullet v(t_n) - (x, \dot{v} + A^* v)_n \\ \underline{C}_n \underline{x} \bullet \underline{v} = X \bullet v(t_{n-1}) \quad \left(= X \bullet \sum_{\ell=1}^{k+2} v_\ell \right) \\ \forall\, \underline{x} = i_{\mathcal{X},n}(x, X),\ \forall\, \underline{v} = i_{V,n}(v). \end{cases}$$

With our choice of bases the matrix $\underline{\Gamma}_n$ has the standard structure which characterizes a one-step method

$$(33) \qquad \underline{\Gamma}_n \underline{x} = \begin{bmatrix} x_1 \\ \vdots \\ x_{k+1} \\ x_{k+2} \end{bmatrix} - h_n \begin{bmatrix} & 0 \\ \underline{A}_n & \vdots \\ & 0 \\ \underline{b}_n^\top & 0 \end{bmatrix} \begin{bmatrix} x_1 \\ \vdots \\ x_{k+1} \\ x_{k+2} \end{bmatrix}, \quad \text{where } \underline{x} = \begin{bmatrix} x_1 \\ \vdots \\ x_{k+1} \\ x_{k+2} \end{bmatrix}$$

where the elements of the matrix $\underline{A}_n$ and the vector $\underline{b}_n$ are given by the expressions

$$(34) \qquad h_n(\underline{A}_n)_{m\ell} = \int_{I_n} A(t) \varphi_{n\ell}(t)\, \psi_{nm}(t)\, dt, \quad 1 \leq \ell, m \leq k+1$$

$$(35) \qquad h_n(\underline{b}_n)_\ell = \int_{I_n} A(t) \varphi_{n\ell}(t)\, \psi_{n,k+2}(t)\, dt, \quad 1 \leq \ell \leq k+1.$$

The elements of the matrix $\underline{A}_n$ and the vector $\underline{b}_n$ are tabulated in DELFOUR–DUBEAU [1] for the first k's and various quadrature formulae (cf. Appendix A for $k = 0, 1, 2$ and the Legendre quadrature formula).

The matrices $\underline{C}_n, 1 \leq n \leq N$, are simpler and constant

$$(36) \qquad \underline{C}_n = \underline{C} = \begin{bmatrix} 0 & \dots & \dots & 0 & I \\ \vdots & & & \vdots & \vdots \\ \vdots & & & \vdots & \vdots \\ 0 & \dots & \dots & 0 & I \end{bmatrix} \qquad \begin{matrix} 0 = d \times d & \text{matrix with zero entries} \\ I = d \times d & \text{identity matrix.} \end{matrix}$$

We shall also need the following $(k+2)d \times (k+2)d$ matrix

$$(37) \qquad \underline{C}_{N+1}\, \underline{x} \bullet \underline{v} = x_{k+2} \bullet v_{k+2} \quad \Rightarrow \quad \underline{C}_{N+1} = \begin{bmatrix} 0 & \dots & \dots & 0 \\ \vdots & \ddots & \ddots & \vdots \\ \vdots & \ddots & 0 & 0 \\ 0 & \dots & 0 & I \end{bmatrix}.$$

Finally associate with B and f the operator B_n and the vector $\underline{f}_n$

(38) $$B_n : L^2(I_n, E) \to E^{k+2}, \quad (B_n u)_\ell = \int_{I_n} B(t)u(t)\psi_{n\ell}(t)dt, \quad 1 \le \ell \le k+2$$

(39) $$\left(\underline{f}_n\right)_\ell = \int_{I_n} f(t)\psi_{n\ell}(t)dt, \quad 1 \le \ell \le k+2$$

or in product form

(40) $$B_n u \bullet \underline{v} = (Bu, v)_n, \quad \forall u \in L^2(I_n, E),\ \forall \underline{v} = i_{V,n}(v), \quad \Rightarrow \quad B_n^* = B^* i_{V,n}^{-1}.$$

With the above definitions the state equation can be written in the form

(41) $$\left\{ \begin{array}{ll} \underline{\Gamma}_n\, \underline{x}_n = \underline{C}_n\, \underline{x}_{n-1} + B_n u + \underline{f}_n, & 1 \le n \le N \\ \underline{x}_0 = \underline{x}^0, & \underline{x}_{N+1} = \underline{C}_{N+1}\, \underline{x}_N, \end{array} \right\}, \text{ where } \underline{x}^0 = \begin{bmatrix} 0 \\ \vdots \\ 0 \\ x^0 \end{bmatrix}.$$

Finally the associated cost functions becomes

(42) $$J_h(u, x^0) = [\underline{F}\, \underline{x}_{N+1} + 2\underline{\ell}] \bullet \underline{x}_{N+1} + \sum_{n=1}^{N} [\underline{Q}\, \underline{x}_n + 2\underline{q}_n] \bullet \underline{x}_n + (Nu, u)$$

where $\underline{F}$ and $\underline{Q}_n$ are the $(k+2)d \times (k+2)d$–matrices and $\underline{\ell}$ and $\underline{q}_n$ are the $(k+2)d$–vectors defined below

(43) $$\underline{F} = \begin{bmatrix} 0 & \dots & \dots & 0 \\ \vdots & \ddots & \ddots & \vdots \\ 0 & \dots & 0 & 0 \\ 0 & \dots & 0 & F \end{bmatrix}, \quad \underline{\ell} = \begin{bmatrix} 0 \\ \vdots \\ 0 \\ \ell \end{bmatrix}$$

and for $n, 1 \le n \le N$,

(44) $$\left\{ \begin{array}{l} \underline{Q}_n\, \underline{x} \bullet \underline{z} = (Q|_{I_n} x, z)_n \\ \underline{q}_n \bullet z \quad = \quad (q|_{I_n}, z)_n \\ \forall \underline{x} = i_{\mathcal{X},n}(x, X), \quad \forall \underline{z} = i_{\mathcal{X},n}(z, Z). \end{array} \right.$$

3.4 CONVERGENCE AND ERROR ESTIMATES FOR THE APPROXIMATION OF Π and ρ.

The details of the constructions to approximate Π and ρ from the solutions $\underline{P}_n$ and $\underline{r}_n$ to the discrete time equations (2.25)-(2.26) using mesh dependent Invariant Embedding and the convergence and asymptotic error estimates can be found in DELFOUR–OUANSAFI [1].

On each interval I_n we construct the matrix function

$$\Pi_n^h(t) = \sum_{i=1}^{k+2} (\underline{P}_n)_{i,k+2}\, \psi_{ni}(t), \quad t \in I_n, \quad 1 \le n \le N, \tag{45}$$

where$(\underline{P}_n)$ is considered as a $(k+2)\times(k+2)$ matrix of $d \times d$ matrices and $(\underline{P}_n)_{i,k+2}$ denotes the $d \times d$ matrix in its i-th row and $(k+2)$-th column. At each node t_n, $0 \le n \le N$, we will use the matrices $\Pi_{n+1}^h(t_n)$, $0 \le n < N$, and at t_N the matrix

$$\Pi_{N+1}^h(t_N) \overset{\text{def}}{=} (\underline{P}_{N+1})_{k+2,k+2} \qquad (= F). \tag{46}$$

Here Π_{N+1}^h is not a function: the notation $\Pi_{N+1}^h(t_N)$ is chosen for uniformity of notation.

Similarily on each interval I_n we construct the vector function

$$\rho_n^h(t) = \sum_{i=1}^{k+2} (\underline{r}_n)_i \psi_{ni}(t), \quad t \in I_n, \quad 1 \le n \le N, \tag{47}$$

where $\underline{r}_n$ is considered as a $(k+2)$-vector of d vectors and $(\underline{r}_n)_i$ denotes the d-vector in its i-th component. At each node t_n, $0 \le n \le N$, we will use the vectors $\rho_{n+1}^h(t_n)$, $0 \le n < N$, and at t_N the vector

$$\rho_{N+1}^h(t_N) \overset{\text{def}}{=} (\underline{r}_{N+1})_{k+2,k+2} \qquad (= \ell). \tag{48}$$

The last step is to establish the convergence and estimate the asymptotic rate of convergence.

THEOREM 3.2. *Assume that the entries of the matrices A, B, Q and N belong to $H^{k+1}(0,T)$ and that f and q belong to $H^{k+1}(0,T;E)$ for some integer $k \ge 1$.*

As h goes to zero, there exists a constant $c > 0$ (independent of h and n) such that

$$\max_{0 \le n \le N} |\Pi_{n+1}^h(t_n) - \Pi(t_n)| \le ch^{2k+2}, \quad \max_{0 \le n \le N} |\rho_{n+1}^h(t_n) - \rho(t_n)| \le ch^{2k+2} \tag{49}$$

$$\left[\sum_{n=1}^{N} \|\Pi_n^h(\cdot) - \Pi(\cdot)\|_n^2\right]^{1/2} \le ch, \quad \left[\sum_{n=1}^{N} \|\rho_n^h(\cdot) - \rho(\cdot)\|_n^2\right]^{1/2} \le ch. \quad \square \tag{50}$$

REMARK 3.1: The nodal convergence rate for Π and ρ is $2k+2$. It coincides will the convergence rate of the chosen one-step method. However the L^2-convergence rate of Π_n^h and ρ_n^h to Π and ρ is 1 instead of the expected $k+1$. Numerical tests indicate that this estimate is optimal. The explanation of this phenomenon will be given in sections 4 and 5. We shall see that Π_n^h and ρ_n^h L^2-converge with a rate of $k+1$ to a slighlty different limit which depends on the partition of $[0,T]$ as its size goes to zero. $\square$

APPENDIX.

The matrices $\underline{\Gamma}_n$ for k equal to $0, 1$ and 2 and a Legendre quadrature formula are quoted from DELFOUR-DUBEAU [1] with $L = 0$.

1) $\underline{k=0}$.

$$a_1 = 1,\ \tau_1 = \frac{1}{2},\ \varphi_1 = 1,\ \psi_1(\tau) = 1 - \tau,\ \psi_2(\tau) = 1$$

$$\underline{\Gamma}_n \underline{x} = \begin{bmatrix} x_1 \\ x_2 \end{bmatrix} - h_n \begin{bmatrix} \frac{1}{2}\underline{A}_{n1} & 0 \\ \underline{A}_{n1} & 0 \end{bmatrix} \begin{bmatrix} x_1 \\ x_2 \end{bmatrix}, \quad \underline{A}_{n1} = A(t_{n-1} + h_n \tau_1)$$

2) $\underline{k=1}$.

$$a_1 = a_2 = \frac{1}{2},\ \tau_1 = \frac{1}{2} - \frac{1}{2\sqrt{3}},\ \tau_2 = \frac{1}{2} + \frac{1}{2\sqrt{3}}$$

$$\underline{\Gamma}_n \underline{x} = \begin{bmatrix} x_1 \\ x_2 \\ x_3 \end{bmatrix} - h_n \begin{bmatrix} \frac{1}{4}\underline{A}_{n1} & \left(\frac{1}{4} - \frac{1}{2\sqrt{3}}\right)\underline{A}_{n2} & 0 \\ \left(\frac{1}{4} - \frac{1}{2\sqrt{3}}\right)\underline{A}_{n1} & \frac{1}{4}\underline{A}_{n2} & 0 \\ \frac{1}{2}\underline{A}_{n1} & \frac{1}{2}\underline{A}_{n2} & 0 \end{bmatrix} \begin{bmatrix} x_1 \\ x_2 \\ x_3 \end{bmatrix}$$

$$\underline{A}_{ni} = A(t_{n-1} + h_n \tau_i),\ i = 1, 2.$$

This is the method of HAMMER and HOLLINGWORTH [1] of order 4.

3) $\underline{k=2}$.

$$a_1 = \frac{5}{18},\ a_2 = \frac{8}{18},\ a_3 = \frac{5}{18},\ \tau_1 = \frac{1}{2} - \frac{\sqrt{3}}{2\sqrt{5}},\ \tau_2 = \frac{1}{2},\ \tau_3 = \frac{1}{2} + \frac{\sqrt{3}}{2\sqrt{5}}$$

$$\underline{\Gamma}_n \underline{x} = \begin{bmatrix} x_1 \\ x_2 \\ x_3 \\ x_4 \end{bmatrix} - \begin{bmatrix} \frac{5}{36}\underline{A}_{n1} & \left(\frac{2}{9} - \frac{1}{\sqrt{15}}\right)\underline{A}_{n2} & \left(\frac{5}{36} - \frac{1}{\sqrt{15}}\right)\underline{A}_{n3} & 0 \\ \left(\frac{5}{36} + \frac{\sqrt{5}}{8\sqrt{3}}\right)\underline{A}_{n1} & \frac{2}{9}\underline{A}_{n2} & \left(\frac{5}{36} - \frac{\sqrt{5}}{8\sqrt{3}}\right)\underline{A}_{n3} & 0 \\ \left(\frac{5}{36} + \frac{1}{2\sqrt{15}}\right)\underline{A}_{n1} & \left(\frac{2}{9} + \frac{1}{\sqrt{15}}\right)\underline{A}_{n2} & \frac{5}{36}\underline{A}_{n3} & 0 \\ \frac{5}{18}\underline{A}_{n1} & \frac{8}{18}\underline{A}_{n2} & \frac{5}{18}\underline{A}_{n3} & 0 \end{bmatrix} \begin{bmatrix} x_1 \\ x_2 \\ x_3 \\ x_4 \end{bmatrix}$$

$$\underline{A}_{ni} = A\left(t_{n-1} + h_n \tau_i\right),\ i = 1, 2, 3.$$

REFERENCES

I. BABUSKA [1], *Error bounds for finite element methods*, Numer. Math. **16** (1971), 322–333.

I. BABUSKA and A.K. AZIZ [1], *Survey lectures on the mathematical foundations of the finite element method*, in "The Mathematical Foundation of the Finite Element Method with Application to Partial Differential Equations," A. K. Aziz, ed., Academic Press, N.Y., 1973.

F. BREZZI [1], *On the existence, uniqueness, and approximation of saddle point problems arising from Lagrangian multipliers*, R.A.I.R.O. **8** (1972), 129–151.

PH.G. CIARLET [1], "The finite element method for elliptic problems," North Holland, Amsterdam, 1978.

M. C. DELFOUR [1], *The linear quadratic optimal control problem for hereditary differential systems: theory and numerical solutions*, J. Appl. Math. and Opt. **3** (1977), 101–162.

M. C. DELFOUR and F. DUBEAU [1], *Discontinuous polynominal approximations in the theory of one step, hybrid and multistep methods for non linear ordinary differential equations*, Mathematics of Computation **47** (1986), 169–189 and S1–S8.

M. C. DELFOUR, W. HAGER and F. TROCHU [1], *Discontinuous Galerkin methods for ordinary differential equations*, Mathematics of Computation **36** (1981), 455–473.

M. C. DELFOUR AND A. OUANSAFI [1], "Noniterative approximations to the solution of the matrix Riccati differential equation," CRM Report 1690, Université de Montréal, Canada, October 1990.

M. C. DELFOUR and T. TROCHU [1], *Discontinuous finite elements methods for the approximation of optimal control problems governed by hereditary differential system*, in "Distributed Parameter Systems: Modelling and identification," A. Ruberti, ed., Springer-Verlag, New York, 1978, pp. 256–271.

F. DUBEAU [1], "Approximation polynomiale par morceaux des équations différentielles.," Thèse de doctorat, Université de Montréal, février 1981.

P. C. HAMMER and J. W. HOLLINGSWORTH [1], *Trapezoidal methods of approximating solutions of differential equations*, Math. Tables Aids Comput. **9** (1955), 92–96.

T. KAILATH [1], *Some Chandrasekhar -type algorithms for quadratic regulators*, in "Proc. IEEE CDC (New Orleans 1972)," IEEE Publications, New York, 1972, pp. 219-223.

V. I. KRYLOV [1], "Approximation calculation of integrals," MacMillan Co., New York, 1962.

D. LEROY and M. SORINE [1], "Schéma d'approximation pour les équations du type Chandrasekhar intervenant en contrôle optimal," INRIA Report, unpublished.

A. LINDQUIST [1], *Some new non-Riccati algorithms for continuous-time Kalman-Bucy filtering*, Appl. Math. and Optim. **3** (1976), 1-13.

J. L. LIONS [1], "Optimal control of systems governed by partial differential equations," Springer-Verlag, New York, Heidelberg-Berlin, 1971.

J. L. LIONS et E. MAGENES [1], "Problèmes aux limites non homogènes et applications," Vol. 1, 2, 3, Dunod, Paris, 1968, 1970.

J. T. ODEN and J. N. REDDY [1], "An introduction to the mathematical theory of finite elements," John Wiley & Sons, New York, 1976.

A. OUANSAFI [1], "Méthodes d'approximation discontinue des problèmes de commande optimale," Ph.D. Thesis, Université de Montréal, 1984.

NUMERICAL APPROACH TO THE EXACT CONTROLLABILITY OF HYPERBOLIC SYSTEMS

A. ELJENDY[1]

Abstract. In this paper we present the numerical implementation of H.U.M. (Hilbert Uniqueness Method, J.L.Lions[1]). We restrict ourselves to the exact boundary controllability of the wave equation, with Dirichlet controls, but the numerical method presented here can be applied to other kinds of controllability. The problem is discretized by a finite elements of first order in space and by a discrete time Galerkin approximation (Dupont [1]). The efficiency of the method is illustrated by numerical results.

Keywords. Distributed systems - Exact controllability - Boundary control - Finite elements - Discrete time Galerkin approximation - Wave equation.

1).INTRODUCTION.

Let Ω be an open set in $R^n, n \geq 1$, assumed to be bounded with a smooth boundary $\Gamma = \partial\Omega$, Γ of class C^2, Γ^0 non empty part of Γ and let $T \in R$, $T > 0$ be given. T large enough. We consider the wave equation:

$$(1\text{–}1) \qquad y'' - \Delta y = 0 \qquad \text{in } Q = \Omega \times]0, T[$$

with the initial conditions:

$$(1\text{–}2) \qquad y(0) = y^0, \qquad y'(0) = y^1 \qquad \text{in } \Omega$$

and nonhomogeneous Dirichlet boundary action:

$$(1\text{–}3) \qquad \begin{cases} y = v & \text{on } \Sigma^0 = \Gamma^0 \times]0, T[\\ y = 0 & \text{on } \Sigma \setminus \Sigma^0 = (\Gamma \setminus \Gamma^0) \times]0, T[\end{cases}$$

Where:
$y'(x,t) = \frac{\partial y}{\partial t}(x,t), \qquad y''(x,t) = \frac{\partial^2 y}{\partial t^2}(x,t)$
$y(0) : x \longmapsto y(x,0), \qquad y'(0) : x \longmapsto \frac{\partial y}{\partial t}(x,0)$
$y^0 : \Omega \longmapsto R, \qquad y^1 : \Omega \longmapsto R$

The functions y^0 and y^1 are chosen in appropriate Hilbert spaces, which will be defined later on. t is the time variable, $t \in]0, T[$.

[1] University of Jyväskylä, Departement of Mathematics, Seminaarinkatu 15, SF. 40100 Jyväskylä 10, Finland.

$v : \Sigma = \Gamma\times]0,T[\longmapsto R$ is a control function. We have imposed the constraint (1-3) to the control function v. The exact controllability problem can be defined as:

Given $T > 0$ and y^0, y^1, is it possible to find v in suitable Hilbert space ($L^2(\Sigma^0)$, for example) such that: if $y = y(v)$ denotes the solution of the system (1-1)(1-2) and (1-3) then y satisfies :

$$y(x,T,v) = y'(x,T,v) = 0 \qquad \text{in } \Omega\ ? \tag{1–4}$$

2).H.U.M.: Hilbert Uniqueness Method.

In this paragraph we present the steps of the method for the wave equation with Dirichlet boundary control. We start with the wave equation:

$$\begin{cases} \phi'' - \Delta\phi = 0 & \text{in } Q \\ \phi(0) = \phi^0, \quad \phi'(0) = \phi^1 & \text{in } \Omega \\ \phi = 0 & \text{on } \Sigma \end{cases} \tag{2–1}$$

With $\{\phi^0, \phi^1\} \in \mathcal{D}(\Omega) \times \mathcal{D}(\Omega)$. Where $\mathcal{D}(\Omega)$, is the space of functions of class C^∞ with compact support in Ω.
In reality $\{\phi^0, \phi^1\}$ will be chosen in appropriate Hilbert space F. F will be defined later. With these conditions the problem (2-1) admits a unique solution ϕ (J.L.Lions and E.Magenes[1]).
We then solve the system:

$$\begin{cases} \psi'' - \Delta\psi = 0 & \text{in } Q \\ \psi(T) = \psi'(T) = 0 & \text{in } \Omega \\ \psi = \dfrac{\partial\phi}{\partial\nu} & \text{on } \Sigma^0 \\ \psi = 0 & \text{on } \Sigma\setminus\Sigma^0 \end{cases} \tag{2–2}$$

This nonhomogeneous boundary value problem admits always at least a weak solution (J.L.Lions and E.Magenes[1]).
We define then a linear operator Λ by:

$$\Lambda\{\phi^0, \phi^1\} = \{\psi'(0), -\psi(0)\}$$

We compute the scalar product

$$< \Lambda\{\phi^0, \phi^1\}, \{\phi^0, \phi^1\} >= \int_\Omega (\psi'(0)\phi^0 - \psi(0)\phi^1)dx$$

We multiply (2-2) by ϕ and we integrate by parts on Q, and use Green's formula to get:

$$< \Lambda\{\phi^0,\phi^1\},\{\phi^0,\phi^1\} >= \int_{\Sigma^0} \left(\frac{\partial \phi}{\partial \nu}\right)^2 d\Gamma dt$$

We introduce then the semi-norm

$$\| \{\phi^0,\phi^1\} \|_F = \left(\int_{\Sigma^0} \left(\frac{\partial \phi}{\partial \nu}\right)^2 d\Gamma dt \right)^{\frac{1}{2}}, \qquad \{\phi^0,\phi^1\} \in \mathcal{D}(\Omega) \times \mathcal{D}(\Omega)$$

By uniqueness theorem $\| \, . \, \|_F$ will define a norm on the space of the initial data $\{\phi^0,\phi^1\}$, denoted F. The space F is defined as the Hilbert space completion of $\mathcal{D}(\Omega) \times \mathcal{D}(\Omega)$ for the norm $\| \, . \, \|_F$.
The operator Λ is then an isomorphism from F onto F'. For any initial data $\{y^0, y^1\}$ such that $\{y^1,y^0\} \in F'$ the equation:

$$\Lambda\{\phi^0,\phi^1\} = \{y^1,-y^0\}$$

admits a unique solution $\{\phi^0,\phi^1\} \in F$.
We then solve (2-1) and (2-2) and we define the control v by :

$$v = \frac{\partial \phi}{\partial \nu} \qquad \text{on } \Sigma^0$$
$$v = 0 \qquad \text{on } \Sigma \setminus \Sigma^0$$

and we define the solution y by:

$$y = \psi \qquad \text{in } Q$$

To characterize F, one uses an equivalence of norms, for T large enough and one has:

$$F = H_0^1(\Omega) \times L^2(\Omega)$$
$$F' = H^{-1}(\Omega) \times L^2(\Omega)$$

The exact controllability problem is then solved. For more details on H.U.M. see J.L. Lions[1].

To solve then, by H.U.M., the exact controllability problem one must:

(1) Find $\{\phi^0,\phi^1\} \in H_0^1(\Omega) \times L^2(\Omega)$ solution of the equation:

$$\Lambda\{\phi^0,\phi^1\} = \{y^1,-y^0\}$$

With $\{y^1,y^0\}$ given in $F' = H^{-1}(\Omega) \times L^2(\Omega)$.

(2) Resolve the system (2-1).

(3) Compute $\left.\frac{\partial \phi}{\partial \nu}\right|_\Gamma$.

(4) Resolve the system (2-2).

The method presented here consists to transform the steps 1 and 2 to one eigenvalues problem. ∎

3). Resolution of the equation $\Lambda\{\phi^0,\phi^1\} = \{y^1,-y^0\}$.

Let $\{y^1,y^0\}$ be given in $H^{-1}(\Omega)\times L^2(\Omega)$. The solving of the equation:

$$\Lambda\{\phi^0,\phi^1\} = \{y^1,-y^0\} \tag{3-1}$$

leads to minimization problems which can be solved by several methods (Glowinski, Li and J.L.Lions[1], A.El jai and A.Gonzalez[1]). Here we shall present a systematic method which is based on harmonic analysis and eigenfunctions calculation. The solving of the equation (3-1) is equivalent to solving the following minimization problem:

$$\inf_{\{\phi^0,\phi^1\}} L(\phi^0,\phi^1) \tag{3-2}$$

Where L is given by:

$$\begin{aligned} L(\phi^0,\phi^1) &= \frac{1}{2} < \Lambda\{\phi^0,\phi^1\},\{\phi^0,\phi^1\} > - \int_\Omega (\phi^0 y^1 - \phi^1 y^0)dx \\ &= \frac{1}{2}\int_{\Sigma^0}\left(\frac{\partial\phi}{\partial\nu}\right)^2 d\Sigma - \int_\Omega(\phi^0 y^1 - \phi^1 y^0)dx \end{aligned}$$

We introduce eigenfunctions ω_j of the operator $-\Delta$, with homogeneous Dirichlet condition:

$$\begin{cases} -\Delta\omega_j = \lambda_j\omega_j & \text{in } \Omega \\ \omega_j = 0 & \text{on } \Gamma \\ \| \omega_j \| = 1 \end{cases}$$

The eigenvalues λ_j are such that: $(0 < \lambda_1 \le \lambda_2 \le ...)$. Functions $\frac{\omega_j}{\sqrt{\lambda_j}}$, ω_j and $\sqrt{\lambda_j}\omega_j$ are respectively orthonormal bases of $H_0^1(\Omega)$, $L^2(\Omega)$ and $H^{-1}(\Omega)$. The initial data can be written as:

$$y^0 = \sum_{j=1}^{\infty} y_j^0\omega_j, \quad \text{and} \quad y^1 = \sum_{j=1}^{\infty} y_j^1\omega_j$$

$$\phi^0 = \sum_{j=1}^{\infty} \phi_j^0\omega_j, \quad \text{and} \quad \phi^1 = \sum_{j=1}^{\infty} \phi_j^1\omega_j$$

with:

$$y_j^0 = (y^0,\omega_j)_{L^2(\Omega)}, \qquad \text{and } y_j^1 = \lambda_j(y^1,\omega_j)_{H^{-1}(\Omega)} = (y^1,\omega_j)_{L^2(\Omega)}$$

$$\phi_j^0 = \frac{1}{\lambda_j}(\phi^0,\omega_j)_{H_0^1(\Omega)} = (\phi^0,\omega_j)_{L^2(\Omega)}, \qquad \text{and } \phi_j^1 = (\phi^1,\omega_j)_{L^2(\Omega)}$$

We get then :

$$\int_\Omega y^1\phi^0 dx = \sum_{j=1}^{\infty}(y^1,\omega_j)(\phi^0,\omega_j)$$

$$\int_\Omega y^0\phi^1 dx = \sum_{j=1}^{\infty}(y^0,\omega_j)(\phi^1,\omega_j)$$

By the same we have:

$$\phi(x,t) = \sum_{j=1}^{\infty} \phi_j(t)\omega_j(x) \tag{3–3}$$

with:

$$\begin{aligned}\phi_j(t) &= (\phi^0,\omega_j)\cos(t\sqrt{\lambda_j}) + \frac{1}{\sqrt{\lambda_j}}(\phi^1,\omega_j)\sin(t\sqrt{\lambda_j}) \\ &= \phi_j^0 \cos(t\sqrt{\lambda_j}) + \frac{1}{\sqrt{\lambda_j}}\phi_j^1 \sin(t\sqrt{\lambda_j})\end{aligned}$$

and then:

$$\begin{aligned}\frac{1}{2}\int_{\Sigma^0}\left(\frac{\partial\phi}{\partial\nu}\right)^2 d\Sigma &= \frac{1}{2}\sum_{j,k=1}^{\infty}\left(\int_0^T \phi_j(t)\phi_k(t)dt \int_{\Gamma^0}\frac{\partial\omega_j}{\partial\nu}\frac{\partial\omega_k}{\partial\nu}d\Gamma\right) \\ &= \sum_{j,k=1}^{\infty}\beta_{jk}\left(\int_0^T \phi_j(t)\phi_k(t)dt\right)\end{aligned}$$

with:

$$\beta_{jk} = \frac{1}{2}\int_{\Gamma^0}\frac{\partial\omega_j}{\partial\nu}\frac{\partial\omega_k}{\partial\nu}d\Gamma$$

Let us now calculate the $\int_0^T \phi_j(t)\phi_k(t)dt$:

$$\begin{aligned}\int_0^T \phi_j(t)\phi_k(t)dt &= \int_0^T \phi_j^0\phi_k^0 \cos(t\sqrt{\lambda_j})\cos(t\sqrt{\lambda_k})dt \\ &+ \int_0^T \frac{1}{\sqrt{\lambda_j}}\phi_j^1\phi_k^0 \sin(t\sqrt{\lambda_j})\cos(t\sqrt{\lambda_k})dt \\ &+ \int_0^T \phi_j^0\phi_k^1\frac{1}{\sqrt{\lambda_k}}\cos(t\sqrt{\lambda_j})\sin(t\sqrt{\lambda_k})dt \\ &+ \int_0^T \frac{1}{\sqrt{\lambda_j}}\frac{1}{\sqrt{\lambda_k}}\phi_j^1\phi_k^1 \sin(t\sqrt{\lambda_j})\sin(t\sqrt{\lambda_k})dt\end{aligned}$$

which implies:

$$\begin{aligned}\frac{1}{2}\int_{\Sigma^0}\left(\frac{\partial\phi}{\partial\nu}\right)^2 d\Sigma &= \sum_{j,k=1}^{\infty}\beta_{jk}\int_0^T \phi_j^0\phi_k^0 \cos(t\sqrt{\lambda_j})\cos(t\sqrt{\lambda_k})dt \\ &+ 2\sum_{j,k=1}^{\infty}\beta_{jk}\int_0^T \frac{1}{\sqrt{\lambda_j}}\phi_j^1\phi_k^0 \sin(t\sqrt{\lambda_j})\cos(t\sqrt{\lambda_k})dt \\ &+ \sum_{j,k=1}^{\infty}\beta_{jk}\int_0^T \frac{1}{\sqrt{\lambda_j}}\frac{1}{\sqrt{\lambda_k}}\phi_j^1\phi_k^1 \sin(t\sqrt{\lambda_j})\sin(t\sqrt{\lambda_k})dt\end{aligned}$$

We write then:

$$\frac{1}{2}\int_{\Sigma^0}\left(\frac{\partial\phi}{\partial\nu}\right)^2 d\Sigma = \sum_{j,k=1}^{\infty} A_{jk}\phi_j^0\phi_k^0 + \sum_{j,k=1}^{\infty} 2B_{jk}\phi_j^1\phi_k^0 + \sum_{j,k=1}^{\infty} C_{jk}\phi_j^1\phi_k^1$$

with:

$$A_{jk} = \beta_{jk}\int_0^T \cos(t\sqrt{\lambda_j})\cos(t\sqrt{\lambda_k})dt$$

$$B_{jk} = \beta_{jk}\int_0^T \frac{1}{\sqrt{\lambda_j}}\sin(t\sqrt{\lambda_j})\cos(t\sqrt{\lambda_k})dt$$

$$C_{jk} = \beta_{jk}\int_0^T \frac{1}{\sqrt{\lambda_j}}\frac{1}{\sqrt{\lambda_k}}\sin(t\sqrt{\lambda_j})\sin(t\sqrt{\lambda_k})dt$$

and

$$A_{jk} = A_{kj} \quad , \quad C_{jk} = C_{kj}$$

Finally we get:

$$L(\phi^0,\phi^1) = \sum_{j,k=1}^{\infty}\left(A_{jk}\phi_j^0\phi_k^0 + 2B_{jk}\phi_j^1\phi_k^0 + C_{jk}\phi_j^1\phi_k^1\right) - \sum_{j=1}^{\infty}\left(\phi_j^0 y_j^1 - \phi_j^1 y_j^0\right)$$

Let us now calculate the gradient of functional L:

$$\frac{\partial L}{\partial\phi^0}(\phi^0,\phi^1) = \sum_{j,k=1}^{\infty}\left(A_{jk}\phi_j^0\omega_k + A_{jk}\phi_k^0\omega_j + 2B_{jk}\phi_j^1\omega_k\right) - \sum_{j=1}^{\infty} y_j^1\omega_j$$

and

$$\frac{\partial L}{\partial\phi^1}(\phi^0,\phi^1) = \sum_{j,k=1}^{\infty}\left(2B_{jk}\phi_k^0\omega_j + C_{jk}\phi_j^1\omega_k + C_{jk}\phi_k^1\omega_j\right) + \sum_{j=1}^{\infty} y_j^0\omega_j$$

which leads to:

$$\frac{\partial L}{\partial\phi^0}(\phi^0,\phi^1) = \sum_{j,k=1}^{\infty}\left(2A_{jk}\phi_j^0 + 2B_{jk}\phi_j^1\right)\omega_k - \sum_{k=1}^{\infty} y_k^1\omega_k$$

and

$$\frac{\partial L}{\partial\phi^1}(\phi^0,\phi^1) = \sum_{j,k=1}^{\infty}\left(2B_{jk}\phi_k^0 + 2C_{jk}\phi_k^1\right)\omega_j + \sum_{j=1}^{\infty} y_j^0\omega_j$$

The minimum is then given by the following infinite linear system:

$$(3\text{-}4)\qquad \begin{cases} \sum_{j=1}^{\infty}(2A_{jk}\phi_j^0 + 2B_{jk}\phi_j^1) = y_k^1 & \text{for } k = 1,2,...,+\infty \\ \sum_{j=1}^{\infty}(2B_{kj}\phi_j^0 + 2C_{jk}\phi_j^1) = -y_k^0 & \text{for } k = 1,2,...,+\infty \end{cases}$$

Numerically, using Galerkin approximation of the Hilbert space $H_0^1(\Omega) \times L^2(\Omega)$, we approximate the system (3-4) by the following finite linear system:

$$(3\text{-}5)\qquad \begin{cases} \sum_{j=1}^{M}(2A_{jk}\phi_j^0 + 2B_{jk}\phi_j^1) = y_k^1 & \text{for } k = 1,2,...,M \\ \sum_{j=1}^{M}(2B_{kj}\phi_j^0 + 2C_{jk}\phi_j^1) = -y_k^0 & \text{for } k = 1,2,...,M \end{cases}$$

Where M is a fixed integer. Here, M can be chosen large enough, without constraints. We approximate then $\phi^0(x)$ and $\phi^1(x)$ by $\left(\sum_{j=1}^{M}\phi_j^0\omega_j(x)\right)$ and $\left(\sum_{j=1}^{M}\phi_j^1\omega_j(x)\right)$, respectively. ϕ_j^0 and ϕ_j^1, $(j = 1,...,M)$, are solutions of the system (3-5).

4). Application to one-dimensional case.

Let Ω be an open domain of R, $\Omega =]0,1[$. Let $T \in R$ be given, T large enough. We shall solve, using H.U.M., the following system:

$$\begin{cases} y'' - \Delta y = 0 & \text{in }]0,1[\times]0,T[\\ y(x,0) = y^0(x), \quad y'(x,0) = y^1(x) & \text{in }]0,1[\\ y(0,t) = v(t), \quad y(1,t) = 0 & \text{in }]0,T[\end{cases}$$

$\{y^0, y^1\}$ given in $H^{-1}(\Omega) \times L^2(\Omega)$. Eigenfunctions ω_j and eigenvalues λ_j, $j = 1,2,...$, of the operator $(-\Delta)$, with homogeneous Dirichlet boundary:

$$\begin{cases} -\Delta\omega_j = \lambda_j\omega_j & \text{in }]0,1[\\ \omega_j(0) = \omega_j(1) = 0 \\ \| \omega_j \| = 1 \end{cases}$$

are given by:

$$\begin{cases} \lambda_j = j^2\pi^2 & \forall j = 1,2,... \\ \omega_j = \sqrt{2}\sin j\pi x & \forall j = 1,2,... \end{cases}$$

One can verify that:

$$
\begin{aligned}
&\frac{\partial \omega_j}{\partial \nu}(0) \neq 0 \qquad \forall j = 1,2,\dots \\
&\beta_{jk} = jk\pi^2 \\
&A_{jk} = \frac{jk\pi}{2}\left(\frac{\sin((j+k)\pi T)}{j+k} + \frac{\sin((j-k)\pi T)}{j-k}\right) \qquad \text{if } j \neq k \\
&A_{jj} = \frac{j^2\pi^2}{2}\left(\frac{\sin(2j\pi T)}{2j\pi} + T\right) \\
&B_{jk} = \frac{k}{2}\left(\frac{1-\cos((j+k)\pi T)}{j+k} + \frac{1-\cos((j-k)\pi T)}{j-k}\right) \qquad \text{if } j \neq k \\
&B_{jj} = \frac{1}{2}\left(\frac{1-\cos(2j\pi T)}{2}\right) \\
&C_{jk} = \frac{1}{2}\left(\frac{\sin((j-k)\pi T)}{(j-k)\pi} - \frac{\sin((j+k)\pi T)}{(j+k)\pi}\right) \qquad \text{if } j \neq k \\
&C_{jj} = \frac{1}{2}\left(T - \frac{\sin(2j\pi T)}{2j\pi}\right)
\end{aligned}
$$

REMARK 4-1.

Using the coefficients above, if we assume that all eigenvalues λ_j are simple (multiplicity equal to 1), then one recovers the following result given in (Glowinski, Li and J.L.Lions[1]):

$$\lim_{T \to +\infty} T\{\phi^0, \phi^1\} = \{\chi^0, \chi^1\}$$

where χ^0 and χ^1 verify:

$$
\begin{cases}
-\Delta\chi^0 = y^1 & \text{in } \Omega \\
\chi^0 = 0 & \text{on } \Gamma \\
\chi^1 = -y^0 & \text{on } \Omega
\end{cases}
\qquad \blacksquare
$$

REMARK 4-2.

In one-dimensional case, if T is even (here $\mathrm{diam}(\Omega) = 1$, since for every j, it exists an integer p_j such that $T = \frac{2p_j\pi}{\sqrt{\lambda_j}}$, all eigenvalues are simple and for all integers j and k, $\sqrt{\frac{\lambda_j}{\lambda_k}} \in Q$) then ϕ^0 and ϕ^1 are given explicitly by:

$$\phi^0 = \frac{\chi^0}{T} \qquad \text{and} \qquad \phi^1 = \frac{\chi^1}{T} \qquad \blacksquare$$

Generally, we have to solve the following linear system:

$$\begin{cases} \sum_{j=1}^{M} (2A_{jk}\phi_j^0 + 2B_{jk}\phi_j^1) = y_k^1 & \text{for } k = 1,2,...,M \\ \sum_{j=1}^{M} (2B_{kj}\phi_j^0 + 2C_{jk}\phi_j^1) = -y_k^0 & \text{for } k = 1,2,...,M \end{cases}$$

Where M is a fixed integer. We approximate ϕ^0 and ϕ^1 by $\left(\sum_{j=1}^{M} \phi_j^0 \omega_j\right)$ and $\left(\sum_{j=1}^{M} \phi_j^1 \omega_j\right)$, respectively. Using formula (3-3), we can then approximate $\frac{\partial \phi}{\partial \nu}(0)$ by:

$$\frac{\partial \phi}{\partial \nu}(0,t) \approx \sum_{j=1}^{M} \phi_j(t) \frac{\partial \omega_j}{\partial \nu}(0) \qquad \text{on}]0,T[$$

REMARK 4-3.

We do not need to solve the homogeneous boundary problem (2-1), but we have to solve an eigenvalue problem. ∎

We solve then the following system:

$$\begin{cases} \psi'' - \Delta\psi = 0 & \text{in }]0,1[\times]0,T[\\ \psi(T) = \psi'(T) = 0 & \text{in }]0,1[\\ \psi(0) = \frac{\partial \phi}{\partial \nu}(0); \quad \psi(1) = 0 & \text{on }]0,T[\end{cases}$$

We define the basis functions $\{v^i\}_{i=0}^m \subset V_h \subset V$ by:

$$\begin{cases} v^i(x) = \frac{x}{h} - i + 1 & \forall x \in K_i = [(i-1)h, ih]; \quad i = 1,2,...,m \\ v^i(x) = -\frac{x}{h} + i + 1 & \forall x \in K_{i+1}; \quad i = 0,1,...,m \\ v^i(x) = 0 & \text{Otherwise} \end{cases}$$

With V given by:

$$V = \{v \in H^1(\Omega) / \quad v = 0 \text{ on } (\Gamma \setminus \Gamma^0)\}$$

For $\epsilon > 0$, we approximate ψ by the solution ψ_ϵ of a problem analogous to (2-2), with the Newton-Robin type boundary conditions (M.Křížek and P.Neittaanmäki[1]). We get then the following scheme:

$$M_\Omega \left(\psi_\epsilon''\right)^r + \left(A_\Omega + \frac{1}{\epsilon} M_{\Gamma^0}\right) \psi_\epsilon^{r\theta} = \frac{1}{\epsilon} G^{r\theta}; \qquad \theta \in [0,1]$$

where:

$$\phi_h^r = \phi_h(r\Delta t), \qquad r = 1, ..., N$$
$$\phi_h^{r\theta} = \theta\phi_h^{r+1} + (1-2\theta)\phi_h^r + \theta\phi_h^{r-1}, \qquad 0 \le \theta \le 1$$
$$(\phi_h'')^r = \frac{\phi_h^{r+1} - 2\phi_h^r + \phi_h^{r-1}}{(\Delta t)^2}$$

and

$$A_\Omega = \frac{1}{h}\begin{pmatrix} 1 & -1 & 0 & 0 & 0 & \dots & 0 \\ -1 & 2 & -1 & 0 & 0 & \dots & 0 \\ 0 & -1 & 2 & -1 & 0 & \dots & 0 \\ \vdots & \ddots & \ddots & \ddots & \ddots & \ddots & \vdots \\ 0 & \dots & 0 & -1 & 2 & -1 & 0 \\ 0 & \dots & 0 & 0 & -1 & 2 & -1 \\ 0 & \dots & 0 & 0 & 0 & -1 & 2 \end{pmatrix}; \qquad M_\Omega = \frac{h}{6}\begin{pmatrix} 2 & 1 & 0 & 0 & 0 & \dots & 0 \\ 1 & 4 & 1 & 0 & 0 & \dots & 0 \\ 0 & 1 & 4 & 1 & 0 & \dots & 0 \\ \vdots & \ddots & \ddots & \ddots & \ddots & \ddots & \vdots \\ 0 & \dots & 0 & 1 & 4 & 1 & 0 \\ 0 & \dots & 0 & 0 & 1 & 4 & 1 \\ 0 & \dots & 0 & 0 & 0 & 1 & 4 \end{pmatrix}$$

and:

$$M_{\Gamma^0} = \begin{pmatrix} 1 & 0 & \dots & 0 \\ 0 & 0 & \dots & 0 \\ \vdots & \vdots & \ddots & \vdots \\ 0 & 0 & \dots & 0 \end{pmatrix}; \qquad G^r = \begin{pmatrix} \frac{\partial\phi^r}{\partial\nu}(0) \\ 0 \\ \vdots \\ 0 \end{pmatrix}$$

A_Ω, M_Ω and M_Γ^0 are $(m+1)\times(m+1)$ matrices. To discretize the problem in time, we use discrete time Galerkin approximation (Dupont[1]). For each r, we have to solve the following tridiagonal system:

$$\left(M_\Omega + \theta(\Delta t)^2\left(A_\Omega + \frac{1}{\epsilon}M_{\Gamma^0}\right)\right)\psi^r = \left(2M_\Omega - (1-2\theta)(\Delta t)^2\left(A_\Omega + \frac{1}{\epsilon}M_{\Gamma^0}\right)\right)\psi^{r+1}$$
$$- \left(M_\Omega + \theta(\Delta t)^2\left(A_\Omega + \frac{1}{\epsilon}M_{\Gamma^0}\right)\right)\psi^{r+2}$$
$$+ \frac{(\Delta t)^2}{\epsilon}\left(\theta G^{r+2} + (1-2\theta)G^{r+1} + \theta G^r\right)$$

5). Numerical results.

For these simulations we fix:

$$y^0(x) = \sin(\pi x) + \sin(2\pi x) + \sin(3\pi x)$$
$$y^1(x) = x(1-x)$$

The result given here, show for different values of T, Δt and Δx, the evolution of the cost function and initial states error. The value of the cost function is given by $\| v \|^2_{L^2(\Omega)}$. Values of the initial states error are given by:

$$E_0 = \| y(0) - y^0 \|_{L^2(\Omega)}$$
$$E_0' = \frac{\| y(0) - y^0 \|_{L^2(\Omega)}}{\| y^0 \|_{L^2(\Omega)}}$$
$$E_1 = \| y'(0) - y^1 \|_{H^{-1}(\Omega)}$$
$$E_1' = \frac{\| y'(o) - y^1 \|_{H^{-1}(\Omega)}}{\| y^1 \|_{H^{-1}(\Omega)}}$$
$$E_0'' = \| \chi^0 - T\phi^0 \|_{H_0^1(\Omega)}$$
$$E_1'' = \| \chi^1 - T\phi^1 \|_{L^2(\Omega)}$$

where χ^0 and χ^1 verify:

$$\begin{cases} -\Delta\chi^0 = y^1 & \text{in } \Omega \\ \chi^0 = 0 & \text{on } \Gamma \\ \chi^1 = -y^0 & \text{on } \Omega \end{cases}$$

E_0'' and E_1'' are a very severe test, when $T \to +\infty$, for the numerical method designed to solve the equation (3-1) (Glowinski, Li and J.L.Lions[1]). The time step Δt and the space step Δx, are chosen such that they satisfy the stability condition: $\frac{(\Delta t)^2}{(\Delta x)^2} \leq \frac{1}{2}$. We fixe $\frac{\Delta t}{\Delta x} = \frac{1}{2}$ and we choose $\Delta x = \frac{1}{128}$.

T	E_0	E_0'	E_1	E_1'	$\| v \|^2$
2.0	$9.01\ 10^{-3}$	$7.36\ 10^{-3}$	$1.71\ 10^{-2}$	$2.94\ 10^{-1}$	$7.51\ 10^{-1}$
4.0	$5.85\ 10^{-3}$	$4.77\ 10^{-3}$	$1.67\ 10^{-2}$	$2.87\ 10^{-1}$	$3.75\ 10^{-1}$
8.0	$3.89\ 10^{-3}$	$3.17\ 10^{-3}$	$1.65\ 10^{-2}$	$2.83\ 10^{-1}$	$1.88\ 10^{-1}$
12.0	$3.10\ 10^{-3}$	$2.53\ 10^{-3}$	$1.64\ 10^{-2}$	$2.82\ 10^{-1}$	$1.25\ 10^{-1}$
14.0	$2.86\ 10^{-3}$	$2.34\ 10^{-3}$	$1.64\ 10^{-2}$	$2.82\ 10^{-1}$	$1.07\ 10^{-1}$
16.0	$2.66\ 10^{-3}$	$2.17\ 10^{-3}$	$1.63\ 10^{-2}$	$2.81\ 10^{-1}$	$9.39\ 10^{-2}$

Table 1: Cost and initial state error with respect to T.

T	E_0''	E_1''
2.0	$3.00\ 10^{-8}$	$4.19\ 10^{-8}$
4.0	$3.00\ 10^{-8}$	$4.19\ 10^{-8}$
8.0	$3.00\ 10^{-8}$	$4.19\ 10^{-8}$
12.0	$3.00\ 10^{-8}$	$4.19\ 10^{-8}$
14.0	$3.00\ 10^{-8}$	$4.19\ 10^{-8}$
16.0	$3.00\ 10^{-8}$	$4.19\ 10^{-8}$

Table 2: Independence of ϕ^0 and ϕ^1 on T.

T	E_0	E'_0	E_1	E'_1	$\parallel v \parallel^2$
1.5	no contro.	no contro.	no contro.	no contro.	no contro.
2.5	$5.69\ 10^{-3}$	$4.65\ 10^{-3}$	$1.66\ 10^{-2}$	$2.86\ 10^{-1}$	$5.65\ 10^{-1}$
4.5	$2.63\ 10^{-3}$	$2.14\ 10^{-3}$	$1.63\ 10^{-2}$	$2.81\ 10^{-1}$	$3.15\ 10^{-1}$
8.5	$1.32\ 10^{-3}$	$1.07\ 10^{-3}$	$1.62\ 10^{-2}$	$2.80\ 10^{-1}$	$1.70\ 10^{-1}$
12.5	$1.04\ 10^{-3}$	$8.49\ 10^{-4}$	$1.62\ 10^{-2}$	$2.79\ 10^{-1}$	$1.17\ 10^{-1}$
14.5	$9.66\ 10^{-4}$	$7.89\ 10^{-4}$	$1.62\ 10^{-2}$	$2.79\ 10^{-1}$	$1.01\ 10^{-1}$

T	E''_0	E''_1
1.5	no contro.	no contro.
2.5	$3.22\ 10^{-1}$	$1.24\ 10^{-1}$
4.5	$1.86\ 10^{-1}$	$8.97\ 10^{-2}$
8.5	$1.02\ 10^{-1}$	$5.60\ 10^{-2}$
12.5	$7.13\ 10^{-2}$	$4.04\ 10^{-2}$
14.5	$6.18\ 10^{-2}$	$3.55\ 10^{-2}$

Table 4: Convergence of $T\phi^0$ and $T\phi^1$ to χ^0 and χ^1 when $T \to +\infty$.

The results given in tables 1 and 2 are obtained for even T. The results of tables 1 and 3 indicat that the convergence of the method is faster for large T. The results of table 2 confirm the result given in remark 4-2, those of table 4, confirm the convergence of $T\phi^0$ and $T\phi^1$ to χ^0 and χ^1, respectively, when T tends to $+\infty$. They seem also indicate that the cost is equivalent to $\frac{C}{T}$, where C is a constant.

In the following tables, we present the result of the convergence of the method with respect to the space step Δx and the time step Δt, when T is fixed equal to 2.5 and $\frac{\Delta t}{\Delta x} = \frac{1}{2}$.

Δx	E_0	E'_0	E_1	E'_1	$\parallel v \parallel^2$
1/64	$7.46\ 10^{-3}$	$6.09\ 10^{-3}$	$3.26\ 10^{-2}$	$5.61\ 10^{-1}$	$5.64\ 10^{-1}$
1/128	$5.69\ 10^{-3}$	$4.65\ 10^{-3}$	$1.66\ 10^{-2}$	$2.86\ 10^{-1}$	$5.65\ 10^{-1}$
1/256	$4.38\ 10^{-3}$	$3.58\ 10^{-3}$	$8.56\ 10^{-3}$	$1.47\ 10^{-1}$	$5.65\ 10^{-1}$
1/512	$4.07\ 10^{-3}$	$3.32\ 10^{-3}$	$4.86\ 10^{-3}$	$8.37\ 10^{-2}$	$5.65\ 10^{-1}$

Table 5: Cost and initial state error with respect to Δx and Δt.

These results confirm the convergence, in $L^2(\Omega)$ and in $H^{-1}(\Omega)$ for the initial conditions $y(0)$ and $y'(0)$,respectively.

In the end of this presentation of the numerical results we show a visualization of the evolution of the system to zero. Taking $t = 0, 0.25, 0.5, ..., 2.5$, we have shown, for fixed t, the computed function $y(x,t)$; the corresponding calculation was done with $T = 2.5$, $\Delta x = \frac{1}{512}$, $\frac{\Delta t}{\Delta x} = \frac{1}{2}$. The computed approximation of y' shows a similar behavior.

To conclude this paper, we would like to point out that we get the same results and the same behaviors, when we use a direct scheme (see resolution of nonhomogeneous problem), ie: we start the scheme using the solution at levels $t = 0$ and $t = \Delta t$ (using datas y^0 and y^1). The direct scheme corresponds more to the real system. The visualization shown here is obtained by a retrograde scheme.

In spite of approximations used to solve the equation $\Lambda\{\phi^0, \phi^1\} = \{y^1, -y^0\}$ (truncation,...), the result given by this method is quite good. More details on this method, with applications to other kind of controls will be introduced in a next paper wich is in preparation. In the end we want to point out the simplicity of this method and its low cost.

References

Dupont[1]. : L^2-Estimates for galerkin methods for second order hyperbolic equations. S.I.A.M. J. on Num. Anal. Vol. 10, no 5 (1973).

A.El jai and A.Gonzalez[1]. : Exacte contrôlabilité ponctuelle des systèmes hyperboliques bidimensionnels : Approche numérique. 5th symposium I.F.A.C. Perpignan, France (1989).

Glowinski, Li and J.L.Lions[1]. : A numerical approach to the exact boundary controllability of the wave equation (I) Dirichlet controls: Description of the numerical methods. Japan J. of App. Math. Vol. 7, no 1 (1990).

M.Křižek and P.Neittaanmäki[1]. : Finite element approximation of variational problems and applications. Pitman Series. Pure and App. Math. 50 (1990).

J.L.Lions[1]. : Exact controllability, stabilization and perturbations, for distributed systems. S.I.A.M. Review. Vol. 30, no 1 (1988).

J.L.Lions and E.Magenes[1]. : Problèmes aux limites non homogènes et applications. Dunod Paris (1968).

SHAPE SENSITIVITY ANALYSIS FOR STOCHASTIC EVOLUTION EQUATIONS

Dariusz Gątarek and Jan Sokołowski
Systems Research Institute
Polish Academy of Sciences
ul. Newelska 6 , 01-447 Warszawa
Poland

ABSTRACT

Shape differential stability of the solutions to stochastic partial differential equations is shown. Parabolic and hyperbolic stochastic PDE's are considered.

Key words : parabolic stochastic equation, hyperbolic stochastic equation, shape sensitivity analysis, shape derivative.

AMS(MOS) subject classification : 49B22, 49A29, 49A22, 93B30

1. Introduction.

The paper is devoted to the shape differential stability analysis of solutions to stochastic partial differential equations. Optimal control problems for such equations are studied in [2],[3]. We use the method of sensitivity analysis described in detail in [12], [13], [14] for elliptic and parabolic partial differential equations. Further results on the sensitivity analysis of related optimal control problems for the wave equation can be found in [11]. Using the material derivative method of shape sensitivity analysis we differentiate, with respect to real parameter, the stochastic partial differential equation. The form of the shape derivative of the solution is obtained.

Outline of the paper is as follows. In section 2 we provide the results for a stochastic parabolic equation. In section 3 the results on the shape sensitivity analysis for a hyperbolic PDE are derived.

2. Shape sensitivity for parabolic equations.

We provide here an example of the shape sensitivity analysis of stochastic parabolic equation.

Let there be given a family $\mathcal{O}_\varepsilon = T_\varepsilon(V)(\mathcal{O})$ of domains constructed in the following way. For a given vector field

$$V(\cdot,\cdot) \in C^1(0,\delta;C^2(R^n;R^n)), \tag{1}$$

where $\delta > 0$ is a given constant , we define a family $\{\mathcal{O}_s\} \in R^n$, $s \in [0,\delta)$, of domains as follows [14]

$$\mathcal{O}_s = T_s(V)(\mathcal{O}) = \{x \in R^n : \exists X \in \mathcal{O} \text{ such that } x(0) = X, x = x(s)\}, \tag{2}$$

where $x(s) \in R^n$, $s \in [0,\delta)$ is given by the unique solution of the following system

$$\frac{dx}{ds}(s) = V(s,x(s)), \;\; s \in (0,\delta) \tag{3}$$

with initial conditions $x(0) = X$.

Let $\Gamma_\varepsilon = \partial\mathcal{O}_\varepsilon$. Let us consider the following family of equations depending on real parameter $\varepsilon \in [0,\delta)$.

Equation (P_ε) :

$$\frac{\partial}{\partial t}y_\varepsilon - \Delta y_\varepsilon = f + \frac{\partial}{\partial t}M_\varepsilon$$

in $\mathcal{O}_\varepsilon \times (0,T)$,

$$y_\varepsilon = 0 \tag{4}$$

on $\Sigma_\varepsilon = \partial\mathcal{O}_\varepsilon \times (0,T)$ and

$$y_\varepsilon(0,x) = y_0(x) \text{ in } \mathcal{O}_\varepsilon.$$

Define the domain functional

$$J(\mathcal{O}_\varepsilon) = E\int_0^T\int_{\mathcal{O}_\varepsilon}(y_\varepsilon - z_d)^2 dxdt, \tag{5}$$

where $z_d \in L^2(\mathcal{O} \times [0,T];R)$.

Here we denote by y_0 the restrictions to $\Sigma_\varepsilon = \mathcal{O}_\varepsilon \times (0,T)$ of a given element $y_0 \in H^1(R^n)$, M denotes the restriction of the martingale

$$M(t) = \sum_1^m \varphi_i(x)W_i(t)$$

to the domain $\mathcal{O}_\varepsilon$, where W_i, $i = 1,\ldots,m$ are independent one-dimensional Wiener processes. The elements $\varphi_i \in H^1(R^n)$, $i = 1,\ldots,m$ and $f \in L^2(R \times [0,T])$ are given.

For problem (4) with homogeneous Dirichlet boundary condition we can apply the regularity results due to Da Prato, Kwapień and Zabczyk [4] and it follows that $y_\varepsilon \in L^2_a(\Omega\times$

$[0,T]; H^2(\mathcal{O}_\varepsilon))$, where $L^p_a(\Omega \times [0,T], Z)$ denotes the closed subspace of $L^p(\Omega \times [0,T], Z)$ of all progressively measurable stochastic processes with values in Z. We use the regularity to obtain the domain derivative of the equation (4).

Denote $\mathcal{O} = \mathcal{O}_0$ and $y = y_0$. Let y' solve the following equation:

$$\frac{\partial}{\partial t} y' - \Delta y' = 0$$

in $\mathcal{O} \times (0,T)$,

$$y' = V \cdot n \frac{\partial y}{\partial n} \tag{6}$$

on $\Sigma = \partial\mathcal{O} \times (0,T)$ and

$$y'(0,x) = 0 \text{ in } \mathcal{O}.$$

Since $\frac{\partial y}{\partial n} \in L^2_a(\Omega \times [0,T]; H^{\frac{1}{2}}(\Gamma))$ then $y' \in L^2_a(\Omega \times [0,T]; H^1(\mathcal{O}))$.

Compute that

$$dJ(\mathcal{O}; V) = \lim_{\varepsilon \to 0} \varepsilon^{-1} [J(\mathcal{O}_\varepsilon) - J(\mathcal{O})] =$$

$$= E \int_0^T \int_{\mathcal{O}} (y - z_d) y' dx dt + +E \int_0^T \int_{\partial\mathcal{O}} (y - z_d)^2 V \cdot n d\Gamma dt.$$

To avoid the nonhomogeneous boundary conditions in (6) introduce the *adjoint state* as a unique solution of

$$-\frac{\partial}{\partial t} p - \Delta p = y - z_d$$

in $\mathcal{O} \times (0,T)$,

$$p = 0 \tag{7}$$

on $\Sigma = \partial\mathcal{O} \times (0,T)$ and

$$p(T,x) = 0 \text{ in } \mathcal{O}.$$

The equation (7) is understood in the sense of transposition i.e. for any $\varphi \in \mathcal{W} = \{\varphi \in L^2_a(\Omega \times [0,T]; H^1_0(\mathcal{O})) : \frac{\partial \varphi}{\partial t} \in L^2_a(\Omega \times [0,T]; H^{-1}(\mathcal{O}))$ and $\varphi(0) = 0\}$ it holds

$$E \int_0^T \int_{\mathcal{O}} (y - z_d) \varphi dx dt = E \int_0^T \int_{\mathcal{O}} p(\frac{\partial \varphi}{\partial t} - \Delta \varphi) dx dt.$$

Taking into consideration (7) we get

$$E \int_0^T \int_{\mathcal{O}} (y - z_d) y' dx dt = E \int_0^T \int_{\Gamma} \frac{\partial p}{\partial n} \frac{\partial y}{\partial n} V \cdot n d\Gamma dt.$$

Hence

$$dJ(\mathcal{O}; V) = E \int_0^T \int_{\Gamma} \Big(\frac{\partial p}{\partial n} \frac{\partial y}{\partial n} + (y - z_d)^2 \Big) V \cdot n d\Gamma dt. \tag{8}$$

3. Shape sensitivity for hyperbolic equations.

We provide here an example of the shape sensitivity analysis of stochastic hyperbolic equation.

Let the families $\mathcal{O}_\varepsilon = T_\varepsilon(V)(\mathcal{O})$ of domains be constructed as in Section 2. Let us consider the following family of hyperbolic equations depending on real parameter $\varepsilon \in [0, \delta)$.

Equation (H_ε) :

$$\frac{\partial^2}{\partial t^2} y_\varepsilon - \Delta y_\varepsilon = f + \frac{\partial}{\partial t} M_\varepsilon$$

in $\mathcal{O}_\varepsilon \times (0, T)$,

$$y_\varepsilon = 0 \tag{9}$$

on $\Sigma_\varepsilon = \partial \mathcal{O}_\varepsilon \times (0, T)$,

$$\frac{\partial}{\partial t} y_\varepsilon(0, x) = y_1(x) \text{ in } \mathcal{O}_\varepsilon$$

and

$$y_\varepsilon(0, x) = y_0(x) \text{ in } \mathcal{O}_\varepsilon.$$

Define the domain functional analogously as in Section 2

$$J(\mathcal{O}_\varepsilon) = E \int_0^T \int_{\mathcal{O}_\varepsilon} (y_\varepsilon - z_d)^2 dx dt, \tag{10}$$

where $z_d \in L^2(\mathcal{O} \times [0, T]; R)$.

Here we denote by y_0 and y_1 the restrictions to $\Sigma_\varepsilon = \mathcal{O}_\varepsilon \times (0, T)$ of given elements $y_0 \in H^2(R^n)$ and $y_1 \in H^1(R^n)$, M denotes the restriction of the martingale

$$M(t) = \sum_1^m \varphi(x) W_i(t)$$

to the domain $\mathcal{O}_\varepsilon$, where W_i, $i = 1, \ldots, m$ are independent one-dimensional Wiener processes. The elements f, $\varphi_i \in H^1(R)$, $i = 1, \ldots, m$ are given. For problem (9) with homogeneous Dirichlet boundary condition we can apply the regularity [4,5] and it follows that $y_\varepsilon \in L^2_a(\Omega; C(0, T; H^2(\mathcal{O}_\varepsilon)) \cap C^1(0, T; H^1(\mathcal{O}_\varepsilon)))$. We use the regularity to obtain the domain derivative of the equation (9).

Let y' solve the following equation:

$$\frac{\partial^2}{\partial t^2} y' - \Delta y' = 0$$

in $\mathcal{O} \times (0, T)$,

$$y' = V \cdot n \frac{\partial y}{\partial n} \tag{11}$$

on $\Sigma = \partial \mathcal{O} \times (0, T)$,

$$y'(0, x) = 0 \text{ in } \mathcal{O}$$

and

$$\frac{\partial}{\partial t}y'(0,x) = 0 \text{ in } \mathcal{O}.$$

Since $\frac{\partial y}{\partial n} \in L^2_a(\Omega; C(0,T; H^{\frac{1}{2}}(\Gamma)))$ then $y' \in L^2_a(\Omega; C(0,T; H^2(\mathcal{O}_\varepsilon)) \cap C^1(0,T; H^1(\mathcal{O}_\varepsilon)))$.

Compute that

$$dJ(\mathcal{O};V) = \lim_{\varepsilon \to 0} \varepsilon^{-1}[J(\mathcal{O}_\varepsilon) - J(\mathcal{O})] =$$

$$= E\int_0^T \int_{\mathcal{O}} (y - z_d)y' dxdt + +E\int_0^T \int_{\partial\mathcal{O}} (y - z_d)^2 V \cdot n d\Gamma dt.$$

To avoid the nonhomogeneous boundary conditions in (11) introduce the *adjoint state* in the following way. Let $\tilde{p}$ be a unique solution of

$$\frac{\partial^2}{\partial t^2}\tilde{p} - \Delta\tilde{p} = y - z_d$$

in $\mathcal{O} \times (0,T)$,

$$\tilde{p} = 0 \tag{12}$$

on $\Sigma = \partial\mathcal{O} \times (0,T)$,

$$\tilde{p}(T,x) = 0 \text{ in } \mathcal{O}$$

and

$$\frac{\partial}{\partial t}\tilde{p}(T,x) = 0 \text{ in } \mathcal{O}.$$

Let $p(t)$ be a regular version of $E(\tilde{p}(t)|\mathcal{F}_t)$. Compute that for any

$$\varphi \in \{\varphi \in L^2_a(\Omega; C(0,T; H^1(\mathcal{O})) \cap C^2(0,T; H^1(\mathcal{O})')) \text{ and } \varphi(0) = \frac{\partial\varphi}{\partial t}(0) = 0\}$$

it holds

$$E\int_0^T \int_{\mathcal{O}} (y - z_d)\varphi dxdt == E\int_0^T \int_{\mathcal{O}} \varphi(\frac{\partial^2 \tilde{p}}{\partial t^2} - \Delta\tilde{p})dxdt =$$

$$= E\int_0^T \int_{\mathcal{O}} \tilde{p}(\frac{\partial\varphi}{\partial t} - \Delta\varphi)dxdt - E\int_0^T \int_\Gamma \frac{\partial\tilde{p}}{\partial n}\varphi d\Gamma dt =$$

$$= E\int_0^T \int_{\mathcal{O}} p(\frac{\partial\varphi}{\partial t} - \Delta\varphi)dxdt - E\int_0^T \int_\Gamma \frac{\partial p}{\partial n}\varphi d\Gamma dt. \tag{13}$$

Taking into consideration (12) we get

$$E\int_0^T \int_{\mathcal{O}} (y - z_d)y' dxdt = E\int_0^T \int_\Gamma \frac{\partial p}{\partial n}\frac{\partial y}{\partial n} V \cdot n d\Gamma dt.$$

Hence

$$dJ(\mathcal{O};V) = E\int_0^T \int_\Gamma \Big(\frac{\partial p}{\partial n}\frac{\partial y}{\partial n} + (y - z_d)^2\Big) V \cdot n d\Gamma dt. \tag{14}$$

4. REFERENCES.

[1] R.A. Adams,*Sobolev Spaces.* Academic Press, New York, 1975

[2] A. Bensoussan and B. Viot, Optimal control of stochastic linear distributed systems, *SIAM J. Control and Optimization* **13** 1975 p.904–926

[3] A. Bensoussan, *Methodes de Perturbation en Controle Stochastique*, Dunod, Paris 1988

[4] G. Da Prato, S. Kwapień and J. Zabczyk, Regularity of solutions of linear stochastic equations in Hilbert spaces, *Stochastics* **23** 1987 p.1–23

[5] D. Gątarek, On convergence of deterministic and stochastic hyperbolic equations, *Bolletino dell Unione Matematica Itatiana* (in press)

[6] D. Gątarek and J. Sokołowski, Shape sensitivity analysis of optimal control problems for stochastic parabolic equations. *Jornadas Hispano-Francesas sobre Control de Sistemas Distribuidos, Octubre 1990, editado por: Grupo de Analisis Matematico Aplicado de la Universidad de Malaga.* 1991 p.103–114.

[7] I. Lasiecka, J. Sokołowski, Sensitivity analysis of control constrained optimal control problem for wave equation. *SIAM Journal on Control and Optimization* **29** 1991 (in press)

[8] J.L. Lions and E. Magenes, *Problemes aux limites non homogenes.* Dunod, Paris, 1968

[9] M. Metivier, *Semimartingales, a Course on Stochastic Processes*, de Gruyter, Berlin 1982

[10] E. Pardoux, *Equations aux Derivees Particlles Stochastiques non Lineaires Monotonnes*, Thesis, Universite Paris IX, 1975

[11] J. Sokołowski, Shape sensitivity analysis of boundary optimal control problems for parabolic systems. *SIAM Journal on Control and Optimization.* **26** 1988 p.763–787.

[12] J. Sokołowski and J.P. Zolesio, Shape sensitivity analysis of unilateral problems. *SIAM J. Math. Anal.* **18** 1987 p.1416–1437

[13] J. Sokołowski and J.P. Zolesio, *Introduction to shape optimization. Shape sensitivity analysis.* Springer 1991 (in press).

[14] J.P. Zolesio, The material derivative (or speed) method for shape optimization. In: *Optimization of Distributed Parameter Structures,* E.J. Haug and J. Cea (Eds.) Sijthoff and Noordhoff, 1981 p.1089–1151

Boundary Controllability Problems for the Wave and Heat Equations

R. GLOWINSKI

University of Houston, Department of Mathematics
4800 Calhoun Road, Houston, TX 77004, USA,
University of Paris VI and INRIA, France

Abstract. The main goal of this note is to discuss in the spirit of the Hilbert Uniqueness Method of J.L. Lions the solution of boundary controllability problems for the wave and heat equations. After a discussion of the exact boundary controllability of the wave equation by Dirichlet controls we shall consider the (formal) application of HUM to the heat equation by Dirichlet controls, and show that some well known approximate boundary controllability techniques for the heat equations can be interpreted as regular or nondifferentable perturbations of the equation provided by HUM in the exact controllability case.

I. Introduction Synopsis.

The main objective of this article is to discuss, fairly briefly, and motivated by the computational aspects, the application of the *Hilbert Uniqueness Method* (HUM) of J.L. Lions to the *boundary controllability* of the *wave* and *heat equations*. Indeed, the *wave equation* is discussed first (in Section 2) because it has been shown in [1], [2], (see also the references therein) that HUM provides a systematic and constructive way to *exactly* control the wave equations, via *Dirichlet boundary controls*, assuming that the time interval on which one controls is sufficiently large. We consider next (in Section 3) a similar exact control problem for the *heat equation*; here HUM (in fact its variant known as RHUM) provides a fundamental equation which is not well posed in general. If instead of exact controllability, we consider *approximate controllability*, we show on two examples that we obtain *regular* or *nondifferentiable perturbations* of the above fundamental equation. The *numerical treatment* of these approximate problems will be discussed elsewhere.

2. Exact Boundary Controllability of the Wave Equation.

We follow essentially the presentation in [1], [2] (see also [3], [4] for a more computationally oriented discussion).

2.1 Formulation of the Exact Controllability Problem.

Let Ω be a bounded domain of $\mathbf{R}^N$; we denote by Γ the boundary of Ω and we assume Γ to be reasonably smooth. We consider then a time interval $(0,T)$ with $T>0$ and use then the following notation

$$Q=\Omega\times(0,T), \Sigma=\Gamma\times(0,T).$$

It is known (from, e.g., [1], [2]) that if one considers

(2.1) $$\Box\, y = 0 \ in\ Q,$$

(2.2) $$y = u \ on\ \Sigma,$$

(2.3) $$y(0) = y^0, y_t(0) = y^1,$$

with $y^0 \in L^2(\Omega), y^1 \in H^{-1}(\Omega)$, then for T *sufficiently large* there exist controls u in $L^2(\Sigma)$ such that

(2.4) $$y(T) = 0, y_t(T) = 0.$$

In (2.1) - (2.4) we have used the following notation

$$\Box = \frac{\partial^2}{\partial t^2} - \Delta, \Delta = \sum_{i=i}^{N} \frac{\partial^2}{\partial x_i^2}, \quad y(t) : x \to y(x,t), y_t = \frac{\partial y}{\partial t}$$

and $H^{-1}(\Omega)$ is the *dual space* of the Sobolev space $H_0^1(\Omega)$. The statement "T is *sufficiently large*" deserves a comment: indeed it follows from [1], [2] that T has to be strictly larger than $T_c \sim$ diam (Ω); this is not surprising since (2.1) describes waves propagating at *velocity one*. The Hilbert Uniqueness Method described in the next paragraph will select a boundary control which is optimal for some criterium.

2.2 The Fundamental Operator Λ.

Let us define space E by $E = H_0^1(\Omega) \times L^2(\Omega)$ and denote by E' the dual space $H^{-1}(\Omega) \times L^2(\Omega)$. Operator Λ is defined as follows:

(i) *Take* $\mathbf{e} = \{e^0, e^1\} \in E$.

(ii) *Integrate from* 0 *to* T.

(2.5.1) $$\Box\, \varphi = 0 \ in\ Q,$$

(2.5.2) $$\varphi = 0 \ on\ \Sigma,$$

(2.5.3) $$\varphi(0) = e^0, \varphi_t(0) = e^1.$$

It follows then from [5] *that* $\left.\dfrac{\partial y}{\partial n}\right|_\Sigma \in L^2(\Sigma)$.

(iii) *Integrate from* T *to* 0

(2.6.1) $$\Box\, \psi = 0 \ in\ Q,$$

(2.6,2) $$\psi = \frac{\partial y}{\partial n} \ on\ \Sigma,$$

(2.6.3) $$\psi(T) = 0, \psi_t(T) = 0.$$

(iv) *Define* Λ *by*

(2.7) $$\Lambda \mathbf{e} = \{\psi_t(0), -\psi(0)\}. \quad \blacksquare$$

It follows from [5] that $\Lambda e \in E'$. In fact, we have more since $\Lambda \in \mathcal{L}(E, E')$ and also (*symmetry*)

(2.8) $$< \Lambda \mathbf{e}, \mathbf{e}' >= \int_{\Sigma} \frac{\partial y}{\partial n} \frac{\partial y'}{\partial n} d\Gamma dt, \forall e, e' \in E,$$

where, in (2.8), $< \cdot, \cdot >$ denotes the *duality pairing* between E' and E. Now, it follows from [1], [2], that if $T > T_c \sim$ diam (Ω) then operator Λ satisfies

(2.9) $$< \Lambda \mathbf{e}, \mathbf{e} \geq \mathrm{c}||\mathbf{e}||_E^2, \forall \mathbf{e} \in E,$$

where, in (2.9), c is a *positive* constant. If (2.9) is satisfied, then Λ is an *isomorphism* from E onto E'.

2.3 Application to the Exact Boundary Controllability of the Wave Equation.

Let us define $\mathbf{f} \in E'$ by

(2.10) $$\mathbf{f} = \{y^1, -y^0\},$$

and solve

(2.11) $$\Lambda \mathbf{e} = \mathbf{f}.$$

If $T > T_c$, then (2.11) has a *unique solution.* Solve now the corresponding wave equation (2.5); we have then $\left.\frac{\partial y}{\partial n}\right|_{\Sigma} \in L^2(\Sigma)$ and the corresponding solution ψ of (2.6), in addition to $\psi(T) = 0$ and $\psi_t(T) = 0$, satisfies also $\psi(0) = y^0$ and $\psi_t(0) = y^1$; *we have thus achieved the exact boundary controllability, assuming that we take* $u = \left.\frac{\partial y}{\partial n}\right|_{\Sigma}$.

Remark 2.1. From the *symmetry* and *positivity* of Λ, equation (2.11) can be solved by a *conjugate gradient algorithm* operating in space E, such an algorithm is described in the next paragraph.

Remark 2.2. The boundary control constructed through HUM is *optimal* in the following sense: it is the *unique* solution of the following *optimal control problem* (with *constraints* on the *state function*):

$$\text{(OCP)} \qquad \min_{v \in \mathcal{U}_f} J(v),$$

where, in (OCP), we have

$$J(v) = \frac{1}{2}\int_{\Sigma} |v|^2 d\Gamma\, dt,$$

and

$$\mathcal{U}_f = \{v | v \in L^2(\Sigma), \Box y = 0 \text{ in } Q, y = v \text{ on } \Sigma, y(0) = y^0, y_t(0) = y^1, y(T) = 0, y_t(T) = 0\}.$$

Actually, problem $\Lambda \mathbf{e} = \mathbf{f}$ can be viewed as a *dual problem of* (OCP).

2.4 Conjugate Gradient Solution of $\Lambda \mathbf{e} = \mathbf{f}$.

Equation (2.11) can also be written in the following *variational form*

$$\text{(2.12)} \qquad \begin{cases} \mathbf{e} \in E, \\ < \Lambda \mathbf{e}, \mathbf{e}' > = < \mathbf{f}, \mathbf{e}' >, \forall \mathbf{e}' \in E. \end{cases}$$

Problem (2.12) is then a particular case of

$$\text{(2.13)} \qquad \begin{cases} u \in V, \\ a(u, v) = L(v), \forall\, v \in V, \end{cases}$$

where:

(i) V is a real *Hilbert space* for the scalar product $(\cdot,\cdot)$ and the corresponding norm $\|\cdot\|$.

(ii) $a : V \times V \to \mathsf{R}$ is *bilinear, continuous, symmetric*, and $V-$*elliptic* (i.e., $\exists \alpha > 0$ such that a $(v, v) \geq \alpha \|v\|^2, \forall v \in V$).

(iii) $L : V \to \mathsf{R}$ is *linear* and *continuous*.

If properties (i), (ii), (iii), hold, then problem (2.13) has a *unique solution* from the *Lax-Milgram theorem* (see [6, Appendix 1]).

Problem (2.13) can be solved by the following *conjugate gradient algorithm:*

$$\text{(2.14)} \qquad u^0 \in V \quad \textit{is given};$$

solve

$$\text{(2.15)} \qquad \begin{cases} g^0 \in V, \\ (g^0, v) = a(u^0, v) - L(v), \forall v \in V, \end{cases}$$

and set

$$\text{(2.16)} \qquad w^0 = g^0. \quad \blacksquare$$

For $n \geq 0, u^n, g^n, w^n$ being known, compute $u^{n+1}, g^{n+1}, w^{n+1}$ as follows
Compute,

(2.17)
$$\rho_n = \frac{||g^n||^2}{a(w^n, w^n)}$$

and take

(2.18)
$$u^{n+1} = u^n - \rho_n w^n.$$

Solve

(2.19)
$$\begin{cases} g^{n+1} \in V, \\ (g^{n+1}, v) = (g^n, v) - \rho_n a(w^n, v), \quad \forall v \in V, \end{cases}$$

and compute

(2.20)
$$\gamma_n = \frac{||g^{n+1}||^2}{||g^n||^2},$$

(2.21)
$$w^{n+1} = g^{n+1} + \gamma_n w^n.$$

Do $n = n + 1$ and go to (2.17). ■

Concerning the convergence of algorithm (2.14) - (2.21), it can be shown that

(2.22)
$$||u^n - u|| \leq c||u^0 - u||\left(\frac{\sqrt{\nu_a} - 1}{\sqrt{\nu_a} + 1}\right)^n,$$

where u is the solution of (2.13), and where the *condition number* ν_a of $a(\cdot, \cdot)$ is defined by

$$\nu_a = ||A|| \, ||A^{-1}||,$$

where A is the *unique* operator in $\mathcal{L}(V, V)$ defined by

$$a(v, w) = (Av, w), \quad \forall v, w \in V.$$

Application to the solution of problem (2.11), (2.12)

Problem (2.12), being a particular case of (2.13), can be solved by the conjugate gradient algorithm (2.14) - (2.21). We think it is worth to describe the resulting algorithm here:

Initialization

(2.23) $$\mathbf{e}_0 = \{e_0^0, e_0^1\} \in E \text{ is given.}$$

Solve from 0 to T

(2.24.1) $$\Box\, \varphi_0 = 0 \text{ in } Q,$$
(2.24.2) $$\varphi_0 = 0 \text{ on } \Sigma,$$
(2.24.3) $$\varphi_0(0) = e_0^0, \quad \frac{\partial \varphi_0}{\partial t}(0) = e_0^1.$$

Solve from T to 0

(2.25.1) $$\Box\, \psi_0 = 0 \text{ in } Q,$$
(2.25.2) $$\psi_0 = \frac{\partial \varphi_0}{\partial n} \text{ on } \Sigma,$$
(2.25.3) $$\psi_0(T) = 0, \quad \frac{\partial \psi_0}{\partial t}(T) = 0.$$

Solve

(2.26) $$-\Delta g_0^0 = \frac{\partial \psi_0}{\partial t}(0) - y^1, g_0^0 \in H_0^1(\Omega),$$

and set

(2.27) $$g_0^1 = y^0 - \psi_0(0),$$

(2.28) $$\mathbf{w}_0 = \{w_0^0, w_0^1\} = \{g_0^0, g_0^1\}. \quad \blacksquare$$

Then, for $0 \geq n$, *assuming that* $\mathbf{e}_n, \varphi_n, \psi_n, \mathbf{g}_n, \mathbf{w}_n$ *are known we compute* $\mathbf{e}_{n+1}, \varphi_{n+1}, \psi_{n+1}, \mathbf{g}_{n+1}, \mathbf{w}_{n+1}$ *as follows:*

Descent:

Solve from 0 *to* T

(2.29.1) $$\Box\, \bar{\varphi}_n = 0 \text{ in } Q,$$
(2.29.2) $$\bar{\varphi}_n = 0 \text{ on } \Sigma,$$
(2.29.3) $$\bar{\varphi}_n(0) = w_n^0, \frac{\partial \bar{\varphi}_n}{\partial t}(0) = w_n^1,$$

and then from T to 0

$$\Box \bar{\psi}_n = 0 \text{ in } Q, \tag{2.30.1}$$

$$\bar{\psi}_n = \frac{\partial \bar{\varphi}_n}{\partial n} \text{ on } \Sigma, \tag{2.30.2}$$

$$\bar{\psi}_n(T) = 0, \frac{\partial \bar{\psi}_n}{\partial t}(T) = 0. \tag{2.30.3}$$

Compute now $\bar{\mathbf{g}}_n \in E$ *by*

$$-\Delta \bar{g}_n^0 = \frac{\partial \bar{\psi}_n}{\partial t}(0), \quad \bar{g}_n^0 \in H_0^1(\Omega), \tag{2.31}$$

$$\bar{g}_n^1 = -\bar{\psi}_n(0), \tag{2.32}$$

and then ρ_n *by*

$$\rho_n = \frac{||\mathbf{g}_n||_E^2}{< \Lambda \mathbf{w}_n, \mathbf{w}_n >} = \frac{\int_\Omega |\nabla g_n^0|^2 dx + \int_\Omega |g_n^1|^2 dx}{\int_\Omega \nabla \bar{g}_n^0 \cdot \nabla w_n^0 dx + \int_\Omega \bar{g}_n^1 w_n^1 dx}. \tag{2.33}$$

Once ρ_n *is known, compute*

$$\mathbf{e}_{n+1} = \mathbf{e}_n - \rho_n \mathbf{w}_n, \tag{2.34}$$

$$\varphi_{n+1} = \varphi_n - \rho_n \bar{\varphi}_n, \tag{2.35}$$

$$\psi_{n+1} = \psi_n - \rho_n \bar{\psi}_n, \tag{2.36}$$

$$\mathbf{g}_{n+1} = \mathbf{g}_n - \rho_n \bar{\mathbf{g}}_n. \tag{2.37}$$

Construction of the new descent direction

Compute

$$\gamma_n = \frac{||\mathbf{g}_{n+1}||_E^2}{||\mathbf{g}_n||_E^2}. \tag{2.38}$$

and set

$$\mathbf{w}_{n+1} = \mathbf{g}_{n+1} + \gamma_n \mathbf{w}_n. \quad \blacksquare \tag{2.39}$$

Do $n = n + 1$ *and go to* (2.29).

Algorithm (2.23) - (2.39) looks complicated but is in fact quite easy to implement. The computer implementation of the above algorithm to solve –approximately– equation $\Lambda \mathbf{e} = \mathbf{f}$ i.e. to achieve exact controllability, has been discussed with many details in [3], [4], [7], [8], (*finite difference* and *conforming finite element* approximations of $\Lambda \mathbf{e} = \mathbf{f}$) and in [9] (*mixed finite element* approximations). The corresponding numerical results are quite good.

3. Boundary Controllability of the Heat Equation.

In this section, we discuss the boundary controllability of the *heat equation*, first *exact controllability* and then two cases of *approximate boundary controllability*. We follow the approach taken in Section 2 where we discussed the boundary controllability of the wave equation.

3.1 Formulation of the Exact Controllability Problem.

Using the notation of Section 2, we look for a *boundary control* u such that he solution y of the *heat equation*

$$y_t - \Delta y = 0 \ in \ Q, \tag{3.1}$$

$$y = u \ on \ \Sigma, \tag{3.2}$$

$$y(0) = y_0 \tag{3.3}$$

will satisfy

$$y(T) = y_T \tag{3.4}$$

with y_0 and y_T given in $L^2(\Omega)$ and $H^{-1}(\Omega)$, respectively.

Among those Dirichlet controls, we look for the one (necessarily *unique* if it exists) which is of *minimal norm* on $L^2(\Sigma)$ i.e., is solution of the minimization problem:

$$\min_{v \in \mathcal{U}_f} J(v) \tag{3.5}$$

with, in (3.5),

$$J(v) = \frac{1}{2}\int_\Sigma |v|^2 d\Gamma dt, \tag{3.6}$$

$$\mathcal{U}_f = \{v | v \in L^2(\Sigma), \ the \ pair \ \{v, y\} \ satisfies \ (3.1) - (3.4)\}. \tag{3.7.}$$

3.2. Derivation of the Optimality Conditions for Problem (3.5) - (3.7).

To derive the *optimality conditions* satisfied by the solution of problem (3.5)-(3.7), we *approximate* this last problem by the following one

$$\min_{v \in L^2(\Sigma)} J_\epsilon(v), \tag{3.8}$$

with $J_\varepsilon(\cdot)$ defined (with $\varepsilon > 0$) by

$$J_\varepsilon(v) = \frac{1}{2}\int_\Sigma |v|^2 d\Gamma dt + \frac{1}{2\varepsilon}||y(T) - y_T||^2_{H^{-1}(\Omega)}, \tag{3.9}$$

where, in (3.9), y is the solution of (3.1)-(3.3). The $H^{-1}(\Omega)$ norm will be denoted by $||\cdot||_{-1}$ in the following and is classically defined by :

$$||f||_{-1} = \left(\int_\Omega |\nabla\varphi|^2 dx\right)^{\frac{1}{2}} \text{ with } \varphi \in H_0^1(\Omega) \text{ solution of } -\Delta\varphi = f.$$

It follows from, e.g., [10] that problem (3–8) has a unique solution, denoted by u_ε and that the *optimal pair* $\{u_\varepsilon, y_\varepsilon\}$ is characterized by the existence of p_ε such that

$$\frac{\partial y_\varepsilon}{\partial t} - \Delta y_\varepsilon = 0 \; in \; Q, \tag{3.10}$$

$$y_\varepsilon = u_\varepsilon \; on \; \Sigma, \tag{3.11}$$

$$y_\varepsilon(0) = y_0, \tag{3.12}$$

$$-\frac{\partial p_\varepsilon}{\partial t} - \Delta p_\varepsilon = 0 \; in \; Q, \tag{3.13}$$

$$p_\varepsilon = 0 \; on \; \Sigma, \tag{3.14}$$

$$p_\varepsilon(T) \in H_0^1(\Omega) \; and \; is \; the \; solution \; of \; -\varepsilon\Delta p_\varepsilon(T) = y_\varepsilon(T) - y_T, \tag{3.15}$$

$$u_\varepsilon = \frac{\partial p_\varepsilon}{\partial n} \; on \; \Sigma. \tag{3.16}$$

We shall go back to the optimality system (3.10) - (3.16) in a while; for the moment we shall use it to derive optimality conditions for problem (3.5) - (3.7). Assuming that problem (3.5) - (3.7) has a solution, it can be shown that this solution will satisfy the optimality conditions obtained by taking the limit in (3.10) -(3.16) as $\varepsilon \to 0$; we obtain then

$$\frac{\partial y}{\partial t} - \Delta y = 0 \text{ in } Q, \tag{3.17}$$

$$y = u \text{ on } \Sigma, \tag{3.18}$$

$$y(0) = y_0, \tag{3.19}$$

$$y(T) = y_T, \tag{3.20}$$

$$-\frac{\partial p}{\partial t} - \Delta p = 0 \text{ in } Q, \tag{3.21}$$

$$p = 0 \text{ on } \Sigma, \tag{3.22}$$

$$u = \frac{\partial p}{\partial n} \text{ on } \Sigma. \tag{3.23}$$

3.3 An HUM Related Formulation of Problem (3.5) - (3.7).

We follow the approach taken in Section 2 for the *wave equation.* We define first an operator $\Lambda \in \mathcal{L}(H_0^1(\Omega), H^{-1}(\Omega))$ as follows:

$$\Lambda e = -\psi(T) \quad , \forall e \in H_0^1(\Omega), \tag{3.24}$$

where $\psi(T)$ is obtained via

$$-\varphi_t - \Delta\varphi = 0 \text{ in } Q, \tag{3.25.1}$$

$$\varphi = 0 \text{ on } \Sigma, \tag{3.25.2}$$

$$\varphi(T) = e, \tag{3.25.3}$$

$$\psi_t - \Delta\psi = 0 \text{ in } Q, \tag{3.26.1}$$

$$\psi = \frac{\partial \varphi}{\partial n} \text{ on } \Sigma, \tag{3.26.2}$$

$$\psi(0) = 0. \tag{3.26.3}$$

We clearly have, $\forall e, e' \in H_0^1(\Omega)$,

$$\begin{aligned} < \Lambda e, e' > &= - < \psi(T), \varphi'(T) > = - \int_Q \frac{\partial}{\partial t}(\psi\varphi') dx\, dt \\ &= - \int_Q \left(\frac{\partial \psi}{\partial t}\varphi' + \frac{\partial \varphi'}{\partial t}\psi\right) dx\, dt = \int_Q (\Delta\varphi'\psi - \Delta\psi\varphi') dx\, dt \\ &= \int_\Sigma \left(\frac{\partial \varphi'}{\partial n}\psi - \frac{\partial \psi}{\partial n}\varphi'\right) d\Gamma\, dt = \int_\Sigma \frac{\partial \varphi}{\partial n}\frac{\partial \varphi'}{\partial n} d\Gamma\, dt, \end{aligned}$$

i.e.,

$$(3.27) \qquad < \Lambda e, e' >= \int_\Sigma \frac{\partial \varphi}{\partial n}\frac{\partial \varphi'}{\partial n} d\Gamma dt, \forall e, e' \in H_0^1(\Omega).$$

Above, $< \cdot, \cdot >$ denotes the duality pairing between $H^{-1}(\Omega)$ and $H_0^1(\Omega)$. It follows from (3.27) that Λ is *self-adjoint* and *positive semi-definite;* in fact, Λ is *positive definite.*
Next, we define $Y_0 \in L^2(0,T;H_0^1(\Omega)) \cap C^0([0,T];L^2(\Omega))$ as the *unique* solution of

$$(3.28.1) \qquad \frac{\partial}{\partial t}Y_0 - \Delta Y_0 = 0 \ in \ Q,$$

$$(3.28.2) \qquad Y_0 = 0 \ on \ \Sigma,$$

$$(3.28.3) \qquad Y_0(0) = y_0.$$

By subtracting the relations (3.28) to the corresponding ones in (3.17) - (3.19), we obtain after comparing to (3.25), (3.26) that $p(T)$ satisfies

$$(3.29) \qquad \Lambda p(T) = Y_0(T) - y_T.$$

Therefore, *from a formal point of view,* we can obtain the exact boundary control of minimal norm in $L^2(\Sigma)$ by solving

$$(3.30) \qquad \Lambda e = Y_0(T) - y_T,$$

then by taking

$$(3.31) \qquad u = \left.\frac{\partial \varphi}{\partial n}\right|_\Sigma,$$

where φ is the solution of (3.25) associated to the solution e of (3.30).
We insist on the fact that problem (3.30) is *not well posed in general.*

Remark 3.1. We observe that to construct the operator Λ defined by (3.24)–(3.26) we have been performing the two time integrations, namely (3.25) and (3.26), from T to 0 and then from 0 to T, respectively. The fact that this is opposite to what has been done in Section 2, to define the operator Λ associated to the wave equation, justifies the terminology *Reverse Hilbert Uniqueness Method* (R.H.U.M.) used in [2].

The various regularized variants of the above controllability problem can be solved by iterative methods closely related to the algorithm below.

Writing (3.30) in variational form, we obtain

$$(3.32) \qquad \begin{cases} e \in H_0^1(\Omega), \\ < \Lambda e, e' >=< Y_0(T) - y_T, e' >, \forall e' \in H_0^1(\Omega). \end{cases}$$

From the *symmetry* and *positive definiteness* of the bilinear form

$$\{e, e'\} \rightarrow < \Lambda e, e' >,$$

the variational problem (3.32) is closely related to the general problem (2.13), whose conjugate gradient solution has been discussed in Section 2.4.

Applying algorithm (2.14) - (2.21) to the particular problem (3.30), (3.32) we obtain the following iterative technique:

$$(3.33) \qquad e^0 \textit{ is given in } H_0^1(\Omega);$$

solve first,

$$(3.34.1) \qquad -\frac{\partial p^0}{\partial t} - \Delta p^0 = 0 \textit{ in } Q,$$

$$(3.34.2) \qquad p^0 = 0 \textit{ on } \Sigma,$$

$$(3.34.3) \qquad p^0(T) = e^0,$$

and set

$$(3.35) \qquad u^0 = \left.\frac{\partial p^0}{\partial n}\right|_{\Sigma}.$$

Solve now

$$(3.36.1) \qquad \frac{\partial y^0}{\partial t} - \Delta y^0 = 0 \textit{ in } Q,$$

$$(3.36.2) \qquad y^0 = u^0 \textit{ on } \Sigma,$$

$$(3.36.3) \qquad y^0(0) = y^0.$$

Solve finally the following Dirichlet problem:

$$(3.37) \qquad \begin{cases} g^0 \in H_0^1(\Omega), \\ \int_\Omega \nabla g^0 \cdot \nabla v dx =< y_T - y^0(T), v >, \forall v \in H_0^1(\Omega), \end{cases}$$

and set

$$(3.38) \qquad w^0 = g^0. \quad \blacksquare$$

Then, for $n \geq 0$, assuming that e^n, g^n, w^n, are known, we compute $e^{n+1}, g^{n+1}, w^{n+1}$ as follows:
Solve

$$-\frac{\partial \overline{p}^n}{\partial t} - \Delta \overline{p}^n = 0 \text{ in } Q, \tag{3.39.1}$$

$$\overline{p}^n = 0 \text{ on } \Sigma, \tag{3.39.2}$$

$$\overline{p}_n(T) = w^n, \tag{3.39.3}$$

and set

$$\overline{u}^n = \left.\frac{\partial \overline{p}^n}{\partial n}\right|_{\Sigma}. \tag{3.40}$$

Solve now

$$\frac{\partial \overline{y}^n}{\partial t} - \Delta \overline{y}^n = 0 \text{ in } Q, \tag{3.41.1}$$

$$\overline{y}^n = \overline{u}^n \text{ on } \Sigma, \tag{3.41.2}$$

$$\overline{y}^n(0) = 0, \tag{3.41.3}$$

and then

$$\begin{cases} \overline{g}^n \in H_0^1(\Omega) \\ \int_\Omega \nabla \overline{g}^n \cdot \nabla v dx = - < \overline{y}^n(T), v >, \forall\, v \in H_0^1(\Omega). \end{cases} \tag{3.42}$$

Compute

$$\rho_n = \frac{\int_\Omega |\nabla g^n|^2 dx}{< \Lambda w^n, w^n >} \left(= - \frac{\int_\Omega |\nabla g^n|^2 dx}{< \overline{y}^n(T), w^n >} = \frac{\int_\Omega |\nabla g^n|^2 dx}{\int_\Omega \nabla \overline{g}^n \cdot \nabla w^n dx}\right), \tag{3.43}$$

and then

$$e^{n+1} = e^n - \rho_n w^n, \tag{3.44}$$

$$g^{n+1} = g^n - \rho_n \overline{g}^n. \tag{3.45}$$

If $||g^{n+1}||_{H_0^1(\Omega)}/||g^0||_{H_0^1(\Omega)} \leq \varepsilon$, *take $e = e^{n+1}$ and solve* (3.25) *to obtain $u = \left.\frac{\partial \varphi}{\partial n}\right|_{\Sigma}$; if the above stopping test is not satisfied, compute*

$$\gamma_n = \frac{\int_\Omega |\nabla g^{n+1}|^2 dx}{\int_\Omega |\nabla g^n|^2 dx}, \tag{3.46}$$

and then

$$w^{n+1} = g^{n+1} + \gamma_n w^n. \quad \blacksquare \tag{3.47}$$

Do $n = n + 1$ and go to (3.39).
Numerical experiments with algorithm (3.33)–(3.47) will be reported elsewhere.

3.5 A First Approximate Boundary Controllability Problem.

The problem to be considered in this section is in fact problem (3.8), (3.9) (see Section 3.2), derived from problem (3.5)–(3.7) by *penalization* of the *exact* final condition (3.4). Consider the optimality system (3.10)–(3.16) and denote by e_ε the function $p_\varepsilon(T)$; it follows then from (3.15), and from the definition of Λ and Y_0 (cf. Section 3.3) that e_ε satisfies the following equation

$$-\varepsilon\Delta e_\varepsilon + \Lambda e_\varepsilon = Y_0(T) - y_T, \tag{3.48}$$

which is definitely a *(regular) perturbation* of equation (3.30). Equation (3.48) can be solved by a *conjugate gradient algorithm* which is nothing but a simple variant of the above algorithm (3.33)–(3.47).

3.6 A Second Approximate Boundary Controllability Problem.

Following J.L. Lions [11], we consider the variation of the regularized problem (3.8), (3.9) defined, for $\varepsilon > 0$, by

$$\min_{v\in K_\varepsilon} J(v), \tag{3.49}$$

where in (3.49) we have

$$J(v) = \frac{1}{2}\int_\Sigma v^2 d\Gamma\, dt, \tag{3.50}$$

$$K_\varepsilon = \{v | v \in L^2(\Sigma), \|y(T) - y_T\|_{H^{-1}(\Omega)} \le \varepsilon\}. \tag{3.51}$$

with, in (3.51), y the solution of

$$y_t - \Delta y = 0 \text{ in } Q, \tag{3.52}$$

$$y = v \text{ on } \Sigma, \tag{3.53}$$

$$y(0) = y_0. \tag{3.54}$$

Problem (3.49) has a *unique* solution, which converges as $\varepsilon \to 0$, to the solution of problem (3.5)–(3.7), if this last solution exists. The major difficulty with problem (3.49) is the fact that the state function y has to satisfy the inequality in (3.51).

Using *convex duality* techniques we can show that the *unique* solution of the control problem (3.49)–(3.51) is given by

$$u_\varepsilon = \frac{\partial p_\varepsilon}{\partial n} \text{ on } \Sigma, \tag{3.55}$$

where p_ε is the solution of the *backward heat equation*.

(3.56.1) $$-\frac{\partial p_\varepsilon}{\partial t} - \Delta p_\varepsilon = 0 \text{ in } Q,$$

(3.56.2) $$p_\varepsilon = 0 \text{ on } \Sigma,$$

(3.56.3) $$p_\varepsilon(T) = e_\varepsilon,$$

where e_ε is the solution of the following *variational inequality*

(3.57) $$\begin{cases} e_\varepsilon \in H_0^1(\Omega); \forall e \in H_0^1(\Omega) \text{ we have} \\ < \Lambda e_\varepsilon, e - e_\varepsilon > + \varepsilon(j(e) - j(e_\varepsilon)) \geq < Y_0(T) - y_T, e - e_\varepsilon >, \end{cases}$$

with, in (3.57), Λ and Y_0 as in Section 3.3, and the *convex functional* $j(\cdot)$ defined by

(3.58) $$j(e) = (\int_\Omega |\nabla e|^2 dx)^{\frac{1}{2}} = ||e||_{H_0^1(\Omega)}.$$

Numerical solution methods for variational inequalities closely related to (3.57) have been discussed in [12] (see also [6]); they include *regularization methods* where $j(\cdot)$ will be approximated by e.g.,

$$j_\gamma = \sqrt{\gamma^2 + \int_\Omega |\nabla e|^2 dx},$$

with γ a small number, and *duality* methods where one uses the fact that

$$j(e) = \max_{\mu \in B_2(0;1)} \int_\Omega \mu \cdot \nabla e \, dx,$$

where

$$B_2(0;1) = \{\mu | \mu \in (L^2(\Omega))^N, ||\mu||_{L^2(\Omega)^N} \leq 1\}.$$

Applications of these methods to the solution of (3.57) will be discussed elsewhere.

Remark 3.2. Equation (3.57) can also be written as

(3.59) $$\Lambda e_\varepsilon + \varepsilon \partial j(e_\varepsilon) = Y_0(T) - y_T,$$

where $\partial j(\cdot)$ denotes the subgradient of $j(\cdot)$. Equation (3.59) appears as a *nondifferentiable perturbation* of equation (3.30).

Remark 3.3. Taking $e = 0$ and $e = 2e_\varepsilon$ in (3.57) we can easily show that

(3.60) $$< \Lambda e_\varepsilon, e_\varepsilon > + \varepsilon(\int_\Omega |\nabla e_\varepsilon|^2 dx)^{\frac{1}{2}} = < Y_0(T) - y_T, e_\varepsilon > .$$

Since

$$(3.61) \qquad < Y_0(T) - y_T, e_\epsilon > \leq ||Y_0(T) - y_T||_{-1} \left(\int_\Omega |\nabla e_\epsilon|^2 dx \right)^{\frac{1}{2}},$$

it follows from (3.60) that

$$(3.62) \qquad < \Lambda e_\epsilon, e_\epsilon > \leq (||Y_0(T) - y_T||_{-1} - \epsilon) \left(\int_\Omega |\nabla e_\epsilon|^2 dx \right)^{\frac{1}{2}}.$$

Since operator Λ is a *positive* one, it follows from (3.62) that if $||Y_0(T) - y_T||_{-1} \leq \epsilon$, then $e_\epsilon = 0$ is the solution of the regularized problem (3.57); this implies in turn that the corresponding boundary control is $u_\epsilon = 0$. This was obvious from the beginning if $||Y_0(T) - y_T||_{-1} \leq \epsilon$.

Acknowledgements: The author would like to thank J.L. Lions for various discussions concerning the control problems addressed in this article. The support of NSF (Grant INT 8612680) is also knowledged. Finally, we would like to thank Vicki Kulik for processing this article.

References

[1] J.L. Lions, Exact controllability, stabilization and perturbations for distributed systems, *SIAM Review*, **30**, (1988), pp. 1–68.

[2] J.L. Lions, *Controlabilité exacte, perturbation et stabilisation des systèmes distribués*, **Vol. 1 and 2**, Masson, Paris, 1988.

[3] E.J. Dean, R. Glowinski, C.H. Li, Supercomputer solution of partial differential equation problems in Computational Fluid Dynamics and in Control, *Computer Physics Communications*, **53**, (1989), pp. 401–439.

[4] R. Glowinski, C.H. Li, J.L. Lions, A numerical approach to the exact controllability of the wave equation (I) Dirichlet controls: Description of the numerical methods, *Japan J. of Applied Mathematics*, **7**, (1990), pp. 1–76.

[5] J.L. Lions, *Controle des Systèmes Distribués Singuliers*, Dunod, Paris, 1983.

[6] R. Glowinski, *Numerical Methods for Nonlinear Variational Problems*, Springer, New York, 1984.

[7] R. Glowinski, C.H. Li, On the numerical implementation of the Hilbert Uniqueness Method for the exact boundary controllability of the wave equation, *C.R. Acad. Sc., Paris*, **T. 311**, Série I, (1990), pp. 135–142.

[8] R. Glowinski, Contribution of the Stokes problem to the exact controllability of the wave equation (to appear).

[9] R. Glowinski, W. Kinton, M.F. Wheeler, A mixed finite element formulation for the boundary controllability of the wave equation, *Int. J. Num. Meth. Eng.*, **27**, (1989), pp. 623–635.

[10] J.L. Lions, *Optimal Control of Systems Governed by Partial Differential Equations*, Springer–Verlag, New York, 1971.

[11] J.L. Lions, Exact Controllability for Distributed Systems. Some Trends and Some Problems, in *Applied and Industrial Mathematics*, R. Spigler, ed., Kluwer, Dordretch, 1991, pp. 59–84.

[12] R. Glowinski, J.L. Lions, R. Tremolieres, *Numerical Analysis of Variational Inequalities*, North–Holland, Amsterdam, 1981.

ABOUT CRITICAL POINTS OF THE ENERGY IN AN ELECTROMAGNETIC SHAPING PROBLEM

by Antoine HENROT and Michel PIERRE

Département de Mathématiques, B.P. 239
Université de Nancy I - 54506 VANDOEUVRE-lès-Nancy
URA CNRS 0750 et INRIA-Lorraine, Projet NUMATH

1. INTRODUCTION

This note presents some remarks about the critical points of the energy functional in some electromagnetic problem. Part of it is motivated by the results in [5], [6].

The energy functional under consideration is given here by

$$(1.1)\qquad E(\Omega) = \frac{1}{2}\int_\Omega |\nabla\psi_\Omega|^2 - \int_\Omega j_0\,\psi_\Omega = -\frac{1}{2}\int_\Omega |\nabla\psi_\Omega|^2 = -\frac{1}{2}\int_\Omega j_0\,\psi_\Omega$$

where ψ_Ω is the solution of the problem

$$(1.2)\qquad \begin{cases} -\Delta\psi_\Omega = j_0 \text{ in } \mathcal{D}'(\Omega) \\ \psi_\Omega = 0 \text{ on } \partial\Omega \\ \psi_\Omega \text{ is bounded in } \Omega \end{cases}$$

and where Ω is the complement of a compact set ${}^c\Omega$ in $\mathbb{R}^2$ such that the area of ${}^c\Omega$ is prescribed and j_0 is a given (regular enough) function with compact support in Ω. In the corresponding electromagnetic problem, j_0 is a given distribution of current, Ω is the complement of the domain occupied by a liquid metal, ψ_Ω is the potential of the total magnetic induction field (see [8], [9]). The shape $\partial\Omega$ assumed by the boundary of the liquid corresponds to a critical point of the energy $E(\Omega)$ with the constraint that the area of the complement ${}^c\Omega$ of Ω (that is to say the surface of the liquid) is prescribed. The equilibrium set of equations is (1.2) plus the following "nonlinear" extra condition defining the boundary $\partial\Omega$

$$(1.3)\qquad |\nabla\psi_\Omega|^2 = \text{constant on } \partial\Omega.$$

Here we assume that the surface tension of the liquid is zero. We refer to [1], [2], [9], [10], [16] for more details and to [15], [5], [20] for variational formulations of (1.2), (1.3).

We are interested here at the same time in the stability (or instability) of equilibrium shapes and in the existence (or non existence) of a global minimum for the functional $E(\Omega)$.

By stability of the equilibrium defined by (1.2), (1.3), we mean that $E(.)$ has a strict local minimum at Ω. More precisely, for $u: \partial\Omega \to \mathbb{R}$, we consider the open set Ω_u defined by perturbation of $\partial\Omega$ in the normal direction $\vec{n}$ to $\partial\Omega$ according to

$$(1.5) \qquad \partial\Omega_u = \{\xi + \vec{n}(\xi)\, u(\xi)\}.$$

Then, formally, the equilibrium shape Ω is said to be stable (following [15], [16], [5]) if

$$(1.6) \qquad E(\Omega) < E(\Omega_u)$$

for u small enough, $u \neq 0$, and measure $({}^c\Omega_u)$ = measure $({}^c\Omega)$. As shown in [6], [5] in a suitable framework, if one sets $K(u) = E(\Omega_u)$, this requires that

$$(1.7) \qquad \begin{cases} K''(0)\,(u, u) \geq 0 \\ \text{for all } u : \partial\Omega \to \mathbb{R} \text{ with } \int_{\partial\Omega} u = 0, \end{cases}$$

the condition $\int_{\partial\Omega} u = 0$ corresponding to the constraint on the surface of ${}^c\Omega_u$. But, as proved in [6] (see also similar computations in [18] in a more general framework)

$$(1.8) \qquad K''(0)\,[u, u] = Q(u)$$

with

$$Q(u) = \int_{\partial\Omega} u\, C_e u - \sigma u^2$$

where C_c is the pseudo-differential operator on $\partial\Omega$ defined by $C_e u = \frac{\partial U}{\partial n}$ on $\partial\Omega$ where U is the bounded harmonic function on Ω such that $U = u$ on $\partial\Omega$.

Our goal here is to first study the positivity of Q on hyperplanes of the form $\int_{\partial\Omega} hu = 0$ where $h: \partial\Omega \to \mathbb{R}$ is given. In particular, we give a slightly more direct proof of the instability result obtained in [6] in the case $h \equiv 1$ (we nevertheless follow ideas introduced in [5], [6]). As a consequence of this instability, no local "regular" strict minimum exists for $E(\Omega)$ (we refer to Section 2 for the precise meaning of "regular" and for remarks about interpretation of this result).

Next, we study the existence of a global minimum for $E(\Omega)$ over all open sets Ω (without regularity) such that measure $({}^c\Omega)$ is given. We prove that the infimum of $E(.)$ is finite if and only if $\int_{\mathbb{R}^2} j_o = 0$. Then we prove that the global infimum is approached by sets Ω_n for which ${}^c\Omega_n$ "disappears" at infinity. This implies that relevant equilibria have to be found among local minima of the energy. It emphasizes the difference with the case of the "interior" shaping problem where Ω is bounded and its area is prescribed (see [3], [13]). Finally let us mention that the latter results carry over to the case of nonzero surface tension (see remarks at the end of Section 3).

2. THE NONPOSITIVITY RESULT

Let Γ be a Jordan curve and Ω its exterior. We assume that Γ is Lipschitz (see e.g. [7] for a definition). We denote by $\vec{n}$ its normal derivative (which is defined a.e. on Γ) directed toward the interior of Γ. We define the "capacity" operator C_e on the Sobolev space $H = H^{1/2}(\Gamma)$ by

(2.1) $\quad \forall\, u \in H,\ C_e u = \nabla U.\vec{n}_{|_\Gamma} \in H^{-1/2}(\Gamma) = H'$

where U is the unique solution of (see [12], [4]).

$$(2.2)\quad \begin{cases} U \in H^1(\Omega \cap B_R) & \text{for all } R,\ U(x) = O(1) \text{ as } |x| \to \infty \\ \Delta U = 0 & \text{on } \Omega \\ U = u & \text{on } \Gamma = \partial\Omega. \end{cases}$$

Let now

(2.3) $\quad \sigma \in L^\infty(\Gamma),\ h \in H'.$

We consider the following quadratic form on H

(2.4) $\quad \forall\, u \in H \quad Q(u) = \langle C_e u, u\rangle_{H' \times H} - \int_\Gamma \sigma u^2.$

Since Γ is assumed to be a Jordan curve, we introduce a conformal mapping ϕ which transforms the unit circle Γ_o into Γ and the exterior Ω_o of Γ_o into Ω. We know that ϕ' is bounded at infinity. We also assume it is bounded around Γ_o:

(2.5) $\quad \phi'$ is bounded on Ω_o.

It is known that (2.5) implies the existence of a radial limit for ϕ' on Γ_o (see e.g. [14]). We still denote it by ϕ', so that

(2.6) $\quad [e^{i\theta} \to \phi'(e^{i\theta})] \in L^\infty(\Gamma_o).$

In the following, for any $w: \Omega \to \mathbb{R}$ (or $w: \Gamma \to \mathbb{R}$) we set

(2.7) $\quad \hat{w}(\xi) = w \circ \phi(\xi)$

its corresponding function on Ω_o (or Γ_o). Then, according to (2.5), we have the following properties:

(2.8) $\quad \nabla w \in L^2(\Omega) \Leftrightarrow \nabla \hat{w} \in L^2(\Omega_o)$

(2.9) $w \in H^{1/2}(\Gamma) \Leftrightarrow \hat{w} \in H^{1/2}(\Gamma_o) = H_o$.

Moreover, by a straightforward change of variable, we have

Lemma 2.1 For all $u \in H$,

(2.10) $Q(u) = \hat{Q}(\hat{u}) = <C_e \hat{u}, \hat{u}>_{H_o' x H_o} - \int_{\Gamma_o} \hat{\sigma} \hat{u}^2$

where C_e denotes the capacity operator on Γ_o defined as in (2.1) with Γ replaced by Γ_o and

(2.11) $\hat{\sigma}(e^{i\theta}) = \sigma(\phi(e^{i\theta}))\, |\phi'(e^{i\theta})|$.

Remark: Here $C_e \hat{u}$ is equal, but for the sign, to the radial derivative of the harmonic extension of $\hat{u}$ to Ω_o.

We now state our nonpositivity result for $\hat{Q}$. Recall that $\hat{h} = h \circ \phi$ where $h \in H^{-1/2}(\Gamma)$ is given.

Theorem 2.2 Assume

(2.12) $\int_{\Gamma_o} \hat{\sigma} \geq 2\pi, \quad \int_{\Gamma_o} \hat{\sigma} \cos\theta = \int_{\Gamma_o} \hat{\sigma} \sin\theta = 0, \quad \hat{\sigma} \not\equiv 1.$

Then, there exists $\hat{u} \in H_o$ such that

(2.13) $<\hat{h}, \hat{u}>_{H_o' x H_o} = 0, \quad \hat{Q}(\hat{u}) < 0.$

Corollary 2.3 Assume Γ is of class C^2 and σ is the curvature of Γ (seen from the interior of Γ). Then, if Γ is not a circle, there exists $u \in H$ such that

(2.14) $<h, u>_{H' x H} = 0, \quad Q(u) < 0.$

Remark 2.4: The condition (2.14) says that the restriction of Q to the hyperplane defined as the kernel of $h \in H'$ ($h \not\equiv 0$) fails to be positive. This was proved in [6], [5] for $h \equiv 1$. Actually, it is true for all h, since it is more a property of Q itself which turns out to have at least two negative eigenvalues under above assumptions. The conditions in (2.12) for $\hat{\sigma}$ are satisfied if σ is the curvature of Γ and if Γ is not a circle (which corresponds to $\hat{\sigma} \equiv 1$). We will see later that almost anything can happen for the two first eigenvalues of $\hat{Q}$ (and therefore of Q) if they do not hold.

Remark 2.5: Corollary 2.3 says that equilibria of the shaping problem (1.2), (1.3) are unstable. However the proof requires that $\partial\Omega$ be a Jordan curve and that C^2- regularity holds. This does not generally happen if $\int j_o = 0$ (which corresponds to a natural physical situation) since cusps then appear in the "natural" equilibrium shapes at the points where the total magnetic field vanishes. This is due to the fact that the surface tension is zero.

Nevertheless, any analytic Jordan curve can be considered as an equilibrium shape for some j_o with $\int j_o \neq 0$ (see [8], [5]). Therefore, as noticed in [5], instability does occur for a significant family of solutions.

Proof of theorem 2.2

We first notice that for $\hat{u} \equiv 1$,

$$(2.15) \quad \hat{Q}(1) = - \int_{\Gamma_o} \hat{\sigma} < 0.$$

In particular, if $<\hat{h}, 1>_{H_o^{'} x H_o} = 0$, then (2.13) holds with $\hat{u} \equiv 1$.

We will now assume that

$$(2.16) \quad <\hat{h}, 1>_{H_o^{'} x H_o} \neq 0.$$

For all $n \geq 1$, the real and imaginary parts of the holomorphic function $z \to \frac{1}{z^n}$ on $\partial\Omega_o$ are respectively equal to $\cos n\theta$ and $-\sin n\theta$ on Γ_o. We deduce that

$$(2.17) \quad C_e (\cos n\theta) = -\frac{\partial}{\partial r}\left(\frac{\cos n\theta}{r^n}\right)\Big|_{r=1} = n \cos n\theta$$

$$(2.18) \quad C_e (\sin n\theta) = -\frac{\partial}{\partial r}\left(\frac{\sin n\theta}{r^n}\right)\Big|_{r=1} = n \sin n\theta.$$

In particular, using the first property in (2.12)

$$(2.19) \quad \hat{Q}(\cos\theta) + \hat{Q}(\sin\theta) = \int_{\Gamma_o} (1 - \hat{\sigma}) \leq 0.$$

Then we have one of the three following situations:

$$(2.20) \quad \hat{Q}(\cos\theta) < 0 \qquad \text{or} \qquad (2.20)\text{bis} \quad \hat{Q}(\sin\theta) < 0$$

or

$$(2.21) \quad \hat{Q}(\cos\theta) = \hat{Q}(\sin\theta) = 0.$$

Let us denote by $\hat{Q}(.,.)$ the bilinear form associated with $\hat{Q}$, namely, since C_e is symmetric

$$\forall \hat{u}, \hat{w} \in H_0, \hat{Q}(\hat{u}, \hat{w}) = <C_e \hat{u}, \hat{w}>_{H_0' x H_0} - \int_{\Gamma_0} \hat{\sigma} \hat{u} \hat{w} = \hat{Q}(\hat{w}, \hat{u}).$$

Since $\hat{Q}(\cos \theta, 1) = -\int \hat{\sigma} \cos \theta = 0$ and $\hat{Q}(1) < 0$, if $\hat{Q}(\cos \theta) < 0$, then the restriction of $\hat{Q}$ to the plane spanned by $\{1, \cos \theta\}$ in H_0 is negative. But its intersection with the hyperplane of function $\hat{u}$ in H_0 such that $<\hat{h}, \hat{u}>_{H_0' x H_0} = 0$ is of positive dimension. Therefore (2.13) holds for some $\hat{u}$ in this intersection.

We argue in the same way in the case (2.20)bis when $\hat{Q}(\sin \theta) < 0$.

Let us now assume (2.21). If $<\hat{h}, \cos \theta>_{H_0' x H_0} \neq 0$, then there exists a real number α such that

$$<\hat{h}, 1 + \alpha \cos \theta>_{H_0' x H_0} = 0.$$

Since $\hat{Q}(1 + \alpha \cos \theta) = \hat{Q}(1) < 0$, we obtain (2.13) with $\hat{u} = 1 + \alpha \cos \theta$.

If now $<\hat{h}, \cos \theta>_{H_0' x H_0} = 0$, we assume by contradiction that the restriction of $\hat{Q}$ to the hyperplane $(\hat{h})^{\perp}$ orthogonal to $\hat{h}$ in H_0 is nonnegative. Then, it reaches its minimum $(\hat{h})^{\perp}$ at $[\theta \to \cos \theta]$. Consequently

$$\hat{Q}(\hat{w}, \cos \theta) = 0 \text{ for all } \hat{w} \in (\hat{h})^{\perp}.$$

Since $\hat{Q}(1, \cos \theta) = 0$ and since $\hat{w} \equiv 1$ does not belong to $(\hat{h})^{\perp}$ by assumption (2.16), we even have

$$\hat{Q}(\hat{w}, \cos \theta) = 0 \text{ for all } \hat{w} \in H_0,$$

that is to say

$$\int_{\Gamma_0} \cos \theta \, (1 - \hat{\sigma}) \, \hat{w} = 0, \forall \hat{w} \in H^{1/2}(\Gamma_0).$$

This implies $\hat{\sigma} \equiv 1$ which contradicts the last assumption in (2.12). Therefore, the restriction of $\hat{Q}$ to $(\hat{h})^{\perp}$ cannot be nonnegative, whence the existence of $\hat{u}$ satisfying (2.13).

Proof of Corollary 2.3

By lemma 2.1, we are reduced to verify that $\hat{\sigma}$ given by (2.11) satisfies (2.12). Since σ is the curvature of Γ (seen from inside), we have

$$(2.22) \quad \int_{\Gamma} \sigma = 2\pi = \int_{\Gamma_0} \hat{\sigma}.$$

For the other conditions, we argue as in [5], [6] using the expression of the curvature in terms of ϕ, namely

(2.23) $\sigma(\phi(e^{i\theta})) = \dfrac{1}{|\phi'(e^{i\theta})|}\{1 + \mathrm{Re}\,(\dfrac{\phi''(e^{i\theta})\,e^{i\theta}}{\phi'(e^{i\theta})})\}$

where Re denotes the real part. Since $\phi'(z) = a_o + \sum\limits_{n\geq 2} \dfrac{a_n}{z^n}$ for $|z| > 1$, we have for some sequence (b_n)

(2.24) $\dfrac{\phi''(z)\,z}{\phi'(z)} = \sum\limits_{n\geq 2} \dfrac{b_n}{z^n}$ for $|z| > 1$

so that, for all $r > 1$

$$\int_0^{2\pi} \mathrm{Re}\,(\frac{\phi''(re^{i\theta})\,re^{i\theta}}{\phi'(re^{i\theta})})\,e^{i\theta}\,d\theta = 0.$$

By letting r tend to 1, this implies by (2.23), (2.11)

$$\int_0^{2\pi} \hat{\sigma}\,(e^{i\theta})\,e^{i\theta}\,d\theta = 0,$$

so that the only condition to be checked in (2.12) is now that $\hat{\sigma} \equiv 1$. But, if $\hat{\sigma} \equiv 1$, from (2.23), (2.24) we easily deduce that $\phi'' \equiv 0$ so that Γ is a circle which is excluded by assumption.

Remark: We can write

$$\hat{Q}(\hat{u}) = \langle A\,\hat{u},\,\hat{u}\rangle_{H^{-1/2}(\Gamma_o)\times H^{1/2}(\Gamma_o)}$$

where $A\,\hat{u} = C_c\,\hat{u} - \hat{\sigma}\,\hat{u}$. If $\lambda_0 > \|\hat{\sigma}^+\|_\infty$, then $(\lambda_0 I + A)^{-1}$ defines a self adjoint compact operator from $L^2(\Gamma)$ into itself. Therefore, there exists an orthonormal basis $\{\psi_i\}_{i=1,\ldots,\infty}$ of $L^2(\Gamma)$ with $A\,\psi_i = \lambda_i\,\psi_i$, $\lambda_1 \leq \lambda_2 \leq \ldots \leq \lambda_i \leq \ldots \leq +\infty$ and

$$\hat{Q}(\hat{u}) = \sum_{i=1}^{\infty} \lambda_i\,\hat{u}_i^2 \quad \text{if } \hat{u} = \sum_{i=1}^{\infty} \hat{u}_i\,\psi_i.$$

Then applying property (2.13) to the hyperplane of functions $\hat{u}$ in $H^{1/2}(\Gamma)$ such that $\hat{u}_1 = 0$ shows that

(2.25) $\lambda_1 \leq \lambda_2 < 0.$

The assumptions (2.12) are natural according to Corollary 2.3. However they might seem very particular with respect to the study of the quadratic form (2.10). The following easy computations show that everything can happen for the sign of λ_1, λ_2 if (2.12) does not hold. It also indicates that a necessary and sufficient condition on $\hat{\sigma}$ for (2.13) to hold is not easy to reach.

Proposition 2.4 Let $\hat{\sigma} = a + b \cos \theta$. Then

(2.26) $a > 1 \Rightarrow \lambda_1 \leq \lambda_2 < 0$

(2.27) $a < 0$, b small enough $\Rightarrow 0 < \lambda_1 \leq \lambda_2$

(2.28) $0 < a < 1$, b small enough $\Rightarrow \lambda_1 < 0 < \lambda_2$.

Remark: In the latter case, one can characterize the hyperplanes on which the restriction of $\hat{Q}$ is positive (see [11] for a systematic study of this kind of properties for quadratic forms and applications to the study of the stability of electromagnetic shaping for non vanishing surface tension).

Proof of Proposition 2.4 For $\hat{u} \in H^{1/2}(\Gamma_o)$, we introduce its Fourier expansion

(2.29) $$\hat{u} = \alpha_o + \sum_{n\geq 1} \alpha_n \cos n\,\theta + \beta_n \sin n\,\theta$$

where

(2.30) $$\sum_{n\geq 1} n\,(\alpha_n^2 + \beta_n^2) < \infty.$$

Using (2.10), (2.17), (2.18), we easily obtain

(2.31) $$\hat{Q}(\hat{u}) = \pi\,[\sum_{n\geq 1} (n - a)\,(\alpha_n^2 + \beta_n^2) - 2a\alpha_o^2 - 2b\alpha_o\alpha_1 - b \sum_{n\geq 1} \alpha_n\alpha_{n+1} + \beta_n\beta_{n+1}].$$

Except for critical values of (a, b) this can be rewritten

(2.32) $$\hat{Q}(\hat{u}) = \pi\,[-2a\,(\alpha_o + \frac{b}{2a}\alpha_1)^2 + \sum_{n\geq 1} b_n\,(\alpha_n - \frac{b}{2b_n}\alpha_{n+1})^2 + c_n\,(\beta_n - \frac{b}{2c_n}\beta_{n+1})^2]$$

where b_n, c_n are sequences defined by

(2.33) $b_1 = 1 - a + \frac{b^2}{2a}$, $\forall\, n \geq 2$ $\quad b_n = n - a - b^2/4b_{n-1}$

(2.34) $c_1 = 1 - a$, $\forall\, n \geq 2$ $\quad c_n = n - a - b^2/4c_{n-1}$.

The restriction of $\hat{Q}$ to the plane of $\hat{u}$ such that $\alpha_i = 0 = \beta_{i+1}$ for all $i \geq 1$ is equal to $\pi(-2a\alpha_o^2 + (1 - a)\,\beta_1^2)$. If $a > 1$, it is negative, whence conclusion (2.26).

If $a < 0$ and $b^2 \leq \min(|a|, 1)$, we easily check that all b_n and c_n are positive (even greater than 1/2) so that $\hat{Q}$ is positive.

If $0 < a < 1$ and $|b| \leq 1 - a$, we check that b_n, c_n are all positive. The conclusion (2.28) follows.

3. ABOUT EXISTENCE OF MINIMA FOR THE ENERGY FUNCTIONAL

To proceed further in the understanding of critical points for the energy (1.2), we now study the existence of a global minimum without a priori regularity or structure assumption on the open sets Ω. It turns out that differences again appear between the two cases $\int j_o = 0$ or $\neq 0$ (see Remark 2.5).

We denote by $\mathcal{O}$ the family of open sets Ω in $\mathbb{R}^2$ such that ${}^c\Omega$ is bounded. The natural "energy" space is not here $H^1_0(\Omega)$ because of the behavior at infinity but $W(\Omega)$ = the closure of $C^\infty_o(\Omega)$ for the norm

$$\|\psi\|^2_{W(\Omega)} = \|\nabla\psi\|^2_{L^2(\Omega)} + \|\rho\psi\|^2_{L^2(\Omega)}$$

where $\rho(x) = \{(1 + |x|) \ln (2 + |x|)\}^{-1}$. Recall (see [12], [4]) that an equivalent norm is given by

$$|||\psi|||^2_{W(\Omega)} = \|\nabla\psi\|^2_{L^2(\Omega)} + \|\psi\|^2_{L^2(K)}$$

where K is a given not empty bounded open set in Ω.

Theorem 3.1 Let $j_o \in L^2(\mathbb{R}^2)$ with compact support. For Ω in $\mathcal{O}$ and $\psi \in W(\Omega)$, we set

$$E(\Omega, \psi) = \frac{1}{2} \int_\Omega |\nabla\psi|^2 - \int_\Omega j_o \psi. \tag{3.1}$$

Then, for $\alpha > 0$, the number $c(\alpha)$ defined by

$$c(\alpha) = \inf \{E(\Omega, \psi) ; \Omega \in \mathcal{O}, \text{measure } ({}^c\Omega) = \alpha, \psi \in W(\Omega)\} \tag{3.2}$$

is finite if and only if

$$\int_{\mathbb{R}^2} j_o = 0. \tag{3.3}$$

Moreover, for all $\alpha > 0$

$$c(\alpha) = \inf \{E(\mathbb{R}^2, \theta) ; \theta \in W(\mathbb{R}^2)\} \tag{3.4}$$

$$c(\alpha) < E(\Omega, \theta)\ \forall\ \Omega \in \mathcal{O} \text{ with } \Omega \neq \mathbb{R}^2, \forall\ \theta \in W(\mathbb{R}^2). \tag{3.5}$$

Remark: The properties (3.4), (3.5) show that equilibrium positions cannot be found by looking for global minima of the energy: the complement ${}^c\Omega$ of Ω (= the domain occupied by the liquid) has a tendency to "disappear" at infinity as shown by the proof below. This situation is thus completely different from the minimization problem where the surface of Ω (instead of the surface of ${}^c\Omega$) is prescribed. This is the case in the corresponding "interior shaping" problem where the liquid metal is confined in a bounded region (see [3], [13]) or in similar fluid problems such as those considered in [19], [20].

Proof of Theorem 3.1 Let us consider

$$A_R = \{x \in \mathbb{R}^2 ; R \leq |x| \leq R + \varepsilon\}$$

where ε is chosen such that the area of A_R is equal to α, i.e.

$$\pi(R + \varepsilon)^2 - \pi R^2 = \alpha = \pi (2R\varepsilon + \varepsilon^2).$$

Let Ω_R be the complement of A_R and let $\psi_R \in H_0^1(\Omega_R)$ be defined by

$$\psi_R(x) = 0 \quad \text{if } |x| \geq R + \varepsilon$$

$$\begin{cases} \psi_R \in H_0^1(B_R) \quad (B_R = \{x : |x| < R\}) \\ -\Delta\, \psi_R = j_o \text{ in } B_R. \end{cases}$$

Then we claim that, if (3.3) does not hold

$$\lim_{R\to\infty} E(\Omega_R, \psi_R) = -\infty.$$

Indeed, if one sets $\varphi_R(y) = \psi_R(Ry)$, $|y| \leq 1$, then φ_R is solution of

(3.6) $\quad \varphi_R \in H_0^1(B_1)$, $-\Delta\, \varphi_R(y) = R^2\, j_o\,(Ry)$ in B_1.

But, $[\, y \to R^2\, j_o\,(Ry)]$ tends to $(\int_{\mathbb{R}^2} j_o)\, \delta$ as R tends to ∞ where δ is the Dirac mass at 0. Since δ does not belong to $H^{-1}(B_R)$, if $\int_{\mathbb{R}^2} j_o \neq 0$,

$$\lim_{R\to\infty} \int_{B_1} |\nabla \varphi_R|^2 = +\infty.$$

On the other hand, by the definition of ψ_R and by the change of variable $x = Ry$, we have

(3.7) $$E(\Omega_R, \psi_R) = -\frac{1}{2} \int_{B_R} |\nabla \psi_R|^2 = -\frac{1}{2} \int_{B_1} |\nabla \varphi_R|^2 \underset{R\to\infty}{\longrightarrow} -\infty \,.$$

This proves $c(\alpha) = -\infty$ if (3.3) does not hold.

Assume now that (3.3) holds. For $\psi \in W(\Omega)$, we denote by $\tilde{\psi}$ its extension by 0 to $\mathbb{R}^2$. It belongs to $W(\mathbb{R}^2)$ and

(3.9) $\quad E(\Omega, \psi) = E(\mathbb{R}^2, \tilde{\psi})$.

This proves that $c(\alpha)$ defined in (3.2) satisfies

(3.10) $\quad c(\alpha) \geq c_o = \inf\{E(\mathbb{R}^2, \theta)\,;\, \theta \in W(\mathbb{R}^2)\}$.

If (3.3) holds, we have for all $\theta \in W(\mathbb{R}^2)$

(3.11) $\quad E(\mathbb{R}^2, \theta) = E(\mathbb{R}^2, \theta - \bar{\theta})$

where we denote by $\bar{\theta}$ the average of θ over a fixed ball B_{R_o} of radius R_o containing the support of j_o. By Poincaré-Wirtinger's inequality, there exists $C = C(R_o)$ such that for all θ in $W(\mathbb{R}^2)$

(3.12) $$\int_{B_{R_o}} (\theta - \bar{\theta})^2 \leq C^2 \int_{B_{R_o}} |\nabla\theta|^2.$$

Therefore, by (3.11), (3.12)

(3.13) $$\begin{aligned} E(\mathbb{R}^2, \theta) &\geq \frac{1}{2} \int_{\mathbb{R}^2} |\nabla\theta|^2 - \|j_o\|_{L_2} \|\theta - \bar{\theta}\|_{L_2(B_{R_o})} \\ &\geq \frac{1}{2} \int_{\mathbb{R}^2} |\nabla\theta|^2 - C \|j_o\|_{L_2} \{ \int_{B_{R_o}} |\nabla\theta|^2\}^{1/2} \\ &\geq (\frac{1}{2} - \varepsilon) \int_{B_{R_o}} |\nabla\theta|^2 - \frac{C^2}{4\varepsilon} \|j_o\|^2_{L^2} \text{ for all } \varepsilon > 0. \end{aligned}$$

This proves that $c_o \geq -\frac{C^2}{2} \|j_o\|^2_{L^2} > -\infty$ $(\varepsilon = \frac{1}{2})$. Whence the first statement of the theorem.

For the (3.4), (3.5), we first notice that it is obvious if (3.3) does not hold. Indeed we then have (see (3.7), (3.10))

$$-\infty = c(\alpha) \geq c_o \Rightarrow -\infty = c_o = c(\alpha).$$

We now assume that (3.3) holds. Recall that $c(\alpha) \geq c_o$. We introduce the complement Ω_n of the ball $B(x_n, r)$ where $|x_n| = R_n$ and

(3.14) $$\lim_{n\to\infty} R_n = +\infty, \qquad \pi r^2 = \alpha,$$

and ψ_n the solution of

(3.15) $$\begin{cases} \psi_n \in W(\Omega_n) \\ -\Delta \psi_n = j_o \text{ on } \Omega_n. \end{cases}$$

We will prove that $E(\Omega_n, \psi_n)$ converges to $E(\mathbb{R}^2, \psi)$ where

(3.16) $$\psi = j_o * (-\frac{1}{2\pi} \ln |x|)$$

and that

(3.17) $$E(\mathbb{R}^2, \psi) = c_o < E(\mathbb{R}^2, \theta) \text{ for all } \theta \in W(\mathbb{R}^2), \theta \neq \psi.$$

According to (3.9), (3.10), the proof of (3.4), (3.5) will then be complete.

For (3.17), we use that $\psi \in W(\mathbb{R}^2)$ (since $\int j_o = 0$) and for all $\theta \in C_0^\infty(\mathbb{R}^2)$

$$\begin{aligned} E(\mathbb{R}^2, \theta) - E(\mathbb{R}^2, \psi) &\geq \int_{\mathbb{R}^2} \nabla\psi (\nabla\theta - \nabla\psi) - \int_{\mathbb{R}^2} j_o (\theta - \psi) + \frac{1}{2} \int_{\mathbb{R}^2} |\nabla(\theta - \psi)|^2 \\ &\geq \frac{1}{2} \int_{\mathbb{R}^2} |\nabla(\theta - \psi)|^2, \end{aligned}$$

the last equality being obtained by integration by parts (allowed since $|\psi| = 0(|x|^{-1})$ and $|\nabla\psi| = 0(|x|^{-2})$ as $|x| \to \infty$) and by using $-\Delta\psi = j_o$. By density of $C_o^\infty(\mathbb{R}^2)$ in $W(\mathbb{R}^2)$, we obtain

$$\forall\, \theta \in W(\mathbb{R}^2) \text{ with } \theta \neq \psi,\ E(\mathbb{R}^2, \theta) > E(\mathbb{R}^2, \psi)$$

whence (3.17).

Now, since $\psi - \psi_n$ is harmonic on Ω_n, we have

$$(3.18)\quad 0 = -\int_{\Omega_n} (\psi - \psi_n)\, \Delta(\psi - \psi_n) = -\int_{\partial\Omega_n} \psi\, \nabla(\psi - \psi_n) \cdot \vec{n} + \int_{\Omega_n} |\nabla(\psi - \psi_n)|^2.$$

Since $|\psi| = 0(|x|^{-1})$ and $|\nabla\psi| = 0(|x|^{-2})$ as $|x| \to \infty$,

$$(3.19)\quad \lim_{n\to\infty} \int_{\partial\Omega_n} \psi\, (\nabla\psi \cdot \vec{n}) = 0.$$

Thus, according to (3.18), to prove that $\nabla\tilde{\psi}_n \to \nabla\psi$ in $L^2(\mathbb{R}^2)$, it is sufficient to prove that $\sup_{x\in\partial\Omega_n} |\nabla\psi_n(x)|$ is uniformly bounded. For this, we remark that $\theta_n(y) = \psi_n(x_n + y)$ is solution of

$$(3.20)\quad \begin{cases} \theta_n \in W(\Omega_r) \quad (\Omega_r = \{x \in \mathbb{R}^2, |x| > r\}) \\ -\Delta\, \theta_n = j_o\,(. + x_n) \text{ in } \Omega_r. \end{cases}$$

Using (3.12) and (3.3), we have for $B_{R_o} \supset \text{Supp}\, j_o$

$$(3.21)\quad \int_{\mathbb{R}^2} |\nabla\tilde{\psi}_n|^2 = \int_{\mathbb{R}^2} \psi_n\, j_o = \int_{B_{R_o}} (\psi_n - \bar{\psi}_n)\, j_o \leq C\, \|\nabla\tilde{\psi}_n\|_{L^2(\mathbb{R}^2)}.$$

This proves that $\nabla\tilde{\psi}_n$ and $\nabla\tilde{\theta}_n$ are bounded in $L^2(\mathbb{R}^2)$. If we now set $C_r = \{x \in \mathbb{R}^2;\ r < |x| < 2r\}$, since θ_n is harmonic on C_r (for n large enough) and vanishes on $\{|x| = r\}$, it is uniformly bounded in $H^1(C_r)$ and by regularity results in $C^1(C_r)$ (at least). This proves that $\sup_{x\in\partial\Omega_r} |\nabla\theta_n(x)| = \sup_{x\in\partial\Omega_n} |\nabla\psi_n(x)|$ is bounded independently of n whence (3.19) and

$$(3.22)\quad \nabla\tilde{\psi}_n \to \nabla\psi \text{ in } L^2(\mathbb{R}^2).$$

This implies that $E(\Omega_n, \psi_n) = -\frac{1}{2}\int_{\Omega_n} |\nabla\psi_n|^2$ converges to $E(\mathbb{R}^2, \psi) = -\frac{1}{2}\int_{\mathbb{R}^2} |\nabla\psi|^2$ and (3.16) is proved.

Remark: The fact that the sets Ω are required to be open is not really a restriction in Theorem 3.1. A similar result could be stated for measurable sets Ω and $W(\Omega) = \{\varphi \in W(\mathbb{R}^2) ; \tilde{\varphi} = 0$ quasi-everywhere outside $\Omega\}$ where $\tilde{\varphi}$ is the usual quasi-continuous representation of φ.

A similar result holds for the case of positive surface tension σ where the energy functional is given by

$$(3.23) \quad E_\sigma(\Omega, \theta) = E(\Omega, \theta) + \sigma P(\Omega)$$

with $P(\Omega)$ the perimeter of Ω. Because of the isoperimetric inequality, if measure $({}^c\Omega) = \alpha$ then $P(\Omega) \geq 2\pi r$ where $\pi r^2 = \alpha$. Now if

$$c_\sigma(\alpha) = \inf \{E_\sigma(\Omega, \theta), \Omega \in \mathcal{O}, \text{measure } ({}^c\Omega) = \alpha, \theta \in W(\Omega)\}$$

and if (Ω_n, ψ_n) and ψ are defined as above (in (3.15), (3.16)) then

$$(3.24) \quad E_\sigma(\Omega_n, \psi_n) \geq c_\sigma(\alpha) \geq E(\mathbb{R}^2, \psi) + \sigma\, 2\pi r.$$

But, if (3.3) holds, $E_\sigma(\Omega_n, \psi_n)$ converges to $E(\mathbb{R}^2, \psi) + \sigma\, 2\pi r$ so that

$$c_\sigma(\alpha) = E(\mathbb{R}^2, \psi) + \sigma\, 2\pi r \; < \; E_\sigma(\Omega, \theta) \;\forall\, \Omega \neq \mathbb{R}^2, \forall\, \psi \in W(\Omega).$$

On the other hand, if $\int j_0 \neq 0$, the construction of (Ω_R, ψ_R) (see (3.17)) does not extend to the case $\sigma \neq 0$ since $P(\Omega_R) \to \infty$ as $R \to \infty$. However one can prove that the energy is still unbounded by proving that $E(\Omega_n, \psi_n)$ tends to $-\infty$ (one checks that $\theta_n(\eta) = \psi_n(R_n(u + \eta))$ where $x_n = R_n u$, $|u| = 1$, converges to a solution of $-\Delta\theta = \delta_{-u} \int j_0$ which is of infinite energy).

REFERENCES

[1] J.P. BRANCHER, O. SERO-GUILLAUME, "Sur l'équilibre des liquides magnétiques, applications à la magnétostatique", J. de Mécanique Théorique et Appliquée, 2, n° 2, (1983), 265-283.

[2] J.P. BRANCHER, J. ETAY, O. SERO-GUILLAUME, "Formage d'une lame", J. de Mécanique Théorique et Appliquée, 2, n° 6, (1983), 976-989.

[3] M. CROUZEIX, "Variational approach of a magnetic shaping problem", European Journal of Mechanics, B, to appear.

[4] M. CROUZEIX, J. DESCLOUX, "A bidimensional electromagnetic problem", SIAM J. Math. Analysis, 21, (1990), 577-592.

[5] J. DESCLOUX, "On the two dimensional magnetic shaping problem without surface tension", Report Nr. 07.90, Ecole Polytechnique Fédérale de Lausanne (1990).

[6] J. DESCLOUX, "Stability of the solutions of the bidimensional magnetic shaping problem in absence of surface tension", European Journal of Mechanics, B, to appear.

[7] P. GRISVARD, Elliptic problems in non smooth domains, Monograph and Studies in Math, 24, Pitman, (1965).

[8] A. HENROT, M. PIERRE, "Un problème inverse en formage des métaux liquides", R.A.I.R.O., M^2AN, 23, n° 1, (1989), 155-177.

[9] A. HENROT, M. PIERRE, "About existence of a free boundary in electromagnetic shaping", in Recent advances in nonlinear elliptic and parabolic problems, vol. 208, Pitman Research Notes Series, (1989).

[10] A. HENROT, M. PIERRE, "About existence of equilibria in electromagnetic casting", Quarterly of Applied Mathematics, (1991).

[11] A. HENROT, M. PIERRE, "Stability of equilibria in electromagnetic shaping", to appear.

[12] J.C. NEDELEC, "Approximation des équations intégrales en mécanique et en physique", Rapport, Centre de Mathématiques Appliquées, Ecole Polytechnique, Palaiseau, (1977).

[13] O. PIRONNEAU, Optimal Shape Design for Elliptic Systems, Springer-Verlag, New-York, (1984).

[14] W. RUDIN, Real and Complex Analysis, MacGraw-Hill Series in higher Mathematics, (1974).

[15] O. SERO-GUILLAUME, "Sur l'équilibre des ferrofluides et des métaux liquides", Thèse, INPL Nancy, France, (1983).

[16] J.A. SHERCLIFF, "Magnetic Shaping of molten metal columns", Proc. Royal Soc. London, A 375, (1981), 455-473.

[17] A.D. SNEYD, H.K. MOFFATT, "Fluid dynamical aspects of the levitation melting process, J. of Fluid Mech., 117, 45-70.

[18] J.P. ZOLESIO, "The material derivative by speed method", Proc. Math. Institute, Iowa, (1981).

[19] J.P. ZOLESIO, "Numerical algorithms and existence result for a Bernoulli-like steady free boundary problem", Large Scale Systems, Th. and Appl., 6, (1984), 163-278.

[20] J.P. ZOLESIO, "Introduction to shape optimisation problems and free boundary problems", Séminaire de Mathématiques Supérieures, Université de Montréal, Canada, (1990).

Finite dimesional compensator for flexible structures[1]

I. Lasiecka

Department of Applied Mathematics
University of Virginia
Charlottesville, VA 22903

Introduction

One of the important problems in control theory is design of feedback control system under the assumption that only partial observation of the state is available. This is typically done by constructing an appropriate observer (estimator) of the original state and then the desired feedback control law — called compensator — is based on the information available from the observed (estimated) state variable. When applying this procedure to flexible structures, stability properties of the observer and of the resulting closed loop system are of obvious importance. Thus, we are naturally interested in observers and compensators which produce/retain uniform asymptotic stability properties of the entire structure. When the original control systems are described by infinite dimensional models (as they arise in modelling of flexible structures) then the control implementability requirements will dictate that both observer and the compensator be finite dimensional. This brings us to the main objectives of this paper which are:

(i) to introduce finite dimensional approximations (based on finite elements) of the original infinite dimensional compensator,

[1]This research was partially sponsored by the National Science Foundation Grant DMS 8902811 and by AFOSR Grant AFOSR-87-0321

(ii) to show that these finite dimensional compensators when applied to the original model produce near optimal performance (this is to say that the approximating closed loop system converges to the original one),

(iii) to prove that the stability properties of the resulting closed loop system are preserved uniformly, in the parameter of discretization (this property will guarantee that the stability properties of the finite dimensional compensator do not deteriorate with the increasing order of approximation).

The problem of consistent and convergent approximations of infinite dimensional compensators has attracted a lot of attention in recent years (see [S.1], [G.1], [C.1] and most recently [G.2] and references therein). However, most of the results deal with the dynamics when the so called "spectrum determined growth" condition is satisfied like in the parabolic (or more generally analytic) case or for delay equations ([C.1], [S.1], [I.1]). When it comes to general C_0 - semigroups (like the ones arising in flexible structures) few results are available (see [I.1], [G.2]). Moreover, these results are confined to the situation when the control operator and observation operators are bounded. Even in this bounded case, the above results do not guarantee, in general, that the finite dimensional compensator produces an uniformly (in the parameter of discretization) stable closed loop system (provided, of course, that the original infinite dimensional compensator was stable to begin with). Indeed, in [I.1] the statement on uniform stability of the closed loop system holds under the assumption of "spectrum determined growth condition," and in [G.2] such conclusions holds (for general C_0 - semigroups) only in a very special case when the selected approximation is modal and certain decoupling condition holds for the original system (see Thm 9.3, 9.4 in [G.2]). In fact, the problem of proving uniform stability of finite dimensional compensators for general approximation schemes, has been recognized in [G.2] as an open problem (as reported in [G.2], numerical experiments with finite elements testify that such result should hold). Thus the main contribution and novelty of this paper is that (i) we consider the case of general

C_0 semigroups with unbounded control operator (as they arise in boundary/point control problems) (ii) our approximations are general approximating schemes with the usual consistency and stability requirement (not necessarily modal etc.) and (iii) for such approximations we prove convergence results for finite dimensional compensators and, most importantly, we prove that the resulting closed loop systems are uniformly exponentially stable. We should mention explicitly that our focus is on the dynamics generated by general C_0 - semigroups and not on analytic semigroups. In fact, in this latter, analytic case, stronger results can be established.

The outline of our paper is as follows. In section 2 we shall formulate the problem and we will state our main results. Section 3 is devoted to a brief sketch of the proofs. In section 4 we apply the results of Sect. 2 to the case of optimal infinite dimensional compensator (i.e when the control feedback is determined as a solution of the appropriate Riccati Equation). In order to obtain the desired convergence properties we shall use recent results of [L.1] on the approximations of Algebraic Riccati Equations with unbounded coefficients. Section 5 provides an example of a wave equation with boundary control and boundary observation.

2. Statement of the problem of and of the main results.

Let A be a generator of a C_0 - semigroup e^{At} defined on a Hilbert space H. Let be given the following operators:

$$B: \; U \rightarrow D(A^*)' , \tag{2.1}$$

$$C \in \mathcal{L}(H; Z) , \tag{2.2}$$

$$K \in \mathcal{L}(Z \rightarrow H) , \tag{2.3}$$

where U and Z are another Hilbert spaces and $\mathcal{D}(A^*)'$ denotes dual to $\mathcal{D}(A^*)$ with respect to H-inner product. In what follows we shall assume that the operator B, generally unbounded is subject to the following hypothesis:

$$\text{(H-1)} \quad \int_0^T |B^* e^{A^* t} x|_U^2 \, dt \le C_T \, |x|_H^2 \; ; \quad x \in \mathcal{D}(A^*),$$

where $(Bu, v)_H \equiv (u, B^* v)_U$ $u \in U$; $v \in \mathcal{D}(B^*) \subset \mathcal{D}(A^*)$. We consider the following control system

$$\begin{cases} x_t = Ax + Bu \quad \text{on } (\mathcal{D}(A^*)' \\ x(t = o) = x_0 \in H; \end{cases} \tag{2.4}$$

with partial observation

$$y = Cx. \tag{2.5}$$

An observer (or equivalently estimator) for system (2.4) is the system given by

$$w_t = Aw + Bu + K[y - Cx]. \tag{2.6}$$

It can be shown that the assumption (H-1) guarantees that the system consisting of (2.4) and (2.5) is well posed on H. This is to say that for any initial data $x_0 \in H$, $w(t =) = w_0 \in H$ and control function $u \in L_2[0T; U]$; the solution (x, w) to (2.4), (2.5) belongs to $C[0T; H \times H]$ and it continuous with respect to the data: x_0, w_0, u. If in addition we assume that

the semigroup generated by A - KC is exponentially stable, (2.7)

then

$$|x(t) - w(t)|_H \le C e^{-wt} |x_0 - w_0|_H \tag{2.8}$$

for some positive constants C, w > 0. In this case we say that the observer (2.5) is uniformly exponentially stable. We will be interested in feedback controls which must be based on an estimate w(t) (because the full state x(t) is not available for direct feedback). Thus, we are led to consider feedback laws of the form

$$u(t) = F\, w(t)\,; \quad F\colon\ H \to U\,. \tag{2.9}$$

The observer (2.6) and the control law (2.9) constitute a compensator for the control system (2.4)-(2.5). We will show that the closed loop system consisting of control system (2.4)-(2.5) and the compensator (2.6); (2.9) generates a strongly continuous semigroup on $H \times H$ (note that this result is well known in the case when the operator B is bounded).

The main goal of this paper is to develop an approximation theory for the infinite dimensional compensator i.e for (2.6) and (2.9). To this end we introduce (i) a family of finite dimensional subspaces V_h of H with orthogonal projections $\pi_h\colon\ H \to V_k$ which are strongly convergent to the identity and (ii) a family of finite dimensional operators

$$A_h\colon\ V_h \to V_h\,, \quad K_h\colon\ Z \to V_h\,, \quad F_h\colon\ V_h \to U\,, \quad B_h\colon\ U \to V_h\,.$$

The finite dimensional approximation of the compensator (2.6), (2.9) is defined by the finite dimensional observer

$w_{t,h} = A_h\, w_h + B_h\, u_h - K_h\, C\, w_h + K_h\, C\, x$ together with the finite dimensional control law (2.10)

$$u_h(t) = F_h\, w_h(t)\,. \tag{2.11}$$

Our aim here is to prove that (i) the finite dimensional control law (2.10), (2.11) once inserted into the original system (2.4) produces solutions which are convergent to the solutions of the original infinite dimensional system, (ii) if the original infinite dimensional closed loop system (2.4) - (2.6) is exponentially stable, that this stability is retained — uniformly in the parameter of discretization h — for the closed loop system consisting of (2.4) - (2.10) and (2.11). To accomplish this goal we need to make several — approximation type — hypothesis.

$$(A-1)\quad \begin{cases} \text{(i)} & |e^{A_h t}|_{\mathcal{L}(H)} \le C\,e^{wt} :\ C, w > 0\,. \\ \text{(ii)} & \int_0^T |B_h^* e^{A_h^* t} x_h|_U^2\, dt \le C_T\, |x_h|_H^2 \end{cases}$$

The following convergence holds as $h \to 0$

(A–2) (i) $|(\hat{A}_h^{-1}\,\pi_h - A^{-1})x|_H \to 0 \quad x \in H$ and where $\hat{A} \equiv \lambda I - A$ for some $\lambda > 0$.

(ii) $|(\hat{A}^{-1} - \hat{A}_h^{-1}\,\pi_h)\,Bu|_H \to 0; \quad u \in U;$

(iii) $|\hat{A}^{-1}\,(B_h - B)u|_H \to 0; \quad u \in U;$

(iv) $|(\hat{A}_h^{-1} - \hat{A}^{-1})\,B_h\,u|_H \to 0; \quad u \in U.$

To state our results, we introduce the operators

$\mathcal{A}$: $H \times H \to H \times H$ given by

$$\mathcal{A} \equiv \begin{bmatrix} A & BF \\ KC & A + BF - KC \end{bmatrix},$$

and $\mathcal{A}_h$: $H \times V_h \to H \times V_h$ defined by

$$\mathcal{A}_h \equiv \begin{bmatrix} A & BF_h \\ K_h C & A_h + B_h F_h - K_h C \end{bmatrix}.$$

It is immediate to see that (2.4), (2.5), (2.6) and (2.9) can be written as

$$\frac{d}{dt}\begin{bmatrix} x \\ w \end{bmatrix} = \mathcal{A}\begin{bmatrix} x \\ w \end{bmatrix}, \tag{2.12}$$

and (2.4), (2.10), (2.11) as

$$\frac{d}{dt}\begin{bmatrix} x \\ w_h \end{bmatrix} = \mathcal{A}_h\begin{bmatrix} x \\ w_h \end{bmatrix}. \tag{2.13}$$

Our main results are:

Theorem 2.1

Assume (H-1), (A-1) and that $|F_h|_{\mathcal{L}(H \to U)} + |K_h|_{\mathcal{L}(Z \to H)} \le M$. Then $\mathcal{A}$ (resp $\mathcal{A}_h$) generate C_0 - semigroups on $H \times H$ (resp $H \times V_h$) and there exist constants C, w >0

$$|e^{\mathcal{A}t}|_{\mathcal{L}(H \times H)} + |e^{\mathcal{A}_h t}|_{\mathcal{L}(H \times H)} \le Ce^{wt}. \tag{2.14}$$

If, in addition, approximation assumptions (A-2) are satisfied and

(A–3) (i) $F_h \to F$ strongly $H \to U$ and (ii) $K_h \to K$ strongly $H \to Z$ then

$$e^{\mathcal{A}_h t} \to e^{\mathcal{A}t} \text{ strongly in } H \times H \text{ and uniformly on compact interval of time } 0 < t < T. \tag{2.15}$$

Theorem 2.2

In addition to the assumptions of Theorem 2.1 we assume that

(B-1) the semigroups generated by A-KC and A + BF are exponentially stable and

(B–2) $|e^{(A_h + B_h F_h)t}|_{\mathcal{L}(H)} \le C e^{-wt}$ where C and w are positive constants independent on h,

(B–3) $|(K_h - K) C|_{\mathcal{L}(H)} \to 0$ and K C is compact in H.

Then

$$|e^{\mathcal{A}_h t}|_{\mathcal{L}(H)} \le C e^{-wt}, \text{ where the constants } C, w > 0 \text{ are uniform in } h > 0, \tag{2.16}$$

$$\sup_{t \ge 0} |(e^{\mathcal{A}_h t}\pi_h - e^{\mathcal{A}t})x|_H \to 0 \quad x \in H, \tag{2.17}$$

$$|u_h - u|_{L_2(0\infty;\, U)} \to 0 \tag{2.18}$$

with u given by (2.9) and u_h by (2.11) ■

Remark 2.1

Notice that Theorem 2.1 provides us with basic convergence result — for finite time intervals — valid for the approximations of the compensator. It generalizes the main result of [G.2] Thm. 9.3 where such convergence is stated in the special case when the control operator is bounded. What we consider as the main contribution of this paper is the result of Theorem 2.2 which states that the approximation of the originally stable infinite dimensional compensator reconstructs these stability properties uniformly in the parameter of discretization.

Remark 2.2

Although the results of Theorems 2.1, 2.2 provide us with finite dimensional compensator, the order of the compensator may be excessively high. A natural issue which arises then is that of reduced order compensators. Since the task of finding the minimal order of the (finite dimensional) compensator can be accomplished by routine procedure in finite dimensional theory (see for instance [G.1], [S.1] and references provided there), this question will not be considered in the paper.

3. Outline of the proof of Theorem 2.2

Because of space limitations, we shall outline only the major steps of the proof of Theorem 2.2. For details, we refer the reader to [L.2]. The proof of Theorem 2.1 is simpler and can be easily reconstructed from the proof of Theorem 2.2.

Step 1.

By using regularity hypothesis (H-1) and its discrete counterpart (A-1) we can show that

$$|e^{\mathcal{A}_h t}|_{\mathcal{L}(H \times H)} \leq C e^{w_0 t} \tag{3.1}$$

for some constants C, $w_0 > 0$ independent on h.

Step 2

Since $|(K - K_h) C|_{\mathcal{L}(H)} \to 0$ one proves by the usual perturbation type of arguments that

$$|e^{(A - K_h C)t}|_{\mathcal{L}(H)} \leq C e^{-wt} . \tag{3.2}$$

Step 3

Using the same transformation as in [S.1] i.e.

$$H \equiv \begin{bmatrix} I & I \\ 0 & I \end{bmatrix}$$

it is immediate to notice that (2.16) and (2.17) are equivalent to the same statements with $\mathcal{A}$ (resp $\mathcal{A}_h$) replaced by $\bar{\mathcal{A}}$ resp ($\bar{\mathcal{A}}_h$) where $\bar{\mathcal{A}} \equiv H^{-1} \mathcal{A} H$ and $\bar{\mathcal{A}}_h \equiv H^{-1} \mathcal{A}_h H$, or explicitly

$$\bar{\mathcal{A}} \equiv \begin{bmatrix} A-KC & 0 \\ KC & A+BF \end{bmatrix} \tag{3.3}$$

$$\bar{\mathcal{A}}_h \equiv \begin{bmatrix} A - K_h C & A - A_h + (B - B_h) F_h \\ K_h C & A_h + B_h F_h \end{bmatrix} \tag{3.4}$$

Step 4

Explicit computations of the resolvent for $\bar{\mathcal{A}}_h$ yield: $R(\lambda, \bar{\mathcal{A}}_h) \equiv (r_1, r_2)$ where

$$r_1(\lambda) = [I + T_h(\lambda)]^{-1} \{R(\lambda, A_c) f + R(\lambda, A_C)(A + BF - \lambda)[R(\lambda, A_{hB}) - R(\lambda, A_B]g \tag{3.5}$$

$$+ R(\lambda, A_C) B(F_h - F) R(\lambda, A_{hB})g$$

$$r_2(\lambda) = R(\lambda, A_{hB})[g - K_h C r_1(\lambda)] \tag{3.6}$$

where we have used the following notation: $A_C \equiv A - K_h C$; $A_B \equiv A + BF$; $A_{hB} \equiv A_h + B_h F_h$ and $T_h(\lambda) \equiv R(\lambda, A_C)[(A + BF - \lambda)(R(\lambda, A_{hB}) - R(\lambda, A_B)) + B(F_h - F) R(\lambda, A_{hB})] K_h C$. The above representation of the resolvent can be easily justified for Re $\lambda > R$ with sufficiently large R. We shall here is to show that representation (3.5), (3.6) is also valid for Re $\lambda < - w$ for some $w > 0$. This will be accomplished in several steps.

<u>Step 5</u> we prove that as $h \to 0$

$$R(\lambda, A_{hB}) - R(\lambda, A_B) \to 0 \text{ strongly in } H \tag{3.7}$$

and the convergence is uniform in Re $\lambda > -w$. Hence, by hypothesis (B-3)

$$([R(\lambda, A_{hB}) - R(\lambda, A_B)]\, K_h\, C|_{\mathcal{L}(H)} \to 0 \text{ as } h \to 0 \text{ for all } \operatorname{Re}\lambda > -w\,. \tag{3.8}$$

The proof of (3.7) is rather tedious, and the difficulty of the proof lies in the fact that the operator B is unbounded. The property that is critically used is the following inequality which can be deduced from the result proved in [W.1]

$$|R(\lambda, A_h)B_h|_{\mathcal{L}(U;H)} \leq \frac{C}{\sqrt{\operatorname{Re}\lambda - w_0}} \tag{3.9}$$

uniformly in Re $\lambda > w_0$.

<u>Step 6:</u>

By using perturbation techniques we prove that

$$\int_0^T |B^* e^{A_C t} x|_U^2\, dt \leq C_T\, |x|_H^2\,. \tag{3.10}$$

(3.10) together with the result of Proposition 3.1 of [W.1] gives

$$|R(\lambda, A_C)\,(A + BF - \lambda)|_{\mathcal{L}(H)} \leq M \tag{3.11}$$

where this estimate holds uniformly in Re $\lambda > -w$.

<u>Step 7:</u>

By collecting the results of Step 5 and of Step 6 and recalling hypothesis (A-2) we obtain:

$$|T_h(\lambda)|_{\mathcal{L}(H)} \to 0 \text{ as } h \to 0 \text{ and the convergence is uniform in } \operatorname{Re}\lambda > -w. \tag{3.12}$$

Step 8

From (3.2), (3.5), (3.6), (3.7) and (3.12) we obtain that

$$|R(\lambda, \bar{\mathcal{A}}_h)| \leq M \text{ uniformly in } \operatorname{Re}\lambda > -w. \tag{3.13}$$

Step 9 (3.13) together with stability result of Monauni [M.1] (see also [G-N]) yields

$$|e^{\bar{\mathcal{A}}_h t}|_{\mathcal{L}(H\times H)} \leq C\, e^{-wt} \tag{3.14}$$

uniformly in h.

Step 10

By using the results of Step 9, Step 5 and Step 4 one shows that for a fixed λ_0

$$R(\lambda_0, \bar{\mathcal{A}}_h) - R(\lambda_0, \mathcal{A}) \to 0 \text{ strongly in } H\times H. \tag{3.15}$$

(3.15) together with (3.14) and Trotter Kato Theorem leads to the desired conclusion in (2.17). (2.18) is a simple consequence of (2.17) and (2.16).

4. The optimal finite dimensional compensator and the optimal closed loop system.

In this section we shall specialize the results of Theorem 2.2 to the situation when the feedback control operators F and F_h as well as the estimator's gains K and K_h are determined via the solutions of the appropriate Riccati equations. In what follows we suppose that the feedback law F and the estimator gain K are chosen as

$$\begin{aligned} F &\equiv B^* P \\ K &\equiv \hat{P} C^* \end{aligned} \tag{4.1}$$

where P and $\hat{P}$ are the solutions (if they exist)of the following Algebraic Riccati Equations (ARE).

$$\begin{aligned} &a)\quad (A^* Px,y)_H + (PA\,x,y)_H + (R^*Rx,y)_H = (B^*Px, B^*Py)_U; \quad (x,y) \in \mathcal{D}(A) \\ &b)\quad (A\hat{P}x,y)_H + (\hat{P}A^*x,y) + (Q^*Q\,x,y)_H = (C\hat{P}x, C\hat{P}Y)_Z; \quad (x,y) \in \mathcal{D}(A^*) \end{aligned} \tag{4.2}$$

where $R \in \mathcal{L}(H)$ and $Q \in \mathcal{L}(H)$ are given operators. The compensator (2.6), (2.9) with F and K given by (4.1) is called optimal infinite dimensional compensator and the corresponding closed loop system consisting of (2.4), (2.5), (2.6), (2.9), (4.1), is referred to as the optimal closed loop system.

In order to guarantee unique solvability of ARE (4.2), we assume the following stabilizability/detectability conditions.

$$
\text{(C-1)} \quad \begin{cases} \text{(i)} \quad (A^*, C) \text{ is stabilizable on } H \\ \text{(ii)} \quad (A^*, Q) \text{ is detectable on } H \end{cases} .
$$

Then from standard LQR theory (see [B.1]) we obtain that there exists unique solution $\hat{P}$ to (4.2b) such that $0 \le P = P^* \in \mathcal{L}(H)$ and

$$
| e^{(A - \hat{P} C^* C)t} |_{\mathcal{L}(H)} \le C e^{-wt} . \tag{4.3}
$$

Situation is more complicated in the case of equation (4.2a) since the operator B is generally unbounded and standard LQR theory does not apply. We shall use however recent results on solvability of Riccati Equation with unbounded coefficients (F-L-T). Indeed, according to the theory of [F-L-T], the hypotheses

$$
\text{(C-2)} \quad \begin{cases} \text{(i)} \quad (A, B) \text{ are stabilizable} \\ \text{(ii)} \quad (A, R) \text{ are detectable} \end{cases}
$$

imply the existence and uniqueness of a unique solution $P \in \mathcal{L}(H)$ to eq. (4.2a) such that

$$
| e^{(A - B B^* P)t} |_{\mathcal{L}(H)} \le C e^{-wt} ; \tag{4.4}
$$

$$
B^* P \in \mathcal{L}(D(A);U) . \tag{4.5}
$$

Thus, in general, the gain operator F is unbounded (though densely defined on $\mathcal{D}(A)$). On the other hand, if the observation R satisfies an additional regularity hypothesis

$$\text{(C-3)} \quad \int_0^T |R^* R e^{At} B u|_H \le C_T |u|_U ,$$

then from [F-L-T]

$$B^* P \in \mathcal{L}(H; U) . \tag{4.6}$$

If the feedback operator F and K are as in (4.1), it is only natural to select for the approximations F_h and K_h - solutions to the appropriate approximations of Algebraic Riccati Equations. This leads us to consider the following discrete approximations of (4.2), in the variable $P_h \in \mathcal{L}(V_h)$ (resp $\hat{P}_h$)

$$A_h^* P_h + P_h A_h + \pi_h R^* R \pi_h = P_h B_h B_h^* P_h \tag{4.7}$$

and

$$A_h \hat{P}_h + \hat{P}_h A_h^* + \pi_h Q^* Q \pi_h = \hat{P}_h C^* C \hat{P}_h . \tag{4.8}$$

To apply the results of Theorem 2.2, we need convergence properties for the corresponding Riccati operators. Indeed, such convergence theory has been recently developed (see [L.1]). Following [L.1] we shall assume:

(C-1-h) (i) (A_h, B_h) and $(A_h^*, \pi_h C)$ are uniformly stabilizable.

(C-2-h) $(A_h^*, \pi_h Q)$ and $(A_h, \pi_h R)$ are uniformly detectable.

Then, according to [L.1] there exist unique solution P_h (resp $\hat{P}_h$) to (4.7) (resp. (4.8)) such that as $h \to 0$

$$\text{(i)} \quad P_h \to P \ \text{ strongly in } H , \quad \text{(ii)} \ \hat{P}_h \to \hat{P} \ \text{ strongly in } H , \tag{4.9}$$

$$|B_h^* P_h e^{(A_h - B_h B_h^* P_h)t} x - B^* P e^{(A - B B^* P)t} x|_{L_2(0 \infty ; U)} \to 0 \quad x \in H , \tag{4.10}$$

$$|e^{(A_h - B_h B_h^* P_h)t}|_{\mathcal{L}(H)} \le C e^{-wt} , \tag{4.11}$$

$$|e^{(A_h - \pi_h C K_h)t}|_{\mathcal{L}(H)} \le C e^{-wt} . \tag{4.12}$$

If, in addition, the discrete counterpart of (C-3) holds i.e

(C–3–h) $\int_0^T |R^* R e^{A_h t} B_h u|_U \, dt \le C_T \, |u|_\Gamma$

then

$$|B_h^* P_h \pi_h x - B^* P x|_U \to 0 ; \quad x \in H . \tag{4.13}$$

Choosing

$$F_h \equiv - B_h^* P_h \tag{4.14}$$

$$K_h \equiv \hat{P}_h C^* \tag{4.15}$$

we shall see that the assumptions of Theorem 2.2 are satisfied. Indeed, the assumption (B.1) holds by the virtue of (4.3) and (4.4). Assumption (B.2) is equivalent to (4.11). If, for instance, the operator C is compact (which we shall assume for simplicity), then (B.3) follows from (4.15) and (4.9) (ii). As for Assumption (A-3) (i) - this follows from (4.14) and (4.13), however under the additional regularity hypothesis (C-3), (C-3h). The above discussion is summarized in the following Theorem.

Theorem 4.1

Suppose that for the continuous problem: (i) the regularity hypotheses (H-1) (C-3) hold (ii) stabilizability/detectability conditions (C-1), (C-2) are satisfied and for the approximating problem: (i) assumptions (A-1), (A-2)and (C-1-h) — (C-3-h), hold. Then, if in addition operator C is compact, the conclusions of Theorem 2.1 and 2.2 hold with the feedback gains (4.1), and their approximations given by (4.14), (4.15) where P_h (resp $\hat{P}_h$) satisfy the approximate ARE (4.7) (resp (4.8)).

Remark 4.1

The result of Theorem 4.1 states that the finite dimensional compensator computed on a basis of approximating Riccati Equation provides a near optimal performance and, moreover, preserves, uniformly in the parameter of discretization, asymptotic stability properties of th original optimal closed loop infinite dimensional system. This conclusion is proved for any consistent approximations (i.e. complying with requirements (A-1), (A-2)) but subject to the regularity property imposed on the observation R i.e. (C-3) (note that (C-3) is automatically satisfied if B is bounded or R is "smoothing"). A natural question one would like to ask is: is it possible to reach the same conclusion without assuming (C-3). The obvious difficulty is, of course, the fact that in this case the feedback law F is unbounded (see (4.5)) and the corresponding F_h are not even uniformly bounded in the topology $H \to U$. What is still true however, is that F_h converges to F on optimal dynamics (see (4.10)). By using this property together with (4.5) we can prove that the statement of Theorem 4.1, without assumption (C-3) is still valid, provided however that the approximations are "special". More precisely we require that

$$|A_h \pi_h - A \pi_h|_{\mathcal{L}(H)} \leq C . \tag{4.16}$$

Assumption (4.16) is satisfied, for instance, in the case of model approximations.

5. Example

We consider an example of lightly damped wave equation with boundary control and boundary observations. We shall show that for this example all the assumptions of Theorem 4.1 are satisfied. Other examples of plates or Kirchoff plates could be considered likewise.

Let Ω be a bounded domain with a boundary $\Gamma \equiv \Gamma_0 \cup \Gamma_1$, Γ_1 is nonempty, and let γ_1, γ_2 be a given positive constants. We consider the following boundary control system

$$\begin{cases} z_{tt} = \Delta z - \gamma_1 z_t & (x,t) \in \Omega \times (0, \infty), \\ \frac{\partial}{\partial\gamma} z - \gamma_2 z_t = u & (x,t) \in \Gamma_1 \times (0, \infty), \\ z = 0 \quad \text{on } \Gamma_0, \\ z(t=0) = z_0 \in H^1(\Omega), \\ z_t(t=0) = z_1 \in L_2(\Omega); \end{cases} \tag{5.1}$$

with boundary observation

$$y = z|_{\Gamma_1}. \tag{5.2}$$

It is well known that eq. (5.1) can be written in a semigroup form as

$$x_t = A\,x + Bu: \text{ where} \tag{5.3}$$

$$x \equiv (z, z_t), \quad H \equiv H^1_{\Gamma_0}(\Omega) \times L_2(\Omega); \quad U = L_2(\Gamma); \quad Z = L_2(\Gamma),$$

$$A \equiv \begin{bmatrix} 0, & I \\ \mathcal{A}, & -\gamma_1 I - \gamma_2 \mathcal{A}NN^* \mathcal{A}^* \end{bmatrix}; \quad \mathcal{D}(A) \equiv \mathcal{D}(\mathcal{A}) \times H^1_{\Gamma_0} \text{ with} \tag{5.4}$$

$$\mathcal{A}y \equiv \Delta y; \quad \mathcal{D}(\mathcal{A}) \equiv \{y \in H^2(\Omega);\ y|\Gamma_0 = 0;\ \frac{\partial y}{\partial\gamma}|_{\Gamma_1} = 0\}. \tag{5.5}$$

N: $L_2(\Gamma) \to L_2(\Omega)$ is defined by $Nu \equiv v$ iff $\Delta v = 0$; $v|_{\Gamma_0} = 0$; $\frac{\partial}{\partial\gamma} v|_{\Gamma_1} = u$.

$$Bu \equiv \begin{bmatrix} 0 \\ \mathcal{A}Nu \end{bmatrix}. \tag{5.6}$$

Since A generates an exponentially stable semigroup, stabilizability/detectability conditions (C-1), (C-2) hold automatically. Also, it can be shown [see sect. 4.2 in [L.2] that the hypothesis (H-1) holds as long as $\gamma_2 > 0$. If $\dim \Omega = 1$ then γ_2 can be taken zero). Since we are interested in optimal compensator we select the operator R in compliance with hypothesis (C-3). We define,

$$R\,z \equiv R\begin{bmatrix} z_1 \\ z_2 \end{bmatrix} = \begin{bmatrix} 0 \\ z_1 \end{bmatrix}, \text{ so } \ |Rz(t)|_H = |z(t)|_{L_2(\Omega)}. \tag{5.7}$$

We can easily verify (see [L.3]) that the inequality of hypothesis (C.3) is satisfied.

As an approximating subspace V_h of H we take $V_h \equiv \tilde{V}_h \times \tilde{V}_h$ where $\tilde{V}_h$ is a space of linear splines defined on a quasi uniform mesh of Ω with $\tilde{\pi}_h$ orthogonal projection of $L_2(\Omega)$ onto $\tilde{V}_h$. Let $\mathcal{A}_h$ be a standard Galerkin approximation of the operator $\mathcal{A}$ i.e.

$$(-\mathcal{A}_h u_h, v_h)_{L_2(\Omega)} \equiv (\nabla u_h, \nabla v_h)_{L_2(\Omega)} ; \quad u_h, v_h \in \tilde{V}_h . \tag{5.8}$$

We define D_h: $\tilde{V}_h \to \tilde{V}_h$ as

$$D_h v_h \equiv \tilde{\pi}_h \mathcal{A} N N^* \mathcal{A} v_h \text{ or}$$

$$(D_h v_h, u_h)_{L_2(\Omega)} = (v_h, u_h)_{L_2(\Gamma)} \quad \text{for } u_h, v_h \in \tilde{V}_h . \tag{5.9}$$

As an approximation of the operator A and B we take

$$A_h \equiv \begin{bmatrix} 0 & \tilde{\pi}_h \\ \mathcal{A}_h & -\gamma_1 \tilde{\pi}_h - \gamma_2 D_h \end{bmatrix} ; \tag{5.10}$$

$$B_h u \equiv \begin{bmatrix} 0 \\ \tilde{\pi}_h \mathcal{A} N u \end{bmatrix} . \tag{5.11}$$

Hence

$$B_h^* v_h = B_h^* (v_{h1}, v_{h2}) \equiv v_{h2} |_{\Gamma} , \tag{5.12}$$

and

$$A_h^* = \begin{bmatrix} 0 & -\tilde{\pi}_h \\ -\mathcal{A}_h & -\gamma_1 \tilde{\pi}_h - \gamma_2 D_h \end{bmatrix} . \tag{5.13}$$

It is shown in [L.3] that approximation hypothesis (A-1)-(A-2) (C-1-h)-(C-3h) are verified and consequently all the assumptions listed and Theorem 4.1 are satisfied (including the compactness of C which, in our case, follows from compactness of Sobolev's imbeddings). Hence the conclusion of Theorem 4.1 applies to the model (5.1), (5.2).

Remark 5.1 Notice that in this example — to reach a final conclusion of Theorem 4.1, it is important that damping occurs in the interior and on the boundary. Indeed, interior damping is needed to prove the property of uniform stability of $e^{A_h t}$ (i.e. hypothesis (C-1-h), (C-1-h), while boundary damping is used to prove the property of "uniform regularity" — i.e. assumption (A-1-ii). It should be noted that in one dimensional case when the solutions of (5.1) with $\gamma_2 = 0$ and $u \in L_2(0T)$ are in the space of finite energy $H^1(\Omega) \times L_2(\Omega)$, assumption (A-1-ii) and (A-1) hold with $\gamma_2 = 0$. Thus the result of Theorem 4.1 is valid, if dim $\Omega = 1$, with $\gamma_1 > 0$ and $\gamma_2 = 0$.

6. References

[B.1] A. V. Balakrishnan, Applied Functional Analysis, Springer Verlag 1975.

[C.1] R. Curtain, Finite dimensional compensators for parabolic distributed systems with unbounded control and observation, *SIAM J. Control* Vol. 22 (1984), pp. 255-277.

[G.1] J. S. Gibson, An analysis of optimal model regulation: convergence and stability, *SIAM J. Control Optimiz.* 19 (1981), pp. 686-707.

[G.2] J. S. Gibson, Approximation theory for linear quadratic Gaussian control of flexible structures. *SIAM J. Control, Optimiz.* 29 (1991), pp. 1-38.

[I.1] K. Ito, Finite dimensional compensators for infinite dimensional systems via Galerkin-type approximations, *J. Control Opt.* 28 (1990), pp. 1251-1269.

[F-L-T] F. Flandoli, I. Lasiecka, R. Triggiani, Algebraic Riccati Equations with nonsmoothing observation arising in hyperbolic and Euler-Bernouli boundary control problems, *Ann. Mat. Pura of Appl.* Vol. CLIII (1988), pp. 307-382.

[L.1] I. Lasiecka, Approximations of solutions to infinite - dimensional Algebraic Riccati equations with unbounded input operators, *Numer. Fund. Anal. and Optimiz.* 11 (304) (1990), pp. 303-378.

[L.2] I. Lasiecka, Stabilization of hyperbolic and parabolic systems with nonlinearly perturbed boundary conditions, *Journal Diff. Eq.* Vol. 75 (1988), pp. 53-87.

[L.3] I. Lasiecka, Galerkin approximations of infinite dimensional compensator for flexible structures with unbounded control action.

[M.1] L. Monauni, On the abstract Cauchy problem and the generator problem for semigroups of bounded operators, Technical Report No. 90 (1980). Control Theory Centre, University of Warwick, England.

[P.1] A. Pazy, Semigroups of operators and applications to Partial Differential Equations, Springer-Verlag (1986).

[S.1] J. M. Schumacher, A direct approach to compensator design for distributed parameter systems, *SIAM J. On Control* Vol. 21 (1983), pp. 823-837.

[W.1] G. Weiss, Two conjectures on the admissibility of control operators. Proceedings of the Conference on Distributed parameter Systems, Vorau (Austria), July 1990.

[G-N] G. Greinier, R. Nagel, On the stability of strongly continuous semigroups of positive operators on L^2 (μ). Annati Scuole Normale Pisa X.2 (1983) pp. 257-262.

REMARKS ON BOUNDARY CONTROL FOR POLYHEDRAL DOMAINS AND RELATED RESULTS

Walter Littman

University of Minnesota

Minneapolis, MN 55455, USA

Abstract: We establish boundary controllability for the Neumann (and other) problems for hyperbolic equations with constant coefficients in polyhedral domains not using multiplier methods. The method, which has the advantage that the "energy" remains bounded, is well suited for computations. Applications are discussed.

1. Introduction.

In 1973 D. Russell [R] introduced a method for boundary control for the wave equation in an odd number of space dimensions n which was initially based on Huygens' principle. The main idea was as follows: Assuming the controls active on the whole boundary $\partial\Omega$ of the bounded spatial domain Ω, extend the given data as smoothly as possible to all of R^n to have compact support in a slight enlargement of Ω, and then solve the pure initial value problem for all of space and $t > 0$. By Huygens' principle there is a value $T > 0$ (easily determined from the geometry of Ω) such that this solution vanishes near Ω for $t \geq T$. One then simply reads off the trace of the solution on $\partial\Omega \times [0, T]$ or the trace of its normal derivative, depending on whether one treats the Dirichlet of Neumann problem.

The main advantage of Russell's method is its simplicity; also the fact that the problem is reduced to that of solving a pure initial value problem makes it a good candidate for computations.

The main disadvantages are: a) It is restricted to the *wave* equation for *n odd.* b) There is approximately an extra one-half degree of smoothness required of the initial data for the controls to be in L^2 than seems reasonable from the one dimensional case. c) The controls must be applied on the whole boundary.

A number of methods (see for example, [Ho], [Lio], [LT], [Lit1] [Lit2], [BLR]) have since been developed to overcome some or all of these difficulties. In particular, the method of multipliers or the "Hilbert Space Uniqueness Method" as in [Ho], [Lio], [LT], as well as the recent work of [BLR] seems to overcome all the mentioned drawbacks of Russell's method of "letting nature take its course." Nevertheless, the newer methods have some disadvantages of their own. In addition, being more complicated, both methods, when used in conjunction with the Neumann controls leave open the question whether the resulting motion stays in space of finite energy—assuming the initial data is, and the control Neumann data is in L^2. The methods of [BLR] requires $\partial\Omega$ to be C^∞—thus preventing the applicability to polyhedral domains. Furthermore, even the results of Grisvard [G] for polygonal domains (using "H.U.M") is restricted to $n = 2, 3$.

Our aim is in a sense to "resurrect" Russell's method. In Section 2 we recall some old estimates for equations with constant coefficients. In Section 3 we will show that *Russell's original method* (for the wave equation in odd dimensions) when applied to the Neumann or Robin controls on the whole boundary *works with the same spaces as the newer methods*: when the initial data has finite energy, the boundary controls may be taken in L^2. In addition, the state of the system continues to have finite energy throughout.

In Section 4, we show how a recent method introduced in [LitTa] in connection with the Schrödinger equation can be used for general hyperbolic equations. In particular, for equations with constant coefficients, for polyhedral domains, it may be considered an extension of Russell's method that overcomes all the previous objections. The method also works for domains "with cracks."

In Section 5 we discuss some applications to computations of the controls as well as to homogenization and make some final remarks.

Let us mention that the restriction to constant coefficients and to polyhedral domains we believe to be not fundamental. We hope to be able to remove these restrictions at some future time. $\partial\Omega$ then should be "piecewise smooth." We also hope to treat the case with Dirichlet data.

2. Some estimates for linear equations with constant coefficients.

It seems curious that the estimates we need go back to Hölmander's thesis [H]. We shall state two estimates: the first one a necessary and sufficient condition for an estimate to hold (but somewhat tedious to apply); the second is a much more easily applied condition which is sufficient only but not necessary. We introduce the notation $D = (D_1, \cdots, D_2)$, where $D_j = i^{-1}(\partial/\partial x_j)$. If $P(\xi)$ is a polynomial in $\xi = (\xi_1, \ldots, \xi)$, we let

$$\tilde{P}(\xi) = (\Sigma |P^{(\alpha)}(\xi)|^2)^{1/2},$$

where $P^{(\alpha)}$ are derivatives of P, and the summation extends over all α.

Let Ω be a bounded domain in R^ν and Σ a linear variety of dimension μ, $1 \leq \mu \leq \nu - 1$. We assume that Σ has points in common with Ω. By Σ' we denote any one of the $\nu - \mu$ dimensional linear varieties in R^ν orthogonal to Σ. The surface element in Σ will be denoted by $d\sigma$, that in Σ' by $d\sigma'$. We are interested in the a priori inequality

$$\int_\Sigma |Q(D)u|^2 d\sigma \leq C(\|P(D)u\|^2 + \|u\|^2), \qquad u \in C_0^\infty(\Omega).$$

If this inequality is valid, then whenever u and $P(D)u$ are elements in L^2 with compact support in Ω, then the restriction of $Q(D)u$ exists (in an appropriate sense) as an element of $L^2(\Sigma)$ with support contained in Ω.

THEOREM 2.8 from [H]. *A necessary and sufficient condition in order that the trace of $Q(D)u$ should exist in $L^2(\Sigma)$ whenever Lu and u are in L^2 with compact support in Ω is that $Q(\xi)/\tilde{P}(\xi)$ be uniformly square integrable in the varieties Σ', i.e.,*

$$\int_{\Sigma'} \frac{|Q(\xi)|^2}{\tilde{P}(\xi)^2} d\sigma' < C,$$

where the constant C does not depend on the choice of the variety Σ' orthogonal to Σ.

(The statement is still true if $\mu = 0$ or ν). Our main interest is in the case where $\mu = \nu - 1$. In that case we obtain the following consequence of the preceding result.

THEOREM 2.9 of [H]. *If Σ is a hyperplane with normal N, we have*

$$\int_\Sigma |P_N^{(\alpha)}(D)u|^2 d\sigma \leq C \int |P(D)u|^2 \, dx, \qquad u \in C_0^\infty(\Omega),$$

where $P_N^{(\alpha)}(\xi) = \Sigma N_k \partial P^{(\alpha)}(\xi)/\partial \xi_k$. Thus the restrictions to Σ of all $P_N^{(\alpha)}(D)u$ can be defined as elements of $L^2(\Sigma)$ whenever u and $P(D)u$ are in L^2 with compact support.

In the case where $P(\xi)$ is a regular quadratic form (i.e., $P(D)$ is of second order), $P_N(D)$ is the usual "conormal" derivative. For operators of higher order than two, the operator P_N appears to be an appropriate generalization of "conormal" derivative.

3. The wave equation with an odd number of space dimensions.

The aim of this section is, as stated in introduction, to show that Russell's construction using Huygen's Principle for Neumann or Robin controls on the whole boundary of the bounded domain Ω yields controls in the same spaces as the newer results using multiplier or "H.U.M." methods or the geometric optics method of [BLR]. The main question reduces to the following.

Suppose we solve the *pure* initial value problem (in all of R^{n+1}) with initial data having finite energy. Does the normal derivative $\frac{\partial u}{\partial N}$ (where N is the normal to $\partial\Omega$) of the solution have an L^2 trace on $\partial\Omega$? Now if $\partial\Omega$ is polyhedral, this reduces to the question, "does $\frac{\partial u}{\partial N}$ exist as a trace in L^2 on any hyperplane in R^{n+1}?" To localize the problem, we let $\varphi(x,t)$ be a function in $C_0^\infty(R^{n+1})$ and consider $\square(u\varphi) = f \in L^2$ and having compact support. Thus by results of the previous section $\frac{\partial u}{\partial N}$ exists as an L^2 trace on any hyperplane, in particular the plane portions of $\partial\Omega$ if Ω is a polyhedron. The procedure then proceeds as described in the beginning of the introduction. The mixed problem with the given initial data and newly determined boundary data solves the boundary control problem with Neumann or Robin ($\frac{\partial u}{\partial N} + au$) controls in L^2 of the lateral boundary. Notice that the resulting solution will continue to have finite energy for all the relevant t. We remark that the well posedness of the mixed problem even with Lipschitz $\partial\Omega$ is well known (see for ex. [Lio]).

Also notice that this method lends itself very well to computations (say by finite differences) since only a pure initial value problem need be computed. The solution of the finite difference problem must be extrapolated to a function $\tilde{u}(x,t)$ for all x,t, and the trace of $\frac{\partial \bar{u}}{\partial N}$ must be suitably interpreted. There remains, of course, the theoretical question of the convergence of the constructed $\frac{\partial \bar{u}}{\partial N}$ to an L^2 function on the given hyperplane. Numerical methods, using "H.U.M." for constructing (approximately) the control functions have been given by Glowinski, Li and Lions [GLL].

4. The general case.

We here make use of a very versatile method of boundary control, introduced in [LitTa] in connection with the Schrödinger equation, can be summarized by the principle

local smoothing + reversibility + uniqueness $\Rightarrow$ controllability.

We first describe the method for control on the whole boundary $\partial\Omega$ of the bounded domain Ω in R^n. We illustrate the method with second order hyperbolic equations with data in the finite energy space $H_1(\Omega) \times L^2(\Omega)$, but it will be clear that the method extends to higher order equations as well, in a variety of spaces under appropriate conditions. We assume that the hyperbolic operator L is defined in all of $R^n \times [0,T]$. Let E be an extension operator which extends initial data of finite energy in Ω with data with compact support in a slight enlargement Ω_ϵ of Ω, so as to maintain finite energy. Let S be operator which solves the pure initial value problem for the linear hyperbolic equation $Lu = 0$ in all of x-space and maps the Cauchy data at time u to the Cauchy data at time T. We assume that S is locally smoothing in the sense that the Cauchy data at time T will be smoother in Ω_ϵ. This certainly will be the case if the coefficients of L are C^∞ and each bicharacteristic curve in $\bar{\Omega} \times (0,T)$ either enters or leaves (and stays out of) the cylinder through the lateral boundary, and Ω_ϵ is sufficiently close to Ω. Secondly, we use the fact that the backward pure initial problem from $t = T$ to $t = 0$ is solvable, and call the operator mapping the Cauchy data at $t = T$ to that at $t = 0$ by S^{-1}. We also assume that S^{-1} is locally smoothing in the same sense as S. (The same assumptions on the bicharacteristics mentioned before will insure this also.) Let φ be a smooth "cut off" functions with $\varphi \equiv 1$ near Ω but zero outside of Ω_ϵ. φ will also denote the operation of multiplying by φ.

Now let g be the given initial data. Consider the equation in Ω

$$f - S^{-1}\varphi SEf = g \tag{$*$}$$

for the unknown f. Denoting $S^{-1}\varphi SE$ by K, it is clear that K is smoothing, which makes the above equation a Fredholm equation. Now suppose a solution f exists. Then we define the extension $\tilde{g}$ of g to all of R^n by

$$\tilde{g} = Ef - S^{-1}\varphi SEf.$$

Now apply S to both sides

$$SEf - \varphi SEf = S\tilde{g} = 0 \ \text{ near } \Omega \ \text{ at } t = T.$$

Thus if we solve the initial value problem in $R^n \times \{t \geq 0\}$, with initial data $\tilde{g}$, it will vanish near Ω at $t = T$. We could then continue the solution to be identically zero past $t = T$ in a slightly enlarged Ω. Thus we could read off the lateral boundary data on $\partial\Omega \times [0, T]$ to obtain the right inhomogeneous boundary data that will steer the solution zero in time T. (One, of course, hopes these will be in as good a space as possible.)

We would, of course, like to conclude that boundary controllability holds for all g with finite energy. This is where the "uniqueness" part comes in. We assume that the following uniqueness properly holds: If $Lu = 0$ in $R^n \times (0, T)$ and $u \equiv 0$ in the complement of Ω then $u \equiv 0$ in $R^n \times (0,T)$. This follows, for example, if L has analytic coefficients near $\Omega \times [0,T]$ (by the propagation of the analytic wave front set) or if the equation has coefficients independent of t (see [BLR], for example).

It is well known that one consequence of this uniqueness property is the *approximate controllability* of the system (see, for example, [LitTa]) and we will now make use of this property. (See also the duality argument in [Lit2].)

Suppose first that there is a one-dimensional null space to $I-K$. Then $I-K^*$ has a one-dimensional *null space* $\{\psi\}$. Pick ψ' (in the finite energy space) so that $\langle \psi', \psi \rangle \neq 0$. We will solve $(*)$ for a slightly different right hand side: $g - az$ (in the finite energy space) where a is a scalar. We require $\langle g - az, \psi \rangle = 0$ which can be achieved by $\langle g, \psi \rangle = a\langle z, \psi \rangle$ or $a = \langle g, \psi \rangle / \langle z, \psi \rangle$. This works provided $\langle z, \psi \rangle \neq 0$. Now pick z sufficiently close to ψ' so that $\langle z, \psi \rangle \neq 0$ and in addition such that z can be steered to zero. (That we can do this follows from approximate controllability.) Then if the initial data $(g - az)$ can also be steered to zero, the same follows for g.

Next suppose the null space of $I-K$ is k dimensional. Then try

$$\left\langle g - \sum_{i=1}^{k} a_i z_i, \psi_i \right\rangle = 0$$

$$\langle g, \psi_j \rangle = \sum_{i=1}^{k} a_i \langle z_i, \psi_j \rangle$$

This can be solved for the a_i, provided $\det\{\langle z_i, \psi_j \rangle\} \neq 0$. The z_i should be chosen so that they can be steered to zero at $t = T$. Then proceed as in the case $k = 1$.

Now in what space are the controls? We notice that there are three components to the solution obtained: a) The solution to the pure initial value problem with data Ef, where f is a solution to the Fredholm equation; b) the solution to the pure initial value problem with data Kf; c) the part due to approximate controllability.

Components b) and c) may be taken to be as smooth as desired if the coefficients are C^∞. Hence the smoothness of the controllers is determined by a) i.e., the traces of the boundary operators acting on the solution of a pure initial value problems. We have seen in the case constant coefficients, and Neumann or Robin operators that, for polyhedral boundaries these traces are in L^2 of the polyhedral boundary $\times(0,T)$ provided the initial data have finite energy. *From the procedure it is clear that if the initial data has finite energy, the system will maintain finite energy as the controls are implemented.*

What if the controls are not on the whole boundary? Suppose the boundary consists of two parts, an external part ∂_E and an internal part ∂_I (which may have several components) which we take to the C^∞. We assume that $\partial_E \cap E_I = \emptyset$ (this is probably not essential). Then we replace the solution of the pure initial value problem in $R^n \times (0,T)$ by the solution of the exterior mixed problem with obstacle boundary ∂_I. Using the propagation of singularities for such problems as governed by broken reflected bicharacteristics, we can still obtain a smoothing operator S as before. The uniqueness theorem now follows from a version of Holmgren's theorem, using the fact that homogeneous boundary conditions are imposed on ∂_I. Otherwise, everything works as before. The value for T must, of course, be increased.

5. The case of higher order operators and systems.

Basically everything said goes over to higher order equations if one has, for example, analytic coefficients, provided one assumes strict hyperbolicity. This has been done by a different method in [Lit2] for controls on the entire boundary. For constant coefficients and again for the controls on the entire boundary, an explicitly formula for the controls was obtained in [Lit1]. There it was assumed that the initial data was C^∞. However, that assumption is not needed if one effects the control in two stages: First one solves the pure initial value problem with initial data in an appropriate direct sum of Sobolev spaces. Then at an appropriate time T, one multiplies by a cut-off function $\varphi(x)$ (thus making the Cauchy data C^∞ there) and applies the explicit control result just described in a backward fashion (giving Cauchy data zero at $t = 0$) and subtracts this from the solution to the first stage. One then reads of the traces of the boundary operators acting on the function just obtained.

For the constant coefficient case and polyhedral boundaries, there is a whole class of boundary operators (described in Section 2) including a natural generalization of the Neuman boundary operator for which the method would work. The only missing detail is a uniqueness theorem (that is all one needs at this stage) for the mixed problem for polyhedral domains—which seems not to be known in the general case. The Timoshenko plate equation (cf. [LaLio] and other systems arising in elastic vibration seem promising candidates for this method.

The methods described carry over to systems. For the case of constant coefficients and polyhedral domains, Hörmander's estimates are replaced by their generalization to systems as in [F]. In all cases a uniqueness theorem for the mixed problem for polyhedral domains is needed.

6. Applications.

The applications to computation of the control functions for the odd dimensional wave equation has already been discussed. It may be inferred that the method of Section 4 using an abstract argument for approximate controllability is not useful for computations. However, this is not the case. First, it can be shown in many cases (where energy is conserved, for example) that the operator $I - K$ is 1:1. But even if that is not the case, all one needs is to construct a set of functions z_i that can be steered to zero in time T and such that $\det\{\langle z_i, \psi_j \rangle\} \neq 0$. Such a set satisfying the first condition can easily be constructed by picking nonzero functions vanishing near Ω and with support in Ω_ϵ and applying S^{-1} to them. Chances are they will satisfy the second condition; otherwise, some experimentation will do the trick. The functions ψ_j must, of course, be computed (approximately). Solving the Fredholm equation can be a formidable task. However, it is probably no more complicated than the computations in [GLL]. At any rate for the wave equation is odd space dimension the first method described seems definitely advantageous.

Another application is to the problem of homogenization as described, for example, in [CDZ]. We will describe the problem in two space dimensions, but the results hold in general. Suppose we have a large square (say the unit square) with a number disjoint smaller squares of sides ϵ deleted yielding a domain Ω_ϵ. We are given initial data with finite energy and would like to find Neumann controls on $\partial\Omega_\epsilon$ which will bring the system to rest in finite time. Can we give a uniform estimate (independent of ϵ and the number of deleted squares) on the L^2 norms of the Neumann control functions? The answer from our procedure is obviously affirmative. Using multiplier techniques the answer is by no means obvious.

Bibliography

[BLR] Bardos, C, Lebeau G., and Rauch, J. *Contrôle et stabilization dans des problèmes hyperboliques*, Appendix 2 in [Lio].

[CDZ] Cioranescu, D, Donato, P., and Zuazua, E., *Exact boundary controllability for the wave equation in domains with small holes*, R.89035, Publications du Labratoire d'Analyse Numerique, University Pierre et Marie Curie, Paris, 1990.

[F] Fuglede, B., *A priori inequalities connected with systems of partial differential equations*, Acta Math., 105 (1961), 177–195.

[GLL] Glowinski, R., Li, C.L., and Lions, J.L, *A numerical approach to the exact boundary controllability of the wave equation*, Japan J. Appl. Math., 1 (1990), 1–76.

[G] Grisvard, P., *Singular solutions of elliptic problems and applications to the exact controllability of hyperbolic problems, function spaces, differential operators and nonlinear analysis*, (L. Päivärinta Ed.) Longman (1989), 169–248.

[Ho] Ho, L.,F., Observabilité frontière de l'equation des ondes, C.R. Acad. Sci. Paris Ser. I, Math 302 (1986), 443–446.

[H] Hörmander, L., *On the theory of general partial differential operators*, Acta Math. 94, (1955), 161–248.

[LT] Lasiecka, I. and Triggiani, R., *Exact controllability for the wave equation with Neumann boundary control*, Appl. Math. and Optimization, 19, (1989), 243–290.

[Lio] Lions, J.L., *Contrôlabilité exacte, pertubations et stabilization de systemes distribués*, vol. 1, Masson, Paris, 1988.

[LaLi] Lagnese, J.E. and Lions, J.L., *Modelling analysis and control of thin plates*, Masson, Paris, 1988.

[Lit1] Littman, W., *Boundary control theory for a class of hyperbolic and parabolic equations with constant coefficients*, Annals Sc. N. Sup Pisa Ser IV, 5 (1978), 567–580.

[Lit2] Littman, W., *Near optimal time boundary controllability for a class of hyperbolic equations*, Lecture Notes in Control and Information Sciences, Springer (1987), 306–312.

[LitTa] Littman W., and Taylor S. *Smoothing evolution equations and boundary control theory*, to appear in Journal d'Analyse (Jerusalem).

[R] Russell, D.L., *A unified boundary controllability theory for hyperbolic and parabolic partial differential equations*, Studies in Applied Mathematics, 52, # 3 (1973), 189–211.

Acknowledgement: This research was partially supported by NSF grant DMS90-02919.

STABILITY OF THE TRAVELLING WAVES IN A CLASS OF FREE BOUNDARY PROBLEMS ARISING IN COMBUSTION THEORY

ALESSANDRA LUNARDI
Dipartimento di Matematica
Via Ospedale 72, 09124 Cagliari, Italy

ABSTRACT
We study the stability of the travelling wave solutions of a class of one-phase and two-phase free boundary value problems arising in Combustion Theory. They are transformed into nonlinear evolution equations in Banach space, in such a way that stability of the travelling wave solution of the free boundary problem is equivalent to orbital stability of the null solution of the abstract problem. Then known stability results for abstract parabolic equations are applied.

1. Introduction

We consider a class of one-phase free boundary problems :

$$\begin{cases} u_t(t,x) = u_{xx}(t,x) + u(t,x)\cdot u_x(t,x)\ ,\ \ t \geq 0\ ,\ x \leq \xi(t)\ , \\ u(t,\xi(t)) = a\ ,\ u_x(t,\xi(t)) = 1\ ,\ \ t \geq 0\ , \\ u(t,-\infty) = 0\ ,\ \ t \geq 0, \end{cases} \tag{1.1}$$

with initial values

$$\xi(0) = 0\ ,\ \ u(0,x) = u_0(x)\ ,\ \ x \leq 0\ . \tag{1.2}$$

Here a is a given positive number, and $u_0 :]-\infty,0] \to \mathbf{R}$ is a given function. The unknown are $\xi : [0,+\infty[\to \mathbf{R}$ and $u : \{ (t,x) : t \geq 0\ , x \leq \xi(t) \} \to \mathbf{R}$. Since we look for regular (up to $t = 0$) solutions, the initial datum should satisfy the compatibility condition $u_0(0) = a$, $u_0'(0) = 1$. Problem (1.1)-(1.2) arises in transition from deflagration to detonation, and it is a simplified version of a two-phase problem :

$$\begin{cases} u_t(t,x) = u_{xx}(t,x) + u(t,x)\cdot u_x(t,x)\ ,\ \ t \geq 0\ ,\ x \neq \xi(t)\ , \\ u(t,\xi(t)) = a\ ,\ u_x(t,\xi(t)^+)-u_x(t,\xi(t)^-) = -1\ ,\ \ t \geq 0\ , \\ u(t,-\infty) = 0\ ,\ u(t,+\infty) = b\ ,\ \ t \geq 0, \end{cases} \tag{1.3}$$

The initial data for (1.3) are

$$\xi(0) = 0\ ,\ \ u(0,x) = u_0(x)\ ,\ \ x \in \mathbf{R}\ . \tag{1.4}$$

Again, a and b are given positive numbers, $u_0 : \mathbf{R} \to \mathbf{R}$ is a given function, and $\xi : [0,+\infty[\to \mathbf{R}$, $u : [0,+\infty[\times \mathbf{R} \to \mathbf{R}$ are unknown. Detailed discussions on the derivation of equations (1.1) and (1.3) can be found in [6], [11], [12], [20].

Both in the one-phase and in the two-phase problem, stability and instability of the travelling waves is of interest. We recall that a travelling wave, with speed $c \in \mathbf{R}$, is a solution $(\bar{\xi},\bar{u})$ such that

$$\bar{u}(t,x) = U(x+ct)\,,\quad \bar{\xi}(t) = -ct \tag{1.5}$$

for some $U : \mathbf{R} \to \mathbf{R}$. Concerning problem (1.1), it is easy to see that for each $a > 0$ there is a (unique up to translations) travelling wave solution, given by

$$c = \frac{2+a^2}{2a}\,,\quad U(y) = 2ca\, e^{cy}\,(2c-a+ae^{cy})^{-1} \tag{1.6}$$

Concerning problem (1.3), there is a (unique up to translations) travelling wave solution, if and only if a and b satisfy the condition

$$b > \sqrt{2},\ \ 2b^{-1} < a < b+2b^{-1}. \tag{1.7}$$

If (1.7) holds, we get again $c = \frac{2+a^2}{2a}$, and

$$U(y) = \begin{cases} 2ca\, e^{cy}(2c-a+ae^{cy})^{-1}\,, & y < 0, \\ [b(a+b-2c)-2b^{-1}(a-b)e^{-(b/2-1/b)y}]\cdot[a+b-2c-(a-b)e^{-(b/2-1/b)y}]\,, & y > 0. \end{cases} \tag{1.8}$$

However, the explicit expression of U is not important, since all we need is to know its asymptotic behavior.

Due to the translation invariance, one cannot expect that the travelling wave solution is stable, but merely orbitally stable with asymptotic phase, that is, if u_0 is close to U, the solution $(\xi(t),u(t,\cdot))$ is expected to converge to $(-ct+\xi_\infty, U(\cdot+ct-\xi_\infty))$ for some $\xi_\infty \in \mathbf{R}$. Arguing heuristically, in the one-phase problem ξ_∞ can be evaluated by integrating both members of the differential equation in (1.1) on the set $\{ (s,x) : 0 \le s \le t, -\infty < x \le \xi(t) \}$, and letting $t \to +\infty$. We find that $\xi(t)$ should approach

$$\gamma(t) = -ct - \frac{1}{a}\int_{-\infty}^{0} (u_0(\sigma)-U(\sigma))d\sigma = -ct + \xi_\infty. \tag{1.9}$$

Our aim is to make precise these statements, giving existence and uniqueness theorems for initial data belonging to suitable Banach spaces, stability results in such spaces, and specifying the rate of convergence.

Existence, uniqueness, and regularity of the solutions of the initial value problem for equations (1.1) and (1.3), can be studied in several ways. We remark that by differentiating with respect to t the identity $u(t,\xi(t)) = a$, we get $u_t(t,\xi(t)) + \dot{\xi}(t)u_x(t,\xi(t)) = 0$, so that $\dot{\xi}(t) = -u_{xx}(t,\xi(t))$: therefore our problems are equivalent to suitable Stefan problems in unbounded intervals . The classical methods for Stefan problems in bounded intervals (see e.g. [9] and the references quoted there) can be adapted to our situation, at least in the case of the one-phase problem, to get local existence, uniqueness and regularity of the solution. Another approach was followed in [10] : they transformed problem (1.1) into an elliptic-parabolic problem, which can be solved by parabolic regularization. Coming back to problem (1.1) - (1.2) , they proved existence in the large and uniqueness of the classical solution for every sufficiently smooth initial value u_0 attaining its maximum at the unique point $x = 0$. At the same time they showed that the travelling wave solution U given in (1.6) is orbitally stable in the L^1 norm, and that the interface $\xi(t)$ approaches $\gamma(t)$ as $t \to \infty$. The rate of convergence was not specified. Asymptotic behavior of $\xi(t)$ was studied in [17] for a simplified version of (1.1) (specifically, for the equation $u_t = u_{xx}$; however the method should work also in the present case), as an application of asymptotic behavior results for Stefan problems. It was shown that for every increasing L^1 initial value u_0, with u_0 and u_0' bounded, the function ξ converges to γ as $t \to \infty$, and if u_0' decays exponentially as $x \to -\infty$, then ξ converges exponentially to γ.

Our method consists in transforming equations (1.1), (1.3) into abstract parabolic equations in a suitable Banach space X, in such a way that orbital stability of the travelling wave is equivalent to stability of the null solution of the abstract equation. The abstract equations take the form

$$u'(t) = F(u(t)), \tag{1.10}$$

where $F : D \to X$ is a smooth function, and D is a Banach space continuously embedded in X. The abstract parabolicity character is given by the fact that for every $u \in D$, the Fréchet derivative $F'(u) : D \to X$ generates an analytic semigroup in X. Geometric theory of such equations has been developed in the last five years (see e.g. [16], [13], [14], [15], [7]). In particular, the principle of linearized stability holds : if $F(0) = 0$ and the spectrum of $F'(0)$ is contained in the half plane $\{ \lambda \in C : \mathrm{Re}\,\lambda < -\varepsilon \}$, for some positive ε, then the null solution of (1.9) is exponentially asymptotically stable in the D-norm; if there are some elements of the spectrum of $F'(0)$ with positive real part, and the spectrum does not intersect the imaginary axis, then the null solution is unstable, and the stable and unstable manifolds can be constructed.

There are several choices for the phase space X which make it possible to get existence and regularity of the solution. However, to treat stability and instability, we need very precise spectral information, and we are forced to choose a space of suitably weighted functions as a phase space. From the physical point of view this is not a restriction, since every solution is expected to decay exponentially as $x \to -\infty$ for every value of t. We refer the reader to [18], [19] for a detailed discussion on the use of weighted spaces in studying stability of travelling wave solutions of parabolic equations.

Since the method used is the same both for problem (1.1) and for problem (1.3), we focus our attention here to problem (1.1). The two-phase problem (1.3) was studied in [2], [3], and the results are summarized in Section 5. The rest of the paper is organized as follows : in Section 2 problem (1.1) is reduced to an abstract evolution equation; in Section 3 we briefly recall the results we need about parabolic equations in Banach spaces; in Section 4 we apply the abstract results to our equation, getting exponential orbital stability of the travelling wave U given by (1.6), and analyticity of the free boundary ξ. Similar results were announced in [1] for a simplified version of problem (1.1), without the Burgers nonlinearity $u \cdot u_x$. See also [4], [5] for a slightly different approach.

2. Reducing problem (1.1) to an abstract evolution equation

Through the whole section we denote differentiation with respect to t by a dot, and differentiation with respect to the space variable y by a prime or a subscript.
First of all, we transform (1.1) into a fixed boundary value problem, by means of the changement of coordinates $y = x - \xi(t)$, so that, setting $z(t,y) = u(t,x) = u(t,y+\xi(t))$, problem (1.1) becomes

$$\begin{cases} z_t(t,y) = z_{yy}(t,y) + \dot{\xi}(t)z_y(t,y) + z(t,y)z_y(t,y), \ t \geq 0 \ , \ y \leq 0 \ , \\ z(t,0) = a, \ z_y(t,0) = 1, \ t \geq 0 \ , \\ z(t,-\infty) = 0, \end{cases} \tag{2.1}$$

Setting then

$$s(t) = \xi(t) - \bar{\xi}(t) = \xi(t) + ct, \ v(t,y) = z(t,y) - U(y) \ , \ v_0(y) = u_0(y) - U(y) \ , \tag{2.2}$$

where (c,U) is defined in (1.6), we get

$$\begin{cases} v_t = v_{yy} + [(U-c)v_y]_y + vv_y + \dot{s}\, v_y + \dot{s}\, U', \ t \geq 0 \ , \ y \leq 0 \ , \\ v(t,0) = v_y(t,0) = 0 \ , \ t \geq 0 \ , \\ v(t,-\infty) = 0, \end{cases} \tag{2.3}$$

with initial data

$$s(0) = 0 \ , \ v(0,y) = u_0(y) - U(y) \ , \ y \leq 0 \ . \tag{2.4}$$

Let $\mathcal{X}$ be the Banach space defined by

$$\begin{cases} \mathcal{X} = \{ v :]-\infty,0] \to \mathbf{R} \mid y \to e^{-cy/2}v(y) \in UCB(]-\infty,0]) \}, \\ \| v \| = \sup_{y \leq 0} | e^{-cy/2}v(y) | \ , \end{cases} \tag{2.5}$$

where UCB stands for uniformly continuous and bounded, and let $D(\mathcal{L})$ be the domain of the linear operator $v \to v'' - \frac{d}{dy}[(U-c)v]$ in $\mathcal{X}$:

$$\begin{cases} D(\mathcal{L}) = \{ v \in C^2(]-\infty,0];\mathbf{R}) : v, v', v'' \in \mathcal{X} \ , \ v'(0) = (c-a)v(0) \} \ , \\ \mathcal{L}v = v'' - \frac{d}{dy}[(U-c)v] \ . \end{cases} \tag{2.6}$$

We choose the boundary condition $v'(0) = (c-a)v(0)$ in $D(\mathcal{L})$ because it is satisfied by U' : then U' belongs to the kernel of $\mathcal{L}$; more precisely, Ker $\mathcal{L}$ is spanned by U'. It is easy to see that, setting

$$(Pv)(y) = \frac{1}{a} \int_{-\infty}^{0} v(\sigma)d\sigma \ U'(y) \ , \ \forall v \in X, \tag{2.7}$$

then $P\mathcal{L} = 0$. Therefore system (2.2) can be decoupled by applying the projections P and 1-P , that is setting

$$v(t,y) = Pv(t,\cdot)(y) + [v(t,y)-Pv(t,\cdot)(y)] = p(t)U'(y) + w(t,y) \ . \tag{2.8}$$

Plugging (2.8) into (2.3) , an recalling that U satisfies $U'' = cU' - UU'$, we get

$$\begin{cases} \dot{p}(t) = \dot{s}(t), \ t \geq 0 \\ w_t = w_{yy} + [(U-c)w_y]_y + (1-P)[vv_y + \dot{s}\, v_y] \ , \ t \geq 0, \ y \leq 0 \\ p(t) = -w(t,0), \ w_y(t,0) = (c-a)w(t,0), \ w(t,-\infty) = 0, \ t \geq 0. \end{cases} \tag{2.9}$$

In particular, we find that $w(t,\cdot) \in D(\mathcal{L})$ for each t, and

$$\dot{p}(t) = \dot{s}(t) = -w_t(t,0) = -\mathcal{L}w(t,0)\ ,\quad t \geq 0. \tag{2.10}$$

Moreover, by (2.7) it follows that $P(vv_y) = Pv_y = 0$, so that

$$\begin{cases} w_t(t,y) = \mathcal{L}w(t,y) + [-w(t,0)U'(y)+w(t,y)]\cdot[-w(t,0)U''(y)+w_y(t,y)] + \\ -\mathcal{L}w(t,0)[-w(t,0)U''(y)+w_y(t,y)]\ ,\ t \geq 0\ ,\ y \leq 0\ , \\ w_y(t,0) = (c-a)w(t,0),\ t \geq 0\ ,\ w(t,-\infty) = 0\ ,\ t \geq 0\ , \end{cases} \tag{2.11}$$

$$w(0,y) = w_0(y) = (1-P)v_0(y),\ y \leq 0\ . \tag{2.12}$$

If w is a solution to (2.11)-(2.12), then, defining p, s by (2.10) (with $p(0) = -\mathcal{L}w_0$ and $s(0) = 0$) and v by (2.8), the couple (s,v) is a solution of (2.3)-(2.4), so that the couple (ξ,u), with $\xi(t) = s(t) - ct$, and $u(t,x) = v(t,x-\xi(t))$, is a solution of (1.1)-(1.2). Moreover, stability of the null solution to problem (2.11) is equivalent to orbital stability of the travelling wave U in problem (1.1).

Problem (2.11) can be written as a nonlinear equation of the form (1.10) in the Banach space $\mathcal{X}$, by setting

$$\begin{cases} F : D(\mathcal{L}) \to \mathcal{X}\ , \\ F(w) = \mathcal{L}w + [-w(0)U'+w]\cdot[-w(0)U''+w'] -\mathcal{L}w(0)[-w(0)U''+w']\ . \end{cases} \tag{2.13}$$

Then F is an analytic function, it maps $D = D(\mathcal{L}) \cap X$ into X, and for each $w, z \in D(\mathcal{L})$ we have

$$F'(w)(z) = \mathcal{L}z + [-w(0)U'+w]\cdot[-z(0)U''+z'] + [-z(0)U'+z]\cdot[-w(0)U''+w'] +$$

$$- \mathcal{L}w(0)[-z(0)U'+z] - \mathcal{L}z(0)[-w(0)U'+w]\ .$$

Therefore, $F'(w)$ is a nonlocal elliptic second order operator with C^1 coefficients and principal part $\mathcal{L}_0$ given by

$$\mathcal{L}_0 z = z'' - z''(0)[-w(0)U'+w]\ .$$

LEMMA 2.1 *For every* $w \in D(\mathcal{L})$, *the operator* $F'(w) : D(\mathcal{L}) \to \mathcal{X}$ *generates an analytic semigroup in* $\mathcal{X}$.

Proof : We recall that a linear operator $A : D(A) \to \mathcal{X}$ generates an analytic semigroup if and only if the resolvent set of its complexification $\tilde{A}$ contains a sector $S = \{ \lambda \in C : \lambda \neq \omega, |\arg \lambda| < \theta \}$, for some $\omega \in R$ and $\theta > \pi/2$, and $\| (\lambda-\tilde{A})^{-1} \|_{L(\tilde{\mathcal{X}})} \leq M |\lambda|^{-1}$ for each $\lambda \in S$. Here $\tilde{\mathcal{X}} = \{ \phi+i\psi : \phi, \psi \in \mathcal{X} \}$ is the complexification of $\mathcal{X}$. It is easy to see that if ϕ and ψ are elements of $\mathcal{X}$, then the operator $A : D(A) = D(\mathcal{L}) \to \mathcal{X}$, $Az = z'' + \phi z' + \psi z$, generates an analytic semigroup in $\mathcal{X}$. If η is another function belonging to $\mathcal{X}$, then the operator $B : D(\mathcal{L}) \to \mathcal{X}$, $(Bz)(y) = z''(0)\eta(y)$, is compact. Since $D(\mathcal{L})$ is dense in $\mathcal{X}$, by a result on compact perturbations of densely defined generators (see e.g. [8]) it follows that $A+B$ generates an analytic semigroup in X. In our case, the operator $F'(w)$ is just of this form : therefore, it generates an analytic semigroup in $\mathcal{X}$. ∎

We remark that $F(0) = 0$, $F'(0) = \mathcal{L}$. Since we are interested in stability, the spectral properties of $\mathcal{L}$ are of crucial importance in our analysis.

PROPOSITION 2.2 *We have*

$$\sigma(L) =]-\infty, -\frac{(a-c)^2}{4} - \frac{1}{2}] \cup \{0\},$$

and 0 *is a simple eigenvalue of* L. *Moreover, if* P *is the projection defined in* (2.7), *for each sufficiently small* $\varepsilon > 0$, *we have*

$$P = \frac{1}{2\pi i} \int_{C(0,\varepsilon)} R(\xi, \tilde{L})d\xi,$$

where $\tilde{L}$ *is the complexification of* L.

Proof : Fix $f \in \tilde{X}$ and $\lambda \in C$. The resolvent equation $\lambda u - \tilde{L}u = f$ is an elliptic boundary value problem with variable coefficient. To reduce it to an equation with constant coefficients, it is convenient to set

$$u = \frac{d}{dy}(\phi\psi). \tag{2.14}$$

Then the resolvent equation is transformed into

$$[\lambda\phi - \phi'' - (U-c)\phi']\psi + [-2\phi' - (U-c)\phi]\psi' - \phi''\psi = \int_0^x f(\sigma)d\sigma + k,$$

where k is an arbitrary constant. We choose $\phi(x) = \exp(-\int_0^x (U(\sigma)-c)/2\, d\sigma)$, so that $-2\phi' - (U-c)\phi = 0$, and the equation becomes

$$(\lambda + \frac{(c-a)^2}{4} + \frac{1}{2})\psi - \psi'' = \frac{1}{\phi}[\int_0^x f(\sigma)d\sigma + k]. \tag{2.15}$$

Since u, u' u" should belong to $\tilde{X}$, and $\phi(x) \sim e^{cx/2}$ as $x \to -\infty$, then ψ, ψ', ψ'' should be uniformly continuous and bounded in $]-\infty,0]$. The right hand side of (2.15) is bounded if and only if

$$k = \int_{-\infty}^0 f(\sigma)d\sigma.$$

Due to the choice of k, (2.15) has infinitely many uniformly continuous and bounded solutions, given by

$$\psi(x) = \exp(\sqrt{\mu}x)[c_1 - (2\sqrt{\mu})^{-1}\int_0^x g(\sigma)\exp(-\sqrt{\mu}\sigma)d\sigma] +$$
$$+ (2\sqrt{\mu})^{-1}\exp(-\sqrt{\mu}x)\int_{-\infty}^x g(\sigma)\exp(\sqrt{\mu}\sigma)d\sigma, \tag{2.16}$$

with

$$\mu = \lambda + \frac{(c-a)^2}{4} + \frac{1}{2}, \quad g(x) = \frac{1}{\phi(x)}\int_{-\infty}^x f(\sigma)d\sigma, \quad c_1 \text{ arbitrary},$$

provided $\mu \notin]-\infty,0]$. Replacing ϕ and ψ in (2.14), we find that u satisfies the boundary condition $u'(0) = (c-a)u(0)$ if and only if

$$(\mu - \frac{(c-a)^2}{4} - \frac{1}{2})c_1 = -(2\sqrt{\mu})^{-1}(\mu - \frac{(c-a)^2}{4} - \frac{1}{2}) \int_{-\infty}^{0} g(\sigma)\exp(\sqrt{\mu}\sigma)d\sigma + g(0). \quad (2.17)$$

If $\mu \neq \frac{(c-a)^2}{4} + \frac{1}{2}$, that is, if $\lambda \neq 0$, we get c_1 from (2.17) and we plug it in (2.16). Then, using (2.14), we find a unique solution $u = u(\lambda)$ of the resolvent equation. Moreover, we obtain easily that the function $\lambda \to u(\lambda)$ is holomorphic in a neighborhood of 0, so that 0 is a semisimple eigenvalue of $\tilde{\mathcal{L}}$. Since the eigenspace is one dimensional, then 0 is a simple eigenvalue. For the same reason, there is a unique projection Q such that $Q\tilde{\mathcal{L}} = 0$ on $D(\tilde{\mathcal{L}})$. Since the projection given by the Dunford integral enjoys such a property, then it coincides with P. ∎

Set now

$$X = (1-P)(\mathcal{X}) = \{ w \in \mathcal{X} : \int_{-\infty}^{0} w(\sigma)d\sigma = 0 \}, \ \| w \|_X = \| w \|_{\mathcal{X}},$$

$$D = D(\mathcal{L}) \cap (1-P)(\mathcal{X}), \ L : D = \to X, \ Lw = \mathcal{L}w.$$

As an immediate consequence of Lemma 2.1 and Proposition 2.2 we get:

COROLLARY 2.3 *The operator* $L : D \to X$ *generates an analytic semigroup in* X. *Its spectrum is the real half line* $]-\infty, -\frac{(c-a)^2}{4} - \frac{1}{2}]$. ∎

3. Some results on abstract nonlinear parabolic problems

Let D and X be real Banach spaces such that D is continuously embedded in X. We denote by $\| \cdot \|$ the norm in X and by $\| \cdot \|_D$ the norm in D. Consider the initial value problem

$$\begin{cases} w'(t) = F(w(t)), \ t \geq 0, \\ w(0) = w_0, \end{cases} \quad (3.1)$$

where $F : \Omega \to X$ is a C^2 function, Ω is an open set in D, and $w_0 \in \Omega$. The fundamental assumption, which gives a parabolic character to problem (3.1), is:

$$\begin{cases} \forall w \in \Omega, \ F'(w) : D \to X \text{ generates an analytic semigroup in } X, \\ \text{and the graph norm of } F'(w) \text{ is equivalent to the D-norm.} \end{cases} \quad (3.2)$$

The next local existence,uniqueness, and regularity theorem follows from [13] - [14].

THEOREM 3.1 *Let* Ω *be an open set in* D, *and let* $F : \Omega \to X$ *be a* C^2 *function, satisfying* (3.2). *Fix* $u_0 \in \Omega$ *such that* $F(u_0) \in \bar{D}$. *Then:*

(i) *There are* $\tau = \tau(u_0) > 0$ *and a solution* $u = u(\cdot, u_0)$ *of* (3.1) *in* $[0,\tau[$, *which belongs to* $C([0,\tau[;D) \cap C^1([0,\tau[;X) \cap C^\alpha([\varepsilon,\tau-\varepsilon];D) \cap C^{\alpha+1}([\varepsilon,\tau-\varepsilon];X)$ *for each* $\alpha \in]0,1[$, $\varepsilon \in]0,\tau/2[$. *Moreover for every* $T < \tau$ *we have*

$$\sup_{0<\varepsilon<T} \varepsilon^\alpha [\, u \,]_{C^\alpha([\varepsilon,T];D)} < +\infty , \tag{3.3}$$

and u *is the unique solution of* (3.1) *enjoying such a property for some* $\alpha \in\,]0,1[$.

(ii) *If, in addition,* F *is an analytic function, then* u *is analytic in* $]0,\tau[$ *with values in* D. ■

The proofs of the following stability and instability results can be found in [15], [3].

THEOREM 3.2 *Let* Ω *be an open neighborhood of* 0 *in* D, *and let* $F \in C^2(\Omega;X)$ *satisfy* (3.2) *and be such that* $F(0) = 0$. *Set* $L = F'(0)$, *and let* $\tilde{L}$ *be the complexification of* L. *Assume that*

$$\sup \{\, \mathrm{Re}\,\lambda : \lambda \in \sigma(\tilde{L}) \,\} = -\omega_0 < 0 . \tag{3.4}$$

Then for every $\omega \in [0,\omega_0[$ *there exist* $r, M > 0$ *such that if* $u_0 \in D$, $F(u_0) \in \bar{D}$, *and* $\| u_0 \|_D \le r$, *then* $\tau(u_0) = +\infty$, *and*

$$\sup_{t\ge 0} \| e^{\omega t} u(t,u_0) \|_D \le M . \;\; ■$$

THEOREM 3.3 *Let* Ω *be an open neighborhood of* 0 *in* D, *and let* $F \in C^2(\Omega;X)$ *satisfy* (3.2) *and be such that* $F(0) = 0$. *Set* $L = F'(0)$, *and let* $\tilde{L}$ *be the complexification of* L. *Assume that*

$$\sigma_+(\tilde{L}) = \sigma(\tilde{L}) \cap \{\lambda \in \mathbb{C} : \mathrm{Re}\,\lambda > 0 \} \neq \emptyset , \; \inf \{\, \mathrm{Re}\,\lambda : \lambda \in \sigma_+(\tilde{L}) \,\} = \omega_1 > 0 \tag{3.5}$$

Then problem (1.10) *has a nontrivial backward solution which converges to* 0 *as* $t \to -\infty$. *In particular, the null solution of* (3.1) *is unstable.* ■

4. Application of the abstract results

Theorem 3.1 can be applied to problem (2.11)-(2.12), and gives an existence and regularity result for problem (2.3)-(2.4), and hence for problem (1.1)-(1.2).

THEOREM 4.1 *Let* $u_0 :]-\infty,0] \to \mathbb{R}$ *be a* C^2 *function, such that* $u_0, u_0', u_0'' \in \mathcal{X}$ *and*

$$u_0(0) = a, \;\; u_0'(0) = 1 . \tag{4.1}$$

Then there are a maximal time interval $[0,\tau[$ *and a solution* $(s,v) : [0,\tau[\times]-\infty,0] \to \mathbb{R}$ *of problem* (2.3)-(2.4), *such that* $s \in C^1([0,\tau[)$, *and* v, v_x, v_{xx} *belong to* $\mathcal{X}$. *In addition, for each* $\alpha \in\,]0,1[$ *and* $T \in\,]0,\tau[$, v *satisfies*

$$\sup_{0<\varepsilon<T} \varepsilon^\alpha [\, v_{xx} \,]_{C^\alpha([\varepsilon,T];\mathcal{X})} < +\infty , \tag{4.2}$$

and (s,v) *is the unique classical solution of* (2.3)-(2.4) *satisfying* (4.2) *for some* $\alpha \in\,]0,1[$. *Moreover,* s *is analytic in* $]0,\tau[$, *and* v, v_x, v_{xx} *are analytic with respect to* t *for* $t > 0$.

Proof : We choose $\mathcal{X}$ (defined in (2.5)) as a phase space X, and $D = D(\mathcal{L})$ (defined in (2.6)). D is endowed with the norm

$$\| w \|_D = \| w \|_{\mathcal{X}} + \| w' \|_{\mathcal{X}} + \| w'' \|_{\mathcal{X}} . \tag{4.3}$$

Then D is densely and continuously embedded in X . The function F defined in (2.13) satisfies the assumptions of Theorem 3.1 , with $\Omega = D$: it is an analytic function, and for each $w \in D$ the linear operator $F'(w) : D \to X$ generates an analytic semigroup in X thanks to Lemma 2.1. Moreover it can be easily seen that the graph norm of F'(w) is equivalent to the norm of D . Therefore, we can apply Theorem 3.1 to problem (2.11)-(2.12) , and we get a solution w belonging to $C([0,\tau[;D) \cap C^1([0,\tau[;X)$, which is in addition analytic in $]0,\tau[$ with values in D . This means that w, w_y, w_{yy} are continuous in $[0,\tau[\times]-\infty,0]$, and analytic with respect to t in $]0,\tau[\times]-\infty,0]$. Going back to problem (2.3)-(2.4) , and recalling (2.8), (2.9), (2.10) , one finds

$$s(t) = p(t) - p(0) = -w(t,0) + w_0(0) = -w(t,0) + [(1-P)v_0](0) =$$

$$= - w(t,0) - \frac{1}{a} \int_{-\infty}^{0} [u_0(\sigma)-U(\sigma)]d\sigma , \tag{4.4}$$

$$v(t,y) = -w(t,0)U'(y) + w(t,y) , \tag{4.5}$$

so that v enjoys the same regularity properties of w . Then the statement follows. ∎

Now, coming back to problem (1.1)-(1.2) , we find that there exists a solution $(\xi,u) : \{ (t,x) \mid 0 \le t < \tau, x \le \xi(t) \} \to R$, with

$$\xi(t) = s(t) - ct , \ u(t,x) = v(t,x-\xi(t)) + U(x-\xi(t)) , \tag{4.6}$$

so that (ξ,u) has the same regularity properties of (s,v). Theorem 4.1 gives uniqueness of the solution in a particular class of functions (see (4.2)). Concerning stability of the travelling wave, the following result holds :

THEOREM 4.2 *Let* c, U *be defined by* (1.6) . *For every* $\omega \in]0,-\frac{(a-c)^2}{4} - \frac{1}{2}[$ *there are* $R', M' > 0$ *such that if* $u_0, u_0', u_0'' \in \mathcal{X}$, $\| u_0 \|_{\mathcal{X}} + \| u_0' \|_{\mathcal{X}} + \| u_0'' \|_{\mathcal{X}} \le R$, *and* (4.1) *holds, then* $\tau = +\infty$, *and, setting*

$$\xi_\infty = - \frac{1}{a} \int_{-\infty}^{0} [u_0(\sigma)-U(\sigma)]d\sigma , \tag{4.7}$$

then

$$| \xi(t) + ct - \xi_\infty | \le M'e^{-\omega t} , \ \forall t \ge 0 , \tag{4.8}$$

$$\sup_{x\le\xi(t)} | u(t,x)-U(x-\xi(t)) | + \sup_{x\le\xi(t)} | u_x(t,x)-U'(x-\xi(t)) | +$$

$$+ \sup_{x\le\xi(t)} | u_{xx}(t,x)-U''(x-\xi(t)) | \le M'e^{-\omega t} , \ \forall t \ge 0 . \tag{4.9}$$

Proof : Let us apply the results of Theorems 3.1 and 3.2 to problem (2.11)-(2.12), choosing now $X = (1-P)(\mathcal{X})$ as a phase space, and $D = (1-P)(D(\mathcal{L}))$. The nonlinear function F defined in (2.13) maps D into X . Moreover, F(0) = 0, $F'(0) = \mathcal{L}_{|D}$. By Corollary 2.3 , F'(0) generates an analytic semigroup in X . By a standard perturbation argument, the same is true for every $w \in D$, with $\| w \|_D = \| w \|_{D(\mathcal{L})}$ sufficiently small. Therefore the assumptions of Theorem 3.1 are satisfied, Ω being any sufficiently small neighborhood of 0 in D. By Corollary 2.3, also assumption (3.4) is satisfied, by any $\omega_0 \in]0, -(a-c)^2/2 - 1/2[$: then, by Theorem 3.2, the null solution of (2.11) is exponentially asymptotically stable, in the sense that, if $\omega \in]0,\omega_0[$ and $\| w_0 \|_D$ is sufficiently small, then

$$\| w(t,\cdot) \|_{\mathcal{X}} + \| w_y(t,\cdot) \|_{\mathcal{X}} + \| w_{yy}(t,\cdot) \|_{\mathcal{X}} \leq Me^{-\omega t}, \text{ for each } t \geq 0.$$

But $w_0 = (1-P)v_0 = (1-P)(u_0 - U)$, so that if $\| u_0 - U \|_{D(\mathcal{L})}$ is small, then also $\| w_0 \|_D$ is small, and by (4.4) and (4.6) we find

$$| \xi(t) + ct - \xi_\infty | = | w(t,0) |, \ \forall t \geq 0,$$

so that (4.8) holds. Due to (4.5), we have also

$$\| v(t,\cdot) \|_{\mathcal{X}} + \| v_y(t,\cdot) \|_{\mathcal{X}} + \| v_{yy}(t,\cdot) \|_{\mathcal{X}} \leq M'e^{-\omega t}, \ \forall t \geq 0,$$

and the last statement follows from (4.6). ∎

5. The two-phase problem

The results of this section have been announced in [2] and have been proved in [3], where a larger class of two-phase problems was studied.

In the two-phase problem (1.3), the asymptotic behavior of U as t goes to $+\infty$ is different from the asymptotic behavior as t goes to $-\infty$: we have

$$\begin{cases} U(y) \sim b + k_+e^{-\alpha_+ y}, & y \to +\infty, \\ U(y) \sim k_-e^{-2\alpha_- y}, & y \to -\infty, \end{cases}$$

where

$$\alpha_+ = c/2 - 1/b > 0, \ \alpha_- = -c/2 < 0. \tag{5.1}$$

Therefore we must choose as a phase space a weighted functional space with different weights at $+\infty$ and at $-\infty$. Precisely, we set $c = (2+a^2)/2a$, and

$$\mathcal{X} = \{ v : \mathbf{R} \to \mathbf{R} \mid y \to e^{-cy/2}v(y)_{|]-\infty,0[}, \ y \to e^{(c/2-1/b)y}v(y)_{|]0,+\infty[} \text{ are uniformly continuous and bounded} \},$$

$$\| v \| = \sup_{y\leq 0} | e^{-cy/2}v(y) | + \sup_{y\geq 0} | e^{(c/2-1/b)y}v(y) |, \tag{5.2}$$

$$D(\mathcal{L}) = \{ v \in \mathcal{X} : \forall y \neq 0 \ \exists \ v'(y), v''(y); \ v, v', v'' \in \mathcal{X},$$

$$v'(0^+)-v'(0^-) = (c-a)(v(0^+)-v(0^-)), \ v(0^-)(ac-a^2/2-1) = v(0^+)(ac-a^2/2) \}, \tag{5.3}$$

$$\mathcal{L} : D(\mathcal{L}) \to \mathcal{X}, \ \mathcal{L}v = v'' - \frac{d}{dy} [(U-c)v].$$

The boundary conditions in $D(\mathcal{L})$ are those satisfied by U'. Also in this case, Ker $\mathcal{L}$ is spanned by U'. It is easy to see that, setting

$$(Pv)(y) = \frac{1}{b} \int_{-\infty}^{+\infty} v(\sigma)d\sigma \ U'(y), \ \forall v \in X, \tag{5.4}$$

then P is a projection on $\mathrm{Ker}\,\mathcal{L}$, and $P\mathcal{L} = 0$. Concerning the spectrum of $\mathcal{L}$, the following proposition holds.

PROPOSITION 5.1 *The spectrum of* $\mathcal{L}$ *consists of three subsets : the half line* $]-\infty, -c^2/4 + 1/2]$, *the simple eigenvalue* 0, *and the unique root* λ *(when it does exist) of the equation*

$$(ac - \frac{a^2}{2})[(\lambda + \frac{c^2}{4} - \frac{1}{2})^{1/2} + (\lambda + \frac{c^2}{4})^{1/2}] = \frac{c-a}{2} + (\lambda + \frac{c^2}{4})^{1/2} \tag{5.5}$$

Setting $a_0 = c - [c^2 - 1 - (2c^2-3)^{1/2}]^{1/2}$, $a_c = c + (c^2-1)^{1/2}$, *the following holds :*

(i) *if* $2/b < a < a_0$, *equation* (5.5) *has a unique solution* λ, *and* $(2-c^2)/2 < \lambda < 0$;

(ii) *if* $a_0 \le a \le a_c$, *equation* (5.5) *has no solution at all ;*

(iii) *if* $a_c < a < 2c$, *equation* (5.5) *has a unique solution* $\lambda > 0$. ∎

Arguing as in the previous section, a stability-instability result can be stated.

THEOREM 5.2 *Let* c, U *be defined by* (1.8) , *and let* $2/b < a \le a_c$. *Then the travelling wave solution to problem* (1.3) *is orbitally stable with respect to perturbations belonging to a suitable weighted space. More precisely, for every* $\omega \in]0, c^2/4 - 1/2[$ *(satisfying in addition* $\omega < -\lambda$ *if* $2/b < a < a_0$*) there are* $R', M' > 0$ *such that if* $u_0, u_0', u_0'' \in \mathcal{X}$, $\| u_0 \|_{\mathcal{X}} + \| u_0' \|_{\mathcal{X}} + \| u_0'' \|_{\mathcal{X}} \le R$, *and*

$$u_0(0^+) = u_0(0^-) = a, \quad u_0'(0^+) - u_0'(0^-) = -1 ,$$

then problem (1.3)-(1.4) *has a globally defined solution* (ξ, u), *and, setting*

$$\xi_\infty = -\frac{1}{b} \int_{-\infty}^{+\infty} [u_0(\sigma) - U(\sigma)]d\sigma , \tag{5.6}$$

then

$$| \xi(t) + ct - \xi_\infty | \le M'e^{-\omega t} , \quad \forall t \ge 0 , \tag{5.7}$$

$$\sup_{x \in \mathbb{R}} | u(t,x) - U(x-\xi(t)) | + \sup_{x \ne \xi(t)} | u_x(t,x) - U'(x-\xi(t)) | +$$

$$+ \sup_{x \ne \xi(t)} | u_{xx}(t,x) - U''(x-\xi(t)) | \le M'e^{-\omega t} , \quad \forall t \ge 0 . \tag{5.8}$$

If $a_c < a < 2c$, *the travelling wave solution* U *is orbitally unstable.* ∎

References

[1] C.-M. BRAUNER, A. LUNARDI, C. SCHMIDT-LAINE : *Une nouvelle formulation de modèles de fronts en problèmes totalement non linéaires,* C.R.A.S. Paris **311**, Série I (1990), 597-602.

[2] C.-M. BRAUNER, A. LUNARDI, C. SCHMIDT-LAINE : *Stabilité et instabilité des ondes stationnaires d'un problème de détonation à deux phases,* C.R.A.S. Paris **311**, Série I (1990), 697-700.

[3] C.-M. BRAUNER, A. LUNARDI, C. SCHMIDT-LAINE : *Stability of travelling waves with interface conditions*, preprint.

[4] C.-M. BRAUNER, S. NOOR EBAD, C. SCHMIDT-LAINE : *Sur la stabilité d'ondes singulières en combustion*, C.R.A.S. Paris **308**, Séric I (1989), 159-162.

[5] C.-M. BRAUNER, S. NOOR EBAD, C. SCHMIDT-LAINE : *Nonlinear stability analysis of singular travelling waves in combustion (I) : a one-phase problem*, Nonlinear Analysis (to appear).

[6] J. BUCKMASTER, G.S.S. LUDFORD: *Lectures in Mathematical Combustion*, CBMS-NSF Conference Series in Applied Mathematics **43** (SIAM, 1983).

[7] G. DA PRATO, A. LUNARDI : *Stability, instability and center manifold theorem for fully nonlinear parabolic equations*, Arch. Rat. Mech. Anal. **101** (1988), 115-141.

[8] W.DESCH, W.SCHAPPACHER : *Some perturbation results for analytic semigroups*, Math. Ann. **281** (1988), 157-162.

[9] A. FASANO, M. PRIMICERIO : *Free boundary problems for nonlinear parabolic equations with nonlinear free boundary conditions*, J. Math. Anal. Appl. **72** (1979), 247-273.

[10] D. HILHORST, J. HULSHOF : *An elliptic-parabolic problem in combustion theory : convergence to travelling waves*, Nonlinear Analysis (to appear).

[11] G.S.S. LUDFORD, D.S. STEWART, *The acceleration of fast deflagration waves*, Z.A.M.M. **63** (1983), 291-302.

[12] G.S.S. LUDFORD, D.S. STEWART, *Evolution near Chapman-Jouguet deflagrations*, Trans. 28th Conf. of Army Math., ARO Report 83-1 (1983), 133-142.

[13] A. LUNARDI, *On the local dynamical system associated to a fully nonlinear parabolic equation in Banach space*, in : Nonlinear Analysis and Applications, V. Lakshmikantham Editor, (Dekker, New York-Basel, 1987), 319-326.

[14] A. LUNARDI : *Time analyticity for solutions of nonlinear abstract parabolic evolution equations*, J. Funct. Anal. **71** (1987), 294-308.

[15] A. LUNARDI : *An introduction to geometric theory of fully nonlinear parabolic equations*, in: "Qualitative Theory of Evolution Equations and Applications", ICTP, Trieste (to appear).

[16] A. LUNARDI, E. SINESTRARI : *Existence in the large and stability for nonlinear Volterra equations in Banach space*, in : Integrodifferential Evolution Equations an Applications, G. Da Prato, M. Iannelli Editors, J. Int. Eqns **10** (1985), Special Conference Issue, 213-239.

[17] R. RICCI, X. WEIQING : *On the stability of some solution of the Stefan problem*, preprint.

[18] D.H. SATTINGER : *On the stability of waves of nonlinear parabolic systems*, Adv. in Math. **22** (1976), 312-355.

[19] D.H. SATTINGER : *Weighted norms for the stability of travelling waves*, J. Diff. Eq. **25** (1977), 130-144.

[20] D.S. STEWART, *Transition to detonation on a model problem*, J. Méc. Théor. Appl. **4** (1985), 103-137.

A Convergent Finite Element Scheme for a Wave Equation with a Moving Boundary

J.P. Marmorat
CMA, ENSMP, Sophia Antipolis 06565 Valbonne France
G. Payre
Université de Sherbrooke , J1K 2R1 Canada
J.P. Zolésio
INLN, Parc Valrose, 06034 Nice Cedex

We wish to consider in this paper the numerical approximation of the solution of a wave equation when the boundaries of the spatial domain are moving. This problem has many practical applications in engineering science. One encounters wave systems in evolving domains in widely disseminated situations, such that rolling or unrolling antennas of space satellites, decoding the sound waves emitted by moving underwater objects or simulating the displacement of crane cables. In order to obtain computer simulations of this situations, one may try to make use of the following idea : a first discretization of the partial differential equation with respect to the space variable leads to a second order ordinary differential system $M(h)\ddot{q}(t)+K(h)q(t)=F(t)$. The discretization parameter h gives typically the size of a cell, the number of such cells being held constant during the simulation. When the domain evolves with the time, the parameter h is allowed to vary, and one has to solve $M(h(t))\ddot{q}(t)+K(h(t))q=F(t)$.

We shall give evidence in this paper that the results given by such methods are false, as opposed to those obtained by using the concept of convected dense family defined in [2]. One may find in this reference a new proof of the existence of solution for the continuous problem which generalizes the Galerkin method on basis convected from Ω_t to Ω_0. This approach gives a practical way to generate the convergent numerical solutions we are looking for.

1 A Wave Equation with a Moving Boundary

Let us consider the time interval $[0,T]$. For each value of t in this interval, we define Ω_t to be the set $]0,\ell(t)[$, $\ell(t)$ belonging to $C^1(0,T;R^+)$. The initial value $\ell(0)$ will be denoted by ℓ_0. The evolution domain Q is the non cylindrical set

$$Q=\bigcup_{0<t<T}\{t\}\times\Omega_t$$

Let $c>0$ be the wave celerity, $u_0\in H_0^1(\Omega_0)$ and $u_1\in L_2(\Omega_0)$ be a pair of initial conditions, and $f(t,x)$ a function belonging to $L^2(Q)$. For $0<t<T$ and $x\in\Omega_t$

we denote by $u(t,x)$ the solution of the equation

$$\partial_{tt}u - c^2\partial_{xx}u = f \quad \text{in } Q \tag{1}$$

with boundary conditions

$$u(t,0) = 0 \quad u(t,\ell(t)) = 0 \tag{2}$$

and initial conditions at $t = 0$

$$u(0,x) = u_0(x) \quad \partial_t u(0,x) = u_1(x) \tag{3}$$

1.1 The Method of Characteristics : An Analytical Solution

In the case where $f \equiv 0$ and $T = \infty$, the method of characteristics applied to (1) gives some insights on the properties of the solution. By introducing $\xi = ct - x$ and $\eta = ct + x$, the equation (1) becomes $\partial_{\xi\eta}u = 0$. The solution of this equation can be written by the mean of two arbitrary functions ϕ et ψ :

$$u(\xi,\eta) = \phi(\xi) + \psi(\eta)$$

One has therefore to find $\phi :]-\ell_0, +\infty[\to R$ and $\psi :]0, +\infty[\to R$ such that $u(t,x) = \phi(ct - x) + \psi(ct + x)$ is a solution of (1).

From the boundary conditions (2) one first deduce that $\phi(ct) + \psi(ct) = 0 \;\; \forall t > 0$ and therefore

$$u(t,x) = \phi(ct - x) - \phi(ct + x) \tag{4}$$

then

$$\phi(ct + \ell(t)) = \phi(ct - \ell(t)) \quad \forall t > 0 \tag{5}$$

The initial conditions (3) give, for $x \in]0, \ell_0[$

$$\begin{aligned} \phi(-x) - \phi(x) &= u_0(x) \\ c\,(\phi'(-x) - \phi'(x)) &= u_1(x) \end{aligned}$$

Denoting $U_1(x) = \frac{1}{c}\int_0^x u_1(s)\,ds$ one finally gets for $x \in]0, \ell_0[$

$$\phi(-x) = \frac{1}{2}\{u_0(x) - U_1(x)\} \tag{6}$$

$$\phi(x) = \frac{1}{2}\{-u_0(x) - U_1(x)\} \tag{7}$$

which defines ϕ on the interval $]-\ell_0, \ell_0[$. If $ct + \ell(t)$ and $ct - \ell(t)$ are two strictly increasing functions of t, that is if $|\dot{\ell}(t)| < c \;\; \forall t > 0$, the relation (5) allows to obtain $\phi(x)$ for $x > \ell_0$. Note that ϕ is defined upto an additive constant, but this constant does not affect the expression of u. Note also that $\phi \in H^1(\Omega_0)$ and $\phi(-\ell_0) = \phi(\ell_0)$ such that the preceeding expressions also hold for $x = 0$ and $x = \ell_0$.

1.2 Variation of the Internal energy

The internal energy of the wave system is given by

$$E(t) = \frac{1}{2}\int_0^{\ell(t)} \left[(\partial_t u(t,x))^2 + c^2(\partial_x u(t,x))^2\right]\, dx \tag{8}$$

In this relation, let us substitute the expression (4) for u as a function of ϕ. One obtains

$$E(t) = c^2 \int_0^{\ell(t)} (\dot{\phi}^2(ct-x) + \dot{\phi}^2(ct+x))dx$$

or

$$E(t) = c^2 \int_{ct-\ell(t)}^{ct+\ell(t)} \dot{\phi}^2(x)dx \tag{9}$$

Let us compute then the variation $\Delta_{s,t}E = E(t) - E(s)$ of the energy on a time interval $[s,t]$ small enough so that $cs + \ell(s) > ct - \ell(t)$:

$$\Delta_{s,t}E = c^2\left(\int_{cs+\ell(s)}^{ct+\ell(t)} \dot{\phi}^2(x)dx - \int_{cs-\ell(s)}^{ct-\ell(t)} \dot{\phi}^2(x)dx\right) = c^2(X_+ - X_-)$$

Introducing the variable τ such that $x = c\tau + \ell(\tau)$, the first integral becomes :

$$X_+ = \int_s^t (c + \dot{\ell}(\tau))\dot{\phi}^2(c\tau + \ell(\tau))\, d\tau$$

the relation (5) giving

$$(c + \dot{\ell}(\tau))\dot{\phi}(c\tau + \ell(\tau)) = (c - \dot{\ell}(\tau))\dot{\phi}(c\tau - \ell(\tau))$$

or

$$X_+ = \int_s^t \frac{(c - \dot{\ell}(\tau))^2}{(c + \dot{\ell}(\tau))}\dot{\phi}^2(c\tau - \ell(\tau))\, d\tau$$

Using now τ such that $x = c\tau - \ell(\tau)$ the second integral rewrites as

$$X_- = \int_s^t (c - \dot{\ell}(\tau))\dot{\phi}^2(c\tau - \ell(\tau))\, d\tau$$

Finally,

$$\Delta_{s,t}E = c^2 \int_s^t -2\dot{\ell}(\tau)\frac{c - \dot{\ell}(\tau)}{c + \dot{\ell}(\tau)}\dot{\phi}^2(c\tau - \ell(\tau))\, d\tau \tag{10}$$

This expression shows that, in the case $f \equiv 0$, E is a monotonic function of time as soon as the sign of $\dot{\ell}$ is fixed. $E(t)$ is increasing in time when $\dot{\ell} < 0$ (contracting domain), and decreasing when $\dot{\ell} > 0$ (expanding domain).

1.3 Global Energy

In the present section, we shall assume that the regularity of $f \not\equiv 0$, u_0, and u_1 results in a solution u smooth enough so that the following relations may be justified. Let us consider the evolution of the internal energy (8) :

$$\begin{aligned}\dot{E}(t) = \frac{dE}{dt} &= \int_0^{\ell(t)} \left(\partial_t u\, \partial_{tt}^2 u + c^2 \partial_x u\, \partial_{xt}^2 u\right) dx \\ &+ \frac{1}{2}\left[(\partial_t u(t,\ell(t)))^2 + c^2(\partial_x u(t,\ell(t)))^2\right] \dot{\ell}(t)\end{aligned}$$

Integrating by parts the second term of the integral, it comes

$$\begin{aligned}\dot{E}(t) &= \int_0^{\ell(t)} \left(\partial_{tt}^2 u - c^2 \partial_{xx}^2 u\right) \partial_t u\, dx \\ &+ c^2(\partial_x u\, \partial_t u)(t,\ell(t)) - c^2(\partial_x u\, \partial_t u)(t,0) \\ &+ \frac{1}{2}\left[(\partial_t u(l,\ell(t)))^2 + c^2(\partial_x u(t,\ell(t)))^2\right] \dot{\ell}(t) \qquad (11)\end{aligned}$$

From the boundary conditions, we get

$$\begin{aligned}u(t,0) = 0 &\Rightarrow \partial_t u(t,0) = 0 \\ u(t,\ell(t)) = 0 &\Rightarrow \partial_t u(t,\ell(t)) + \partial_x u(t,\ell(t))\, \dot{\ell}(t) = 0\end{aligned}$$

Multiplying this last relation by $\partial_x u(t,\ell(t))$, we obtain that, at the right boundary, $(\partial_x u\, \partial_t u)(t,\ell(t)) = -(\partial_x u(t,\ell(t)))^2\, \dot{\ell}(t)$. Using this expression in (11) results in

$$\dot{E}(t) = \frac{1}{2}(\partial_x u(t,\ell(t)))^2\, \dot{\ell}(t) \left[\dot{\ell}^2(t) - c^2\right] + \int_0^{\ell(t)} f(t,x)\, \partial_t u(t,x)\, dx \qquad (12)$$

Let us define then, at times s and t, the quantities

$$\begin{aligned}b(s) &= -\frac{1}{2}(\partial_x u(s,\ell(s)))^2 \dot{\ell}(s) \left[\dot{\ell}^2(s) - c^2\right] &(13)\\ B(t) &= \int_0^t b(s)\, ds &(14)\\ g(s) &= -\int_0^{\ell(s)} f(s,x)\, \partial_t u(s,x)\, dx &(15)\\ G(t) &= \int_0^t g(s)\, ds &(16)\end{aligned}$$

The relations (13),(14) and (15),(16) stand for the power and the work applied respectively by the moving boundary $\ell(t)$ and by the forcing fonction f to the wave system. The global energy contained in the system is given by the quantity

$$E_g(t) = E(t) + G(t) + B(t) \qquad (17)$$

The following relations hold :

$$\begin{aligned} E_g(0) &= E(0) = \int_0^{\ell_0} \left[u_1^2(x) + c^2(\partial_x u_0(x))^2\right] dx \\ \dot{E}_g(t) &= \dot{E}(t) + g(t) + b(t) \equiv 0 \\ E_g(t) &= E_g(0) = E_0 \quad \forall t > 0 \end{aligned}$$

2 A Convergence Result

The problem that we are addressing in this work is a specific case of a situation studied in [2]. In this reference, the author introduces the concept of convected dense family in order to obtain convergent approximations of solutions of the wave equation on moving domains in $\mathbf{R}^n$, when Dirichlet conditions hold on the moving parts of the boundary. The following results are proved to hold :

let T_t be a smooth isomorphism from Ω_0 onto Ω_t, and $\{W_{i,m}(x)\}$ a family in $H_0^1(\Omega_0)$ satisfying to the following density property :

$$\forall y \in H_0^1(\Omega_0),\ \forall \epsilon > 0,\ \exists M\ :\ \forall m > M\ \exists\{\alpha_{i,m}\} \text{ such that}$$
$$\| \sum_{i=1}^{m} \alpha_{i,m}\, W_{i,m}(x) - y \|_{H_0^1(\Omega_0)} \leq \epsilon \tag{18}$$

This set of functions is chosen and fixed once for all. We introduce then the *convected family* $\{w_{i,m}(t,.)\}$ at time t by the relation

$$w_{i,m}(t,x) = W_{i,m} \circ T_t^{-1}(x) \tag{19}$$

Then, for m fixed, we look for an approximation $u_m(t,x)$ of the solution $u(t,x)$ under the form :

$$u_m(t,x) = \sum_{i=1}^{m} q_{i,m}(t)\, w_{i,m}(x) \tag{20}$$

where the vector $q(t)$ is solution of the second order ordinary differential linear system :

$$\int_{\Omega_t} [\,\partial_{tt}^2 \;-\; c^2\partial_{xx}^2\,] u_m w_{i,m}\, dx = \int_{\Omega_t} f\, w_{i,m}\, dx \quad 1 \leq i \leq m \tag{21}$$

with the following initial data :

$$\begin{aligned} q_{i,m}(0) &= a_i \\ \dot{q}_{i,m}(0) &= b_i \end{aligned}$$

and

$$\lim_{m\to\infty} \sum_{i=1}^{m} a_i W_{i,m}(x) \;=\; u_0(x) \quad \text{in } H_0^1(\Omega_0) \tag{22}$$

$$\lim_{m\to\infty}\sum_{i=1}^{m} b_i W_{i,m}(x) \quad = \quad u_1(x) \quad \text{in } L^2(\Omega_0)) \tag{23}$$

It is proved in [2] that when using such convected families, a convergence result holds provided that some modifications are done in the linear system (21). That system can be written as

$$M\,\ddot{q}(t) + K\,q(t) = F(t) \qquad 0 < t < T \tag{24}$$

where M and K are the classical symmetric mass and stiffness matrices :

$$M_{i,j}(t) = \int_{\Omega_t} w_i(t,.)\,w_j(t,.)\,dx \quad K_{i,j}(t) = c^2\int_{\Omega_t} \partial_x w_i(t,.)\,\partial_x w_j(t,.)\,dx \tag{25}$$

(in the following, we omit the second subscript m for the sake of clarity) and the vector $F(t)$ is given by $F_i(t) = \int_{\Omega_t} f(t,.)\,w_i(t,.)\,dx$. When the domain is moving, we have to introduce the two non symmetric matrices :

$$C_{i,j}(t) \quad = \quad \int_{\Omega_t} \partial_t w_i(t,.)\,w_j(t,.)dx \tag{26}$$

$$K_{1_{i,j}}(t) \quad = \quad \int_{\Omega_t} \partial^2_{tt} w_i(t,.)\;w_j(t,.)\,dx \tag{27}$$

Then, we have the

Theorem 2.1 *Let $q(t)$ be the solution of the following $m\times m$ linear differential system :*

$$M(t)\ddot{q}(t) + 2C(t)\dot{q}(t) + (K(t) + K_1(t))q(t) = F(t) \tag{28}$$

with the initial data

$$q_i(0) = a_i \qquad \dot{q}_i(0) = c_i \tag{29}$$

with a_i verifying (22), and c_i obeying to

$$\lim_{m\to\infty}\sum_{i=1}^{m} c_i W_i(x) = u_1(x) + (\partial_t T_t)_{t=0}\cdot\nabla u_0(x) \quad \textit{in } L^2(\Omega_0) \tag{30}$$

Then $u_m(t,x) = \sum_{i=1}^{m} q_i(t) w_i(t,x)$ converges in $L^2(0,T;L^2(\Omega_t))$ to the solution $u(t,x)$ of the wave equation (1)-(3) when $m\to\infty$.

3 Numerical Approximation

3.1 Spatial Discretization

The partial differential equation (1) will be approximated by the following scheme : let us consider a fixed integer n, and $n+1$ real functions $x_i(t)$

$(0 \leq i \leq n)$ verifying the relation :

$$0 = x_0(t) < x_1(t) < \ldots < x_i(t) < \ldots < x_n(t) = \ell(t) \quad \forall t \tag{31}$$

This time-varying space discretization may be equivalently defined by the parameters

$$\begin{aligned} h_i(t) &= x_{i+1}(t) - x_i(t) \qquad 0 \leq i <\leq n-1 \\ h(t) &= \max_{0 \leq i \leq n-1} h_i(t) \\ h &= \sup_{t \in [0,T]} h(t) \end{aligned}$$

For each time t and integer i, $0 < i < n$, the function $N_i(t,x)$ is defined by

$$N_i(t,x) = \begin{cases} \frac{x - x_{i-1}(t)}{h_{i-1}(t)} & \text{if } x_{i-1}(t) < x < x_i(t) \\ \frac{x_i(t) - x}{h_i(t)} & \text{if } x_i(t) \leq x < x_{i+1}(t) \\ 0 & \text{elsewhere} \end{cases} \tag{32}$$

These $n-1$ linearly independent functions span (for t fixed) a finite dimensional subspace in which an approximated solution $u_h(t,x)$ of the partial differential equation (1) is considered :

$$u_h(t,x) = \sum_{j=1}^{n-1} q_j(t) N_j(t,x) \tag{33}$$

The $n-1$ functions $q_j(t)$ must be computed in order to obtain a numerical solution of (1).

Using (33) in (1) results in :

$$\partial^2_{tt} \sum_{j=1}^{n-1} q_j(t) N_j(t,x) - c^2 \partial^2_{xx} \sum_{j=1}^{n-1} q_j(t) N_j(t,x) = f(t,x) \tag{34}$$

We may multiply (34) by a test function $N_i(t,x)$ and integrate with respect to x on $[0, \ell(t)]$. Integrating by parts the second term and using the boundary conditions (2), one obtains the second order linear differential system :

$$\begin{aligned} &\sum_{j=1}^{n-1} \int_0^{\ell(t)} \left[\ddot{q}_j N_j + 2 \dot{q}_j \dot{N}_j + q_j \ddot{N}_j \right] N_i \, dx + \\ c^2 &\sum_{j=1}^{n-1} \int_0^{\ell(t)} q_j \, \partial_x N_j \, \partial_x N_i \, dx = \int_0^{\ell(t)} f \, N_i \, dx \qquad 1 \leq i \leq n-1 \end{aligned} \tag{35}$$

This differential system may be written exactly under the form (28) where the respective vectors and matrices are defined by

$$q(t) = (q_i(t)) \tag{36}$$

$$M(t) = (m_{ij}(t)) = \int_0^{\ell(t)} N_i N_j \, dx \tag{37}$$

$$C(t) = (c_{ij}(t)) = \int_0^{\ell(t)} \dot{N}_i N_j \, dx \tag{38}$$

$$K(t) = (k_{ij}(t)) = c^2 \int_0^{\ell(t)} \partial_x N_i \partial_x N_j \, dx \tag{39}$$

$$K_1(t) = (k_{1,ij}(t)) = \int_0^{\ell(t)} \ddot{N}_i N_j \, dx \tag{40}$$

$$F(t) = (f_i(t)) = \int_0^{\ell(t)} f \, N_i \, dx \tag{41}$$

Consequently, in order to use the framework of Theorem 2.1, it remains to define the functions $x_i(t)$ such that the $N_i(t,x)$ are convected from a dense set $W_i(x)$ of functions on $[0, \ell_0]$ by a family of transformations T_t from $[0, \ell(t)]$ to $[0, \ell_0]$. The following choices fulfill these conditions :

$$x_i(t) = \frac{i}{n}\,\ell(t) \quad 0 \leq i \leq n$$

$$W_i(x) = N_i(0, x)$$

It is then a simple exercise to show that

$$N_i(t,x) = W_i(\frac{\ell_0}{\ell(t)}) = W_i \circ T_t^{-1}(x)$$

where T_t is given by

$$T_t(x) = \frac{\ell(t)}{\ell_0}\, x$$

3.2 Time Discretization

The relations (3) will be used in order to obtain the initial conditions needed to integrate the second order differential system (28). The first relation obviously translates in

$$q(0) = q_0 \qquad \text{with} \quad (q_0)_i = u_0(x_i(0)) \tag{42}$$

To obtain the second initial condition is not so straightfull. One must recall that the function $q_i(t)$ approximates $u(t,x)$ *along the curve* $x_i(t)$. By the differentiation chain rule, it comes

$$\dot{q}_i = \frac{dq_i}{dt} = \partial_t u_i + \frac{dx_i}{dt} \partial_x u_i$$

This relation results in the second initial condition :

$$\dot{q}(0) = q_1 \qquad \text{with} \quad (q_1)_i = u_1(x_i(0)) + \dot{x}_i(0)\partial_x u_0(x_i(0)) \tag{43}$$

which is the one dimensional case of expression (30).

The differential system (28)(42)(43) is now completely specified, and may be integrated in time by any standard numerical scheme. However, this scheme must be A-stable in order to correctly handle the hyperbolic nature of the source equation (1). One of the simplest of such schemes is the second order trapezoidal method, which is briefly described below.

In a first step, the second order differential system is transformed in a first order one. Setting

$$\begin{aligned} Y(t) &= \begin{pmatrix} Y_1(t) \\ Y_2(t) \end{pmatrix} = \begin{pmatrix} q(t) \\ \dot{q}(t) \end{pmatrix} \\ G(t) &= \begin{pmatrix} F(t) \\ 0 \end{pmatrix} \end{aligned}$$

the differential system (28) may be written as

$$\begin{aligned} \frac{dY}{dt} &= A(t)Y + B(t) \\ Y(0) &= Y_0 \end{aligned} \tag{44}$$

with

$$\begin{aligned} A(t) &= \begin{pmatrix} 0 & I \\ M^{-1}(t)(K(t) + K_1(t)) & 2M^{-1}(t)C(t) \end{pmatrix} \\ B(t) &= M^{-1}(t)G(t) \\ Y_0 &= \begin{pmatrix} q_0 \\ q_1 \end{pmatrix} \end{aligned}$$

Let us consider now a sequence of time steps $\{\tau_k\}$, $k \geq 0$, bounded above and below ($0 < \tau_m \leq \tau_k \leq \tau_M$), and the associated sequence $\{t_k\}$ defined by

$$t_{k+1} = t_k + \tau_k \qquad k \geq 0 \quad t_0 = 0$$

A numerical approximation of (44) will be defined as the sequence $\{Y_k\}$ obtained by the relation

$$\frac{Y_{k+1} - Y_k}{\tau_k} = A_{k+\frac{1}{2}} \frac{Y_{k+1} + Y_k}{2} + B_{k+\frac{1}{2}} \tag{45}$$

4 A Numerical Example

The sections (2) and (3.1) have established that the solution $q(t)$ of the differential system (28) approaches the solution $u(t, x)$ of (1) along the lines $x_i(t)$

as the discretization parameter n goes to infinity, the matrices $C(t)$ and $K_1(t)$ playing a key role for this to hold. But it has not been established yet if the practical results obtained using (28) are or not substantially better that those given by the usual approach which neglects this two matrices.

Consequently, this section describes a testbed for the numerical solution of equation (1), and the results obtained when using the method exposed in the section 3. In this test, numerical results may be compared in some way to their theoretical values, and two main facts may be derived from these numerical experiences. The first is simply that the claimed convergence of the scheme may be observed as expected, which comes of course as no great surprise. The second, and perhaps more important one, is that the "truncated" scheme (sometimes used in industrial codes) results in a *non convergent* process, leading to unrealistic results whatever the space discretization or time step may be.

In this numerical approximation, the domain has been spatially discretized in the way described in section 3.1. The time steps τ_k are kept constant and small enough to ensure that the numerical errors introduced by the scheme (45) are much smaller than those coming from the spatial approximation. This example has been run twice, with and without the matrices C and K_1, all other parameters keeping the same values. The two sets of results are compared finally against some exact values pertaining to the theoretical solution.

As it has been felt useful to observe the behaviour of solutions when the size of the domain is not monotonic with respect to time, the displacement $\ell(t)$ of the boundary of the domain is a periodical function given by the expression

$$\ell(t) = \ell_0 + a\sin(\omega t) \tag{46}$$

When the forcing function $f(t,x)$ has the value

$$\begin{aligned}
f(t,x) &= 2\cos(\frac{\pi x}{\ell(t)})\pi x a^2 \cos(\omega t)^2 \omega^2 \cos(\frac{\pi t}{\ell(t)})\ell(t)^{-3} \\
&- \sin(\frac{\pi x}{\ell(t)})\pi^2 x^2 a^2 \cos(\omega t)^2 \omega^2 \cos(\frac{\pi t}{\ell(t)})\ell(t)^{-4} \\
&+ \cos(\frac{\pi x}{\ell(t)})\pi x a \sin(\omega t)\omega^2 \cos(\frac{\pi t}{\ell(t)})\ell(t)^{-2} \\
&+ 2\cos(\frac{\pi x}{\ell(t)})\pi x a \cos(\omega t)\omega \sin(\frac{\pi t}{\ell(t)}) \left(\frac{\pi}{\ell(t)} - \frac{\pi t a \cos(\omega t)\omega}{\ell(t)^2}\right) \ell(t)^{-2} \\
&- \sin(\frac{\pi x}{\ell(t)}) \cos(\frac{\pi t}{\ell(t)}) \left(\frac{\pi}{\ell(t)} - \frac{\pi t a \cos(\omega t)\omega}{\ell(t)^2}\right)^2 \\
&- \sin(\frac{\pi x}{\ell(t)}) \sin(\frac{\pi t}{\ell(t)}) \left(\frac{2\pi t a^2 \cos(\omega t)^2\omega^2}{\ell(t)^3} - \frac{2\pi a \cos(\omega t)\omega}{\ell(t)^2}\right. \\
&+ \left.\frac{\pi t a \sin(\omega t)\omega^2}{\ell(t)^2}\right) + \sin(\frac{\pi x}{\ell(t)})\pi^2 \cos(\frac{\pi t}{\ell(t)})\ell(t)^{-2}
\end{aligned}$$

and using the initial conditions

$$u_0(x) = \sin(\frac{\pi x}{\ell_0}) \qquad u_1(x) = 0 \tag{47}$$

the solution $u(t,x)$ of (1) is given by

$$u(t,x) = \cos(\frac{\pi t}{\ell(t)}) \sin(\frac{\pi x}{\ell(t)}) \tag{48}$$

(In practice, the expressions (46) and (48) have been chosen first, and put afterwards into (1) in order to obtain the forcing function f. This step has been done with the help of Maple).

Starting from its initial position, the solution is let to evolve during a full period of the boundary displacement : $0 < t < T = \frac{2\pi}{\omega}$. The accumulated error $\mathcal{E}(t)$ at time t is defined to be

$$\mathcal{E}(t) = \left\{ \int_0^t \int_0^{\ell(t)} (u(t,x) - u_h(t,x))^2 \, dx \, dt \right\}^{\frac{1}{2}} \tag{49}$$

The values of the parameters in this example are the following : $\ell_0 = 1$, $a = 0.5$, $c = 1$ and $\omega = 0.1$. It may be seen clearly from the figures 1 and 2 which show the curves $\mathcal{E}(t)$ for different values of h, that including or neglecting the matrices C and K_1 makes the difference between a convergent and a non convergent process.

The curves describing the evolution of the global energy of the system are plotted on the figure 3, which shows definitely that dropping the matrices C and K_1 leads to unrealistic numerical results. We recall that this quantity is not time dependent, its theoretical value in this particular case being given by $E_g = \frac{\pi^2}{4} \approx 2.47$.

References

[1] J.P.Zolésio : Approximation for Wave Equation in Non Cylindrical Domain. Proceedings of the Comcon Conf. on "Stabilization of Flexible Structures", Montpellier, 1989 to appear in Lectures Notes in Control and Information Sciences, Springer Verlag.

[2] J.P. Zolésio : Eulerian Galerkin Approximation for Wave Equation in Moving Domain. Metz Survey, M. Chipot and J. Saint Jean Paulin eds., pp.112-130, Longman, 1991.

[3] J.P. Zolésio, C. Truchi : Shape stabilization of wave equation. Proc. IFIP Conference on Boundary Control and Boundary Variations, Lectures Notes in Control and Information Sciences, vol. 100, pp. 372–398 Springer Verlag, 1987.

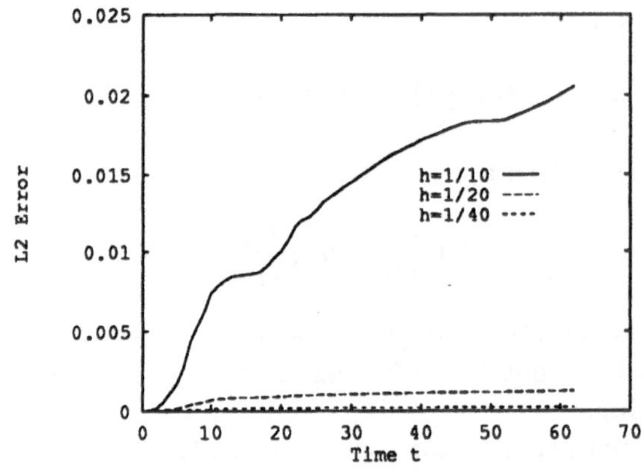

Figure 1: Convected Basis on a Periodical Domain

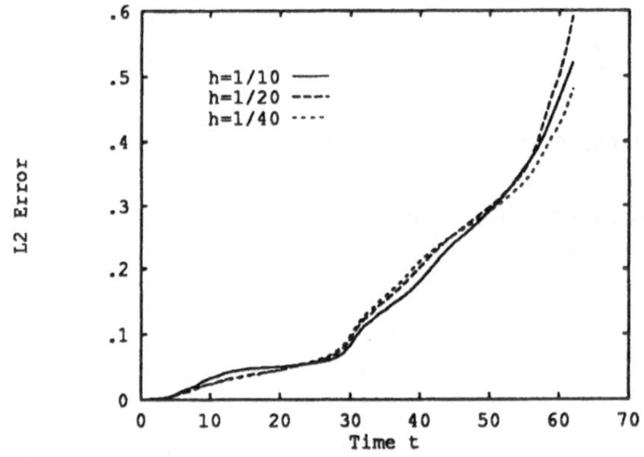

Figure 2: Non-Convected Basis on a Periodical Domain

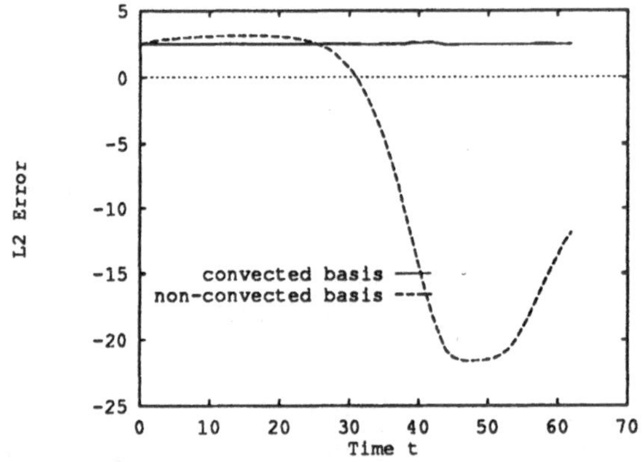

Figure 3: Evolution of the Global Energy $E_g(t)$ on a periodical Domain

A BOUNDARY CONTROLLABILITY APPROACH IN OPTIMAL SHAPE DESIGN PROBLEMS

[1]R. MÄKINEN, [2]P. NEITTAANMÄKI, [3]D. TIBA

[1,2]University of Jyväskylä, Finland
[3]Institute of Mathematics, Bucuresti, Romania

Abstract. We indicate a formulation of optimal shape design problems as boundary control problems, based on some approximate controllability-type results. Numerical examples and a comparison with the standard method are included.

1. INTRODUCTION

We consider the following optimal shape design problem:

$$(\mathsf{P}) \qquad \underset{D}{\text{minimize}}\, J(y_D)$$

subject to

$$(1.1) \qquad \begin{cases} -\Delta y_D = f & \text{in } D, \\ y_D = 0 & \text{on } \partial D. \end{cases}$$

The optimization parameter is the variable domain $D \subset \Omega$, where Ω is a fixed domain in $\mathbb{R}^N$, J is a real valued functional and $f \in H^{m-2}(\Omega)$.

It is our aim to present a fixed domain approach to the problem (P), which is simpler than the usual boundary variation techniques. The key point is a controllability-type result which may be compared mainly with the work of Lions [5], p. 78. We also need and prove approximate controllability properties when constraints are imposed on the control. In the parabolic case, similar problems were studied in [4], but our argument is different and may be also used for nonlinear equations (variational inequalities) as recently shown in [1] and [2]. An announcement of some of the results proved here was published in [8].

For other applications in optimal design and other numerical methods for optimal shape design problems we refer to the monograph [3] and to the paper [9] which is closer to our approach.

---

[1] Research supported by the Academy of Finland

2. Controllability

Let us fix now the domains $D \subset \Omega \subset \mathbf{R}^N$. We assume that $\partial\Omega$ is analytic and D has Lipschitz boundary. Let y be the solution of the boundary value problem

$$-\Delta y = f \quad \text{in } \Omega, \tag{2.1}$$

$$y = u \quad \text{on } \partial\Omega, \tag{2.2}$$

with $f \in H^{m-2}(\Omega)$ and $u \in H^{m-1/2}(\partial\Omega)$, $m > N/2$.

The controllability-type problem we discuss in this section is:

$$(C) \qquad \begin{cases} \text{Find } u \in H^{m-1/2}(\partial\Omega) \text{ such that} \\ y = 0 \text{ on } \partial D \subset \Omega. \end{cases}$$

We approximate it by the following boundary control problem:

$$(\mathbf{P}_n) \qquad \underset{u}{\text{minimize}} \left\{ \frac{1}{2} \int_{\partial D} y^2 \, ds + \frac{1}{2n} \|u\|^2_{H^{m-1/2}(\partial\Omega)} \right\},$$

subject to (2.1)–(2.2), $n \in \mathbb{N}$.

The existence of a unique optimal pair $[y_n, u_n]$ of $(\mathbf{P}_n)$ is quite obvious. Let us denote by $p_n \in H^1(\Omega \setminus \bar{D})$, $q_n \in H^1(D)$ the solution of the adjoint system:

$$-\Delta p_n = 0 \quad \text{in } \Omega \setminus \bar{D}, \tag{2.3}$$

$$-\Delta q_n = 0 \quad \text{in } D, \tag{2.4}$$

$$p_n = 0 \quad \text{on } \partial\Omega, \tag{2.5}$$

$$p_n = q_n \quad \text{on } \partial D, \tag{2.6}$$

$$\frac{\partial p_n}{\partial n} + \frac{\partial q_n}{\partial \nu} = y_n \quad \text{on } \partial D. \tag{2.7}$$

Here $\frac{\partial}{\partial n}$ and $\frac{\partial}{\partial \nu}$ denote exterior normal derivatives on ∂D to $\Omega \setminus \bar{D}$ and D, respectively. They have a sense in $H^{-1/2}(\partial D)$ since Δp_n and Δq_n are square integrable.

We remark that transmission conditions similar to (2.6), (2.7) appear in the study of elliptic equations with discontinuous coefficients. The existence for the system (2.3)–(2.7) may be obtained by a variational argument, minimizing the functional

$$\frac{1}{2} \int_\Omega |\nabla z|^2 \, dx - \int_{\partial D} y_n z \, ds,$$

on the space $H_0^1(\Omega)$, [6] Ch. II.

Proposition 2.1. *The solution of the problem $(\mathbf{P}_n)$ satisfies the optimality condition (maximum principle)*

$$\frac{\partial p_n}{\partial n} = \frac{1}{n} F u_n \quad \text{on } \partial\Omega. \tag{2.8}$$

Here $\frac{\partial}{\partial n}$ *is the exterior normal derivative to* Ω *and* $F : H^{m-1/2}(\partial\Omega) \to H^{1/2-m}(\partial\Omega)$ *is the canonical isomorfism.*

PROOF: Let y_λ be the solution of (2.1) corresponding to $u_n+\lambda v$, $v \in H^{m-1/2}(\partial\Omega)$ arbitrary. By the optimality of u_n, we have

$$0 = \lim_{\lambda\to 0} \frac{1}{\lambda}\left\{\frac{1}{2}\int_{\partial D}(y_\lambda^2 - y_n^2)\,ds + \frac{1}{2n}\left(\|u_n+\lambda v\|^2_{H^{m-1/2}(\partial\Omega)} - \|u\|^2_{H^{m-1/2}(\partial\Omega)}\right)\right\}$$
$$= \frac{1}{n}(u_n, v)_{H^{m-1/2}(\partial\Omega)} + \int_{\partial D} r_n y_n\,ds,$$

where r_n satisfies

$$\begin{cases} -\Delta r_n = 0 & \text{in } \Omega \\ r_n = v & \text{on } \partial\Omega \end{cases}$$

and denotes the weak limit of $(y_\lambda - y_n)/\lambda$ in $H^m(\Omega)$, for $\lambda \to 0$. By (2.7), we get

$$0 = \frac{1}{n}(u_n, v)_{H^{m-1/2}(\partial\Omega)} + \int_{\partial D} r_n\left(\frac{\partial p_n}{\partial n} + \frac{\partial q_n}{\partial \nu}\right) ds$$
$$= \frac{1}{n}(u_n, v)_{H^{m-1/2}(\partial\Omega)} - \int_{\partial\Omega}\frac{\partial p_n}{\partial n} v\,ds,$$

integrating by parts in $\Omega\setminus\bar{D}$ and D respectively. Since v is arbitrary, we obtain (2.8).

THEOREM 2.2. *The problem* (C) *is approximately solvable in the sense that*

(2.9) $$\tilde{y}_n \to 0 \quad \textit{strongly in } L^2(\partial D),$$

where $\tilde{y}_n$ *denotes a convex combination of elements of the sequence* $\{y_n\}$.

PROOF: By taking $u = 0$ in (P_n), we see that the sequence $\{\frac{1}{\sqrt{n}}u_n\}$ is bounded in $H^{m-1/2}(\partial\Omega)$ and $\{y_n\}$ has a bounded trace in $L^2(\partial D)$. Then, from (2.3)-(2.7) it is easy to see that $\{p_n\}$, $\{q_n\}$ are bounded in $H^1(\Omega\setminus\bar{D})$, $H^1(D)$.

We denote by p, q, y the weak limits of $\{p_n\}$, $\{q_n\}$, $\{y_n\}$ in $H^1(\Omega\setminus\bar{D})$, $H^1(D)$, $L^2(\partial\Omega)$ respectively. Passing to the limit in (2.3)-(2.7), we get

$$-\Delta p = 0 \text{ in } \Omega\setminus\bar{D},$$
$$-\Delta q = 0 \text{ in } D,$$
$$p = 0 \text{ on } \partial\Omega,$$
$$p = q \text{ on } \partial D,$$
$$\frac{\partial p}{\partial n} + \frac{\partial q}{\partial \nu} = y \text{ on } \partial D.$$

Moreover, we may take limits in $H^{-1/2}(\partial\Omega)$ in (2.8) and get

$$\frac{\partial p}{\partial n} = 0 \text{ on } \partial\Omega.$$

By the analyticity assumption, the Holmgren uniqueness theorem shows that $p \equiv 0$ in $\Omega\setminus\bar{D}$, so $p = q$ on ∂D and $q = 0$ in D. We conclude that $y = 0$ on ∂D, that is $y_n \rightharpoonup 0$ in $L^2(\partial D)$. According to the Mazur theorem, we may take a sequence of convex combinations $\tilde{y}_n$ (corresponding to a sequence of convex combinations $\tilde{u}_n$) such that $\tilde{y}_n \to 0$ in $L^2(\partial D)$.

REMARK. The analyticity assumption on $\partial\Omega$ may be relaxed according to [7]. However, in optimal design problems, Ω is, in general, at our choice.

3. CONTROL CONSTRAINTS

For the sake of subsequent application, we study the problem (C) under the additional condition

$$u \leq 0 \text{ on } \partial\Omega. \tag{3.1}$$

We prove the following approximate controllability result:

THEOREM 3.1. *For any $\varepsilon > 0$ and any $\Gamma_\varepsilon \subset \partial\Omega$, $\text{meas}\Gamma_\varepsilon < \varepsilon$, there exists a sequence $u_n \in H^{m-1/2}(\partial\Omega)$ such that*

$$u_n \leq 0 \text{ on } \partial\Omega \setminus \Gamma_\varepsilon, \tag{3.2}$$

$$y_n \to 0 \text{ in } L^2(\partial D), \tag{3.3}$$

where y_n is the solution of (2.1)–(2.2) corresponding to u_n.

PROOF: We consider the family of optimal control problems

$$\text{minimize} \left\{ \frac{1}{2} \int_{\partial D} y^2 \, ds + \frac{1}{2n} \|u\|^2_{H^{m-1/2}(\partial\Omega)} \right\} \tag{3.4}$$

over all $[u, y] \in H^{m-1/2}(\partial\Omega) \times H^m(\Omega)$ satisfying (2.1), (2.2) and (3.2).

We approximate the problem (3.4) by the following penalized problem

$$\text{minimize} \left\{ \frac{1}{2} \int_{\partial D} y^2 \, ds + \frac{1}{2n} \|u\|^2_{H^{m-1/2}(\partial\Omega)} + \frac{1}{2\lambda} \int_{\partial\Omega} \chi_\varepsilon u_+^2 \, ds \right\} \tag{3.5}$$

subject to (2.1), (2.2).

Here u_+ is the positive part of u and χ_ε is the indicator function of $\partial\Omega \setminus \Gamma_\varepsilon$ in $\partial\Omega$.

Let $[u_\lambda, y_\lambda]$ denote the unique optimal pair for the problem (3.5), which obviously exists (n is fixed now). Arguing as the previous section, we see that there exist $p_\lambda \in H^1(\Omega \setminus \bar{D})$, $q_\lambda \in H^1(D)$ satisfying together with u_λ and y_λ the optimality system:

$$-\Delta p_\lambda = 0 \quad \text{in } \Omega \setminus \bar{D}, \tag{3.6}$$

$$-\Delta q_\lambda = 0 \quad \text{in } D, \tag{3.7}$$

$$p_\lambda = 0 \quad \text{on } \partial\Omega, \tag{3.8}$$

$$p_\lambda = q_\lambda \quad \text{on } \partial D, \tag{3.9}$$

$$\frac{\partial p_\lambda}{\partial n} + \frac{\partial q_\lambda}{\partial \nu} = y_\lambda \quad \text{on } \partial D, \tag{3.10}$$

$$\frac{\partial p_\lambda}{\partial n} = \frac{1}{n} F u_\lambda + \frac{1}{\lambda} \chi_\varepsilon (u_\lambda)_+ \quad \text{on } \partial\Omega \tag{3.11}$$

If $[u_n, y_n]$ is the optimal pair for the problem (3.4), we have

$$\frac{1}{2}\int_{\partial D} y_\lambda^2\, ds + \frac{1}{2n}\|u_\lambda\|^2_{H^{m-1/2}(\partial\Omega)} + \frac{1}{2\lambda}\int_{\partial\Omega} \chi_\varepsilon(u_\lambda)_+^2\, ds \tag{3.12}$$
$$\leq \frac{1}{2}\int_{\partial D} y_n^2\, ds + \frac{1}{2n}\|u\|^2_{H^{m-1/2}(\partial\Omega)}, \quad \lambda > 0.$$

This shows that $\{y_\lambda\}$ has a bounded trace in $L^2(\partial D)$, $\{u_\lambda\}$ is bounded in $H^{m-1/2}(\partial\Omega)$. Then, by (3.6)–(3.10) we see that $\{p_\lambda\}$ is bounded in $H^1(\Omega\backslash\bar{D})$, $\{q_\lambda\}$ is bounded in $H^1(D)$ and $\{\frac{\partial p_\lambda}{\partial n}\}$ is bounded in $H^{-1/2}(\partial\Omega)$. We may pass to the limit, $\lambda \to 0$, in (3.12) and (2.1), (2.2) to obtain that $y_\lambda \rightharpoonup y_n$ in $H^m(\Omega)$, $u_\lambda \rightharpoonup u_n$ in $H^{m-1/2}(\partial\Omega)$.

Denoting by p_n, q_n the weak limits of p_λ, q_λ we pass to the limit in (3.6)–(3.11) to reobtain the adjoint system of the form (2.3)–(2.7) for the problem (3.4) and the maximum principle

$$\frac{\partial p_n}{\partial n} = \frac{1}{n}Fu_n + \mu_n, \tag{3.13}$$

where μ_n is the weak limit in $H^{1/2-m}(\partial\Omega)$ of $\frac{1}{\lambda}\chi_\varepsilon(u_\lambda)_+$ for $\lambda \to 0$.

All the above estimates, except that for the control, remain valid with respect to n. Besides, we know that $\{\frac{1}{\sqrt{n}}u_n\}$ is bounded in $H^{m-1/2}(\partial\Omega)$. Therefore, (3.13) gives

$$\frac{\partial p}{\partial n} = \mu, \tag{3.14}$$

where μ is the weak limit in $H^{1/2-m}(\partial\Omega)$ of μ_n and p, q denote as well the weak limits of $\{p_n\}$, $\{q_n\}$ respectively.

By the above process, as ε is fixed, it is clear that $\frac{\partial p}{\partial n} = 0$ in Γ_ε. Then, the analyticity assumption on $\partial\Omega$ and the Holmgren theorem gives that $p \equiv 0$ in $\Omega\setminus\bar{D}$ and we get the conclusion as in the previous section.

REMARK. The above argument is general enough to work in nonlinear problems or under state constrants [1], [2]. For the linear unconstrained case a simpler approach is provided in Lions [5], p.78.

4. Application to optimal shape design problems

Let us choose $J(y_D) = \frac{1}{2}\|Cy_D - z\|^2_{H^1(E)}$ in (P). Here $z \in H^1(E)$, $C \in \mathcal{L}(V, H^1(E))$ and $E \subset D \subset \Omega$ is a fixed domain. Moreover, we impose the supplementary hypothesis

$$f \geq 0 \quad \text{in } \Omega.$$

We associate to the problem (P) the following boundary control problem with constraints:

$$\text{minimize}\, \frac{1}{2}\|Cy - z\|^2_{H^1(E)} \tag{4.1}$$

subject to

$$-\Delta y = f \quad \text{in } \Omega, \tag{4.2}$$

$$y = u \quad \text{on } \partial\Omega, \tag{4.3}$$

$$y \geq 0 \quad \text{in } E \tag{4.4}$$

$$u \leq 0 \quad \text{on } \partial\Omega. \tag{4.5}$$

Let $\overline{u}$ be any admissible control for the problem (4.1)–(4.5), from $H^{m-1/2}(\partial\Omega)$. Then $\overline{y}$, the corresponding solution of (4.2), (4.3) is continuous by the Sobolev embedding theorem. The set where $\overline{y} > 0$, $\overline{y} < 0$ are nonvoid open subsets of Ω due to (4.4), (4.5). (We assume that $\overline{y} \neq 0$). We define the subdomain D as the connected component of the first set which contains E. Then $\tilde{y} = \overline{y}|_D$ will satisfy

$$\begin{aligned} -\Delta\tilde{y} &= f \quad \text{in } D, \\ \tilde{y} &= 0 \quad \text{in } \partial D\ . \end{aligned}$$

The regularity of ∂D depends on the regularity of $\overline{y}$, that is of $\overline{u}$. In our case it is at least continuous.

We may view D as admissible for the problem (**P**) with the same value of the cost as $\overline{u}$ in (4.1). Therefore any $\overline{u}$ admissible for the problem (4.1)–(4.5) generates a domain D admissible for (**P**) with the same cost. The converse is more involved and it is based on the controllability results.

For any D admissible in the problem (**P**) one may find a sequence u_n in $H^{m-1/2}(\partial\Omega)$ satisfying the constraint (4.5) in the approximating sense (3.2) (with some $\varepsilon_n \to 0$) and such that the corresponding solution y_n of (2.1), (2.2) satisfies

$$y_n|_D \to 0 \quad \text{strongly in } L^2(\partial D)\ .$$

This is the result of Thm. 3.1. Taking into account the continuity with respect to the boundary data for the equation

$$\begin{aligned} -\Delta z &= f \quad \text{in } D, \\ z &= v \quad \text{in } \partial D, \end{aligned}$$

we see that $y_n \to y$ strongly in $H^{1/2}(D)$, that is $y_n \to y$ strongly in $L^2(E)$. Here y is given by (1.1). Moreover, by the maximum principle y is positive in D since $f \geq 0$, so y_n will approximately satisfy (4.4) in this sense, for n sufficiently large.

We consider the pair $[y_n, u_n]$ as approximately admissible for the problem (4.1). Obviously, the cost associated to it in (4.1) approaches, as $n \to \infty$, the cost associated to D in (**P**).

The above discussion clarifies the relationship between the problems (**P**) and (4.1)–(4.5).

In order to solve the problem (**P**), we propose to study the problem (4.1)–(4.5) which is obtained simply by replacing the variable domain D by the fixed domain Ω and the homogeneous Dirichlet condition, by u. No modification in the cost functional or the state equation is required. Moreover, it is known that the problem (**P**) is nonconvex, while the problem (4.1)–(4.5) is convex.

Due to the state constraint (4.4), for the solution of the problem (4.1) we use a standard penalization technique. We define the approximating problem

$$\text{minimize}\left\{\frac{1}{2}\|Cy - z\|^2_{H^1(E)} + \frac{1}{2\lambda}\int_E y_-^2\,dx\right\} \tag{4.6}$$

subject to (4.2), (4.3), (4.5), $\lambda > 0$.

Since no coercivity properties are given in the problems (4.1), (4.6), the existence of optimal pairs may fail (and in (**P**) too). However the above argument is valid for minimizing sequences or δ-solutions too.

Finally, we point out that solving the problem (4.1) is entirely equivalent with solving (**P**) on the associated domains D defined above from $\overline{y} > 0$. This is a rich class of subdomains of Ω as it may be viewed as indexed by $\overline{u} \in L^2(\partial\Omega)$

5. Numerical approximation

In this Section we describe briefly a numerical procedure to obtain numerical approximations for the problem (**P**). We discretize the state problem using the Finite Element Method (FEM) and formulate the approximation of (**P**) as a mathematical programming problem.

Let us divide a polygonal approximation Ω_h of set Ω into triangles in the standard way of FEM. We denote the triangulation by $\mathcal{T}_h$. Let us consider the standard space V_h of finite elements

$$V_h = \{v_h \in C(\overline{\Omega_h}) \mid v_h|_T \in P_1(T)\ \forall T \in \mathcal{T}_h,\ v_h|_{\partial\Omega_h} = 0\}.$$

We denote by $I = I_\Omega \cup I_\Gamma$ the set of nodes of $\mathcal{T}_h$, where I_Ω and I_Γ denote the sets of nodes in Ω_h and $\partial\Omega_h$ respectively.

Let $\hat{u}_h \in C(\overline{\Omega}_h)$ be a function which is linear within each triangle of Ω_h and

$$\hat{u}_h(x_j) = \begin{cases} 0, & j \in I_\Omega, \\ u(x_j), & j \in I_\Gamma. \end{cases} \tag{5.1}$$

Then the approximation of problem (4.2)–(4.3) is obtained as $y_h = y_{0h} + \hat{u}_h$, where y_{0h} is the solution of the variational equation

$$y_{0h} \in V_h : \int_{\Omega_h} \nabla y_{0h} \cdot \nabla v_h\,dx = \int_{\Omega_h} f v_h\,dx - \int_{\Omega_h} \nabla \hat{u}_h \cdot \nabla v_h\,dx \quad \forall v_h \in V_h, \tag{5.2}$$

The approximation of the solution y is given by

$$y_h(x) = y_{0h}(x) + \hat{u}_h(x) = \sum_{i \notin I_\Gamma} q_{0i}\varphi_i(x) + \sum_{i \in I_\Gamma} u_i\varphi_i(x), \tag{5.3}$$

where $\{\varphi_i(x)\}_{i=1}^{n(h)}$ are the Courant's basis functions. With these notations, (5.2) can be written equivalently as the linear system of equations

$$\mathbf{K}\mathbf{q}_0 = \mathbf{f} - \mathbf{B}\mathbf{u}. \tag{5.4}$$

$\mathbf{K}$ is the "stiffness" matrix of our problem and $\mathbf{f}$ and $\mathbf{Bu}$ are vectors arising from the discretization of the right hand side of (5.2).

Discretizing the cost functional we get the mathematical programming problem

$$\underset{\mathbf{u}}{\text{minimize}}\ J_h^\lambda(y_h) \tag{5.5}$$

subject to $\mathbf{u} \leq 0$ and (5.4).

The variation of J_h^λ with respect to $\mathbf{u}$ is given by

$$\delta_u J_h^\lambda(y_h) = -\mathbf{p}^{\mathrm{T}}\left(\mathbf{B}\,\delta\mathbf{u}\right) + \nabla_u J_h^\lambda(y_h)^{\mathrm{T}}\,\delta\mathbf{u}, \tag{5.6}$$

where $\mathbf{p}$ is the solution of the adjoint equation

$$\mathbf{K}\mathbf{p} = \nabla_{q_0} J_h^\lambda(y_h). \tag{5.7}$$

REMARK. As $\mathbf{K}$ is independent of $\mathbf{u}$, it is beneficial to use direct method (Cholesky, frontal, etc) in solving the linear systems of equations (5.4) and (5.7).

To show that the proposed method works also in practise, we present some numerical results. All computations were made using HP 9000/370-workstation with double precision arithmetic.

In optimization we have used Sequential Quadratic Programming algorithm E04VDF from NAG-library. This code is not intended for large and sparse problems as it uses dense approximation for the Hessian of the cost function. Despite this the code performed well in the authors' tests below where the number of unknown nodal values was up to 160. One could also use conjugate gradient algorithm, but the penalty term arising from the state constraint sometimes leads to ill-conditioning.

In all examples we took $\Omega =]-\frac{3}{2}, \frac{3}{2}[\times]-\frac{3}{2}, \frac{3}{2}[$ for the sake of simplicity, although the analyticity assumption is not satisfied. A uniform triangular finite element mesh with $2(\frac{1}{h})^2$ elements and $(\frac{1}{h}+1)^2$ nodes was used.

EXAMPLE 5.1. We choose in (4.1)–(4.5) $f = 10$, $E =]-\frac{3}{10}, \frac{3}{10}[\times]-\frac{7}{10}, \frac{7}{10}[$, $H(E) = H^1(E)$, C = canonical injection, $z = \frac{3}{2} - 4x_1^2 - x_2^2$ and $h = \frac{1}{30}$. We took $u_i^0 = -4\cdot 5\ \forall i$ as the initial guess. The corresponding cost was $0\cdot 917$. After 4 SQP-iterations and 70 CPU-seconds we obtained a control with the cost $7\cdot 51 \times 10^{-4}$. In Figure 5.1 the initial (- - -) and final (—) domains are shown. No penalty term corresponding to the state constraint (4.4) was needed in this example as the

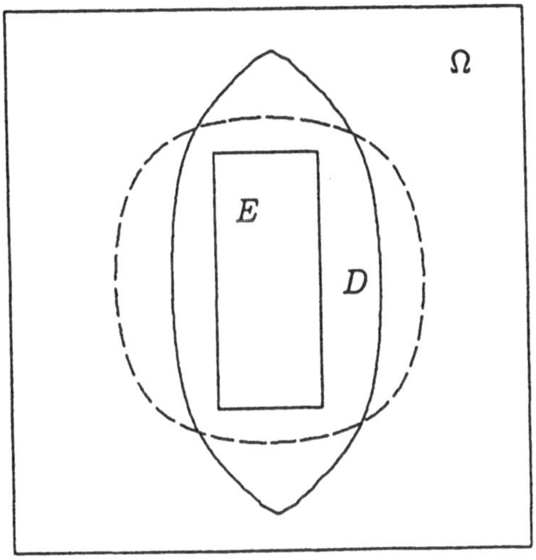

Figure 5.1

computed optimal pair was admissible. Linear systems (5.4) and (5.7) were solved using band-Cholesky method.

The same problem was also solved with conventional moving mesh technique. The radial coordinates of the boundary nodes of the finite element mesh shown in Figure 5.2 were chosen as the control variables. Jacobi conjugate gradient method was used to solve equations (5.4) and (5.7). The unit disc was chosen as the initial guess. This choice does not give exactly same initial domain as above but a one which is close to it as the corresponding cost was 0·947. After 3 SQP-iterations and 195 CPU-seconds we obtained the shape shown in Figure 5.3 together with initial guess. The final cost was $3{\cdot}90 \times 10^{-3}$.

Although the moving mesh approach used less SQP-iterations, the total computational burden is much heavier as the finite element mesh have to be updated at the beginning of each iteration. As the coefficient matrix of equations (5.4) and (5.7) depends now on control, one cannot utilize the same factorization of the coefficient matrix as is the case in the boundary control approach.

EXAMPLE 5.2. Let us consider the following problem

$$\text{(5.1)} \qquad \underset{D}{\text{minimize}} \left\{ J(y_D) = -\int_D 2y_D\,dx \right\}$$

subject to

$$\text{meas}(D) = A,$$
$$\begin{cases} -\Delta y_D = 2 & \text{in } D, \\ \quad y_D = 0 & \text{on } \partial D. \end{cases}$$

This is the well known problem of maximizing the torsional rigidity of a bar with cross-section D (see [3], for example). The amount of material available per cross-

Figure 5.2

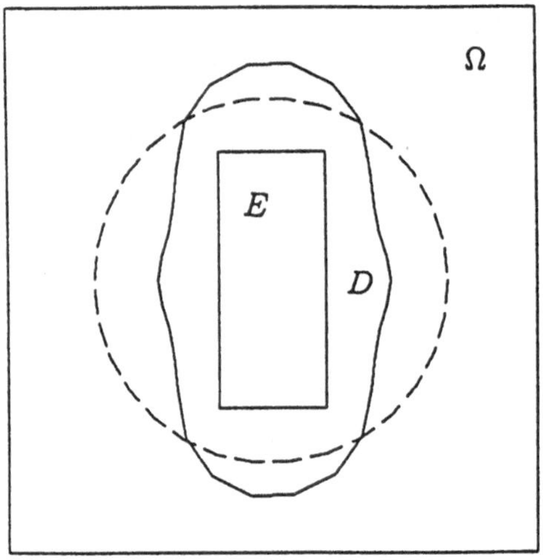

Figure 5.3

section is A. The optimum is known to be circle with radius $r = \sqrt{A/\pi}$ which gives the cost $-\pi/2$.

We note that this problem is slightly more difficult than the problem considered in Section 4 and in the previous example. This is because the variable domain D is explicitly referenced in the definition of the cost functional and area constraint. We can approximate the problem (5.1) by the following boundary control problem

$$\text{(5.2)} \qquad \text{minimize}\left\{ J^{\lambda}(y) = \int_{\Omega} 2\psi^{\varepsilon}(y)\,dx + \frac{1}{2\lambda}\left[A - \int_{\Omega}\chi^{\delta}(y)\,dx\right]^{2}\right\},$$

subject to (4.2),(4.3), (4.5), $\lambda, \varepsilon, \delta > 0$.

Here

$$\psi^{\varepsilon}(s) = \begin{cases} 0, & \text{if } s \le 0 \\ s^2/(2\varepsilon), & \text{if } 0 \le s \le \varepsilon \\ s - \varepsilon/2, & \text{if } s > \varepsilon \end{cases}$$

and

$$\chi^{\delta}(s) = \begin{cases} 0, & \text{if } s \le -\delta \\ 1/2 + 3s/(4\delta) - s^3/(4\delta^3), & \text{if } -\delta < s < \delta \\ 1, & \text{if } s \ge \delta \end{cases}$$

are smooth approximations of y_+ and $\chi_{\{x\in\Omega | y>0\}}$ respectively.

We took in (5.2) $\Omega =]-\frac{3}{2},\frac{3}{2}[\times]-\frac{3}{2},\frac{3}{2}[$, $A = \pi$, $\delta = 10^{-1}$, $\varepsilon = \lambda = 10^{-2}$ and $h = \frac{1}{40}$. As the initial guess for $\mathbf{u}$ we took $u_i = -0{\cdot}8292\ \forall i$. With this choice the (relaxed) area constraint is satisfied. The initial cost was $-1{\cdot}523$. After 8 SQP-iterations and 290 CPU-seconds we obtained a solution shown in Figure 5.4 together with initial guess (dashed line). The final cost was $-1{\cdot}543$. Although quite large regularization parameters ε and δ were used, the error in optimal cost and domain is under 2%.

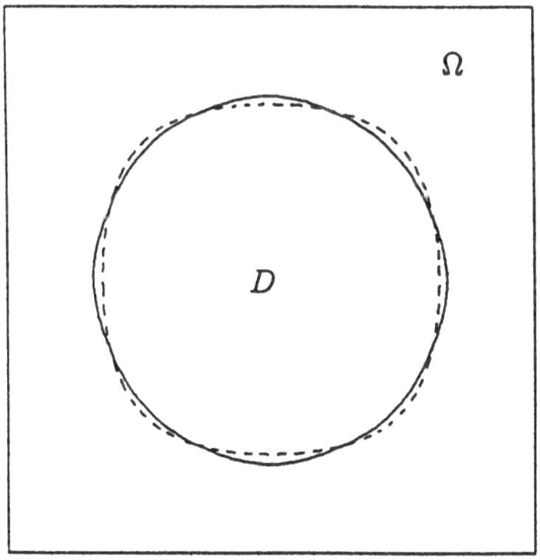

Figure 5.4

6. Conclusions

When conventional boundary variation techniques are used in solving problem (**P**) one must perform a parametrization of the unknown boundary. This parametrization, however, fixes the basic topology of the domain to be found and when (P) is solved numerically, sometimes quite complicated moving finite element meshes have to be used.

The present method does not have the above mentioned difficulties, but it seems limitted to problems with Dirichlet boundary conditions. In the case of Neumann boundary conditions it is possible to obtain similar controllability results, however it is not clear how to define the solution domain D from the solution of the boundary control problem.

References

1. V. Barbu, *Controlling the free boundary for Stefan problem*, in "Proceedings of the symposium "Control of distributed parameter systems"," A. El Jai and M. Amoroux (eds), pp. 255–260. Perpignan, 1989
2. V. Barbu and D. Tiba, *Boundary controllability of the coincidence set in the obstacle problem*, (in print), SIAM J. Control and Optimization.
3. J. Haslinger and P. Neittaanmäki, "Finite element approximation for optimal shape design: theory and applications," Wiley, Chichester, 1988.
4. J. Henry, Thèse, Universitè de Paris VI, 1978.
5. J.-L. Lions, "Optimal control of systems governed by partial differential equations," Springer, Berlin, 1971.
6. J.-L. Lions and E. Magenes, "Non-homogeneous boundary value problems and applications," Springer, Berlin, 1972.
7. E. J. P. G. Schmidt and N. Weck, *On the boundary behaviour of solutions to elliptic and parabolic equations with application to boundary control for parabolic equations*, SIAM J. Control and Optimization **16(4)** (1978).
8. D. Tiba, *Une approche par contrôlabilité dans les problèmes de design optimal*, C.R.A.S. Paris, t. 310, Série I (1990).
9. D. Tiba, P. Neittaanmäki, T. Tiihonen and R. Mäkinen, *Boundary control approach to an optimal shape design problem*, in "Proceedings of the symposium "Control of distributed parameter systems"," A. El Jai and M. Amoroux (eds), pp. 274–277. Perpignan, 1989

Keywords. shape optimization, elliptic problem, finite elements
1980 *Mathematics subject classifications*: 49E15, 35J05, 65K10

[1,2]University of Jyväskylä, Department of Mathematics, Seminaarinkatu 15, SF–40100 Jyväskylä, Finland
[3]Institute of Mathematics, Romanian Academy of Science, Bd. Păcii 220, 79622 Bucharest, Romania

Regularity with interior point control.
Part I: Wave and Euler-Bernoulli equations

R. Triggiani

Department of Applied Mathematics
Thornton Hall
University of Virginia
Charlottesville, VA 22903

Abstract

We study the regularity of wave equations and Euler-Bernoulli equations defined on an open bounded domain Ω, $\dim \Omega = 1, 2, 3$, and subject to the action of point control (through the Dirac distribution δ) at an interior point of Ω. A general approach is used which applies to other dynamics as well (e.g., Kirchhoff equations as in Part II [T.1]). For waves and Kirchhoff equations the results are "$\frac{1}{2}+\epsilon$" sharper in space Sobolev regularity over those that can be obtained by simply using that $\delta \in [H^{\alpha}(\Omega)]'$, $\alpha = \frac{3}{2}+\epsilon$, $1+\epsilon$, $\frac{1}{2}+\epsilon$, for $N = 3, 2, 1$. Instead, in the case of Euler-Bernoulli or Schrödinger equations, the sharp approach produces only an "ϵ-improvement."

1. Introduction, statement of the problem, preliminaries

1.1. Statement of the problem and literature

Let Ω be an open, bounded domain in R^N ($N = 1,2,3$) with sufficiently smooth boundary $\Gamma = \partial\Omega$. In this paper we study the regularity of wave, Euler-Bernoulli, and Schrödinger equations defined on Ω with homogeneous boundary conditions (of various type) on Γ, and subject to the action of point control exercised at an interior point of Ω. As a matter of fact, the specific case of the wave equation with Dirichlet (homogeneous) boundary conditions in the case of $N = \dim \Omega = 3$ has already been studied, with three different proofs reported in J. L. Lions [L.2]: one proof, due to Y. Meyer [M.1], uses harmonic analysis; another proof, due to L. Nirenberg uses propagation properties of waves in R^3, in particular the explicit classical Kirchhoff formula for the solution of the Cauchy problem in R^3 (essentially the same proof appears in [P-S.1]); a third due to J. L. Lions uses a recent trace property for the normal derivative of the corresponding homogeneous problem [L.3], [L-T.1-2], [L-L-T.1]. Some proofs (e.g., Nirenberg's) refer to the dual problem z in (2.7)). Instead, our proof centers on the original problem w in (2.1). Our proof follows a very general approach, which applies to any dynamics [in particular, waves, Euler-Bernoulli (plate) equations, Kirchhoff (plate) equations, Schrödinger equations, etc., regardless of whether or not they possess finite speed of propagation] and in principle to any dimension $N = \dim \Omega$ (we confine our attention to $N = 1,2,3$). The key steps are:

(i) the analysis of (sharp) regularity of the corresponding free space problem by means of Laplace-Fourier transform techniques, which hinges on a sharp *a-priori* estimate (in the present case of wave problems this estimate, Lemma 1.1, is essentially the one established in [L-T.1, Appendix] in the study of regularity of *boundary* forcing terms);

(ii) suitable changes of variables which lead to a corresponding problem on Ω, homogeneous on the boundary but with non-homogeneous forcing term on Ω which does *not*, however, involve the Dirac distribution δ.

In the case of wave equations of this paper (as well as in the case of Kirchhoff (plate) equations to be treated separately in Part II [T.1]), this sharp approach produces regularity results which are "$\frac{1}{2}+\varepsilon$" *stronger* in space regularity, measured in Sobolev space order, over those that can be directly obtained by simply using that $\delta \in [H^{\alpha}(\Omega)]'$, $\alpha = \frac{3}{2}+\varepsilon,\ 1+\varepsilon,\ \frac{1}{2}+\varepsilon$, for $N = 3,2,1$, respectively (see Remark 2.1). In particular, we recover the results of the literature mentioned above in the case of the wave equation with Dirichlet B.C. and $N = \dim \Omega = 3$. In the case of Euler-Bernoulli and Schrödinger equations, the pay-off is smaller, however, as the sharp approach produces only an "ε-improvement" (see Remark 4.1).

In closing, we remark that the *abstract Differential Riccati theory for optimal quadratic cost problems as in* [DaP-L-T.1], [L-T.4] *is applicable now to point control problems on the explicitly identified, sharp regularity spaces of the present paper, and its successor* (Part II [T.1]). This was one of the main motivations of the present work.

Instead, we shall point out in a forthcoming work that exact controllability, as well as uniform stabilization *in the explicitly identified, sharp regularity spaces* of this paper, as well as of [T.2], are out of the question for all these problems with finitely many interior point controls. Thus the abstract Riccati theory as in, say, [F-L-T.1], [L-T.4] is not applicable, since the Finite Cost Condition is not possible. Thus approximate controllability and strong stabilization on these spaces are, by necessity, alternative substitutes which may indeed be achieved.

We shall use freely the Sobolev spaces $H^s(\Omega)$, $H_0^s(\Omega)$, $H_{00}^s(\Omega)$ from [L-M.1], as well as domains of fractional power as e.g. in [G.1].

1.2. A preliminary estimate for wave and Euler-Bernoulli problems

In the study of wave and Euler-Bernoulli equations in free space with point control, we shall need the following preliminary estimate, which is, in fact, contained in the arguments of [L-T.1, Appendix A, pp. 88-90] in connection with the study of the regularity problem of the wave equation with $L_2(0,T;L_2(\Gamma))$-Dirichlet *boundary* control. The proof is repeated here for completeness.

Lemma 1.1. Let $\gamma > 0$ be a fixed constant and let $\omega \in R^1$ be a parameter. Then

$$I(\omega) \equiv \int_0^\infty \frac{y^2 dy}{(y^2+\gamma^2-\omega^2)^2+4\gamma^2\omega^2} \le C < \infty, \tag{1.1}$$

where C is a constant independent of $\omega \in R^1$. ∎

Proof. (i) We first take $\gamma^2-\omega^2 < 0$. Set $a_\omega^2 = \omega^2-\gamma^2 > 0$, and split the integral $I(\omega)$ in (1.1) as

$$I(\omega) = \int_0^{\frac{a_\omega}{2}} + \int_{\frac{a_\omega}{2}}^{2a_\omega} + \int_{2a_\omega}^\infty = I_1(\omega)+I_2(\omega)+I_3(\omega). \tag{1.2}$$

For $I_1(\omega)$ we compute

$$I_1(\omega) = \int_0^{\frac{a_\omega}{2}} \frac{y^2 dy}{(y^2-a_\omega^2)^2+4\gamma^2\omega^2} \leq \int_0^{\frac{a_\omega}{2}} \frac{y^2 dy}{(\frac{3}{4}a_\omega^2)^2+4\gamma^2\omega^2}$$

$$= \frac{1}{3} \frac{(\frac{a_\omega}{2})^3}{(\frac{3}{4}a_\omega^2)^2+4\gamma^2\omega^2} \leq \text{Const.} \tag{1.3}$$

For $I_3(\omega)$ we compute with $-a_\omega^2 \geq -\frac{y^2}{4}$; $\gamma^2 < \omega^2$:

$$I_3(\omega) = \int_{2a_\omega}^{\infty} \frac{y^2 dy}{(y^2-a_\omega^2)^2+4\gamma^2\omega^2} \leq \int_{2a_\omega}^{\infty} \frac{y^2 dy}{(\frac{3}{4}y^2)^2+4\gamma^4} \leq \text{Const.} \tag{1.4}$$

Finally, for $I_2(\omega)$, we set $t = y^2-a_\omega^2$ and compute

$$I_2(\omega) = \int_{\frac{a_\omega}{2}}^{2a_\omega} \frac{y^2 dy}{(y^2-a_\omega^2)^2+4\gamma^2\omega^2} = \frac{1}{2}\int_{-\frac{3}{4}a_\omega^2}^{3a_\omega^2} \frac{\sqrt{t+a_\omega^2}\, dt}{t^2+4\gamma^2\omega^2}$$

(by $t \leq 3a_\omega^2$)
$$\leq a_\omega \int_{-\frac{3}{4}a_\omega^2}^{3a_\omega^2} \frac{1\, dt}{t^2+4\gamma^2\omega^2} = \frac{a_\omega}{2\gamma\omega} \left[\text{arc tan } \frac{t}{2\gamma\omega}\right]_{t=-\frac{3}{4}a_\omega^2}^{t=3a_\omega^2}$$

(by $a_\omega \leq \omega$)
$$\leq \frac{\omega}{2\gamma\omega} \left(\frac{\pi}{2} + \frac{\pi}{2}\right) \leq \text{Const.} \tag{1.5}$$

Thus, estimates (1.3)-(1.5) used in (1.2) yield

$$I(\omega) \leq \text{Const} \quad \text{for all } \omega^2 > \gamma^2. \tag{1.6}$$

(ii) Next, we take $\gamma^2-\omega^2 \geq 0$ and split $I(\omega)$ as

$$I(\omega) = \int_0^1 + \int_1^\infty = J_1(\omega)+J_2(\omega). \tag{1.7}$$

For $J_2(\omega)$ we drop $\gamma^2-\omega^2 \geq 0$ and obtain

$$J_2(\omega) = \int_1^\infty \frac{y^2 dy}{(y^2+\gamma^2-\omega^2)^2+4\gamma^2\omega^2} \leq \int_1^\infty \frac{1}{y^2}\, dy = 1. \tag{1.8}$$

For $J_1(\omega)$ we first restrict to $0 \leq \omega^2 \leq \gamma^2-\varepsilon$, $\varepsilon > 0$ and obtain

$$J_1(\omega) = \int_0^1 \frac{y^2dy}{(y^2+\gamma^2-\omega^2)^2+4\gamma^2\omega^2} \leq \int_0^1 \frac{y^2dy}{(y^2+\varepsilon)^2} \leq \frac{1}{\varepsilon^2}\int_0^1 y^2dy = \frac{1}{3\varepsilon^2}, \tag{1.9}$$

and next consider $\gamma^2-\varepsilon < \omega^2 \leq \gamma^2$ and obtain

$$J_1(\omega) = \int_0^1 \frac{y^2dy}{(y^2+\gamma^2-\omega^2)^2+4\gamma^2\omega^2} \leq \int_0^1 \frac{y^2dy}{4\gamma^2\omega^2}$$

$$\leq \int_0^1 \frac{y^2dy}{4\gamma^2(\gamma^2-\varepsilon)} = \frac{1}{12\gamma^2(\gamma^2-\varepsilon)}. \tag{1.10}$$

Using estimates (1.8), (1.9), and (1.10) in (1.7) yields

$$I(\omega) \leq \text{Const} \quad \text{for all } \omega^2 \leq \gamma^2. \tag{1.11}$$

Thus (1.6) and (1.11) prove (1.1). ∎

Remark 1.1. Estimate (1.1) is sharp (optimal) in the sense that, say for $\omega^2 > \gamma^2$ and with $\varepsilon > 0$, we have instead

$$\int_0^\infty \frac{y^{2+\varepsilon}dy}{(y^2+\gamma^2-\omega^2)^2+4\gamma^2\omega^2} \geq \int_{\frac{a_\omega}{2}}^{2a_\omega} \frac{y^{2+\varepsilon}dy}{(y^2+\gamma^2-\omega^2)^2+4\gamma^2\omega^2}$$

$$= \tfrac{1}{2}\int_{-\frac{3}{4}a_\omega^2}^{3a_\omega^2} \frac{(t+a_\omega^2)^{\frac{1+\varepsilon}{2}}dt}{t^2+4\gamma^2\omega^2} \geq \tfrac{1}{2}\left(\frac{a_\omega^2}{4}\right)^{\frac{1+\varepsilon}{2}} \int_{-\frac{3}{4}a_\omega^2}^{3a_\omega^2} \frac{dt}{t^2+4\gamma^2\omega^2}$$

$$= \tfrac{1}{2}\left(\frac{1}{4}\right)^{\frac{1+\varepsilon}{2}} \frac{a_\omega^{1+\varepsilon}}{2\gamma\omega}\left[\text{arc tan}\,\frac{t}{2\gamma\omega}\right]_{t=-\frac{3}{4}a_\omega^2}^{t=3a_\omega^2} \longrightarrow \infty \tag{1.12}$$

since $\frac{a_\omega^{1+\varepsilon}}{2\gamma\omega} \sim \frac{\omega^{1+\varepsilon}}{\omega} \to +\infty$ as $\omega \to +\infty$, while the "arc tan" term tends to $\frac{\pi}{2}+\frac{\pi}{2}$ as $\omega \to \infty$, since $\frac{a_\omega^2}{\omega} \sim \omega$. ∎

2. Wave equation with homogeneous Dirichlet B.C. and interior point control

2.1. Statement of results

In this section we consider the following wave equation problem in the unknown $w(t,x)$:

$$\begin{cases} w_{tt} = \Delta w + \delta(x)v(t) & \text{in } (0,T]\times\Omega \equiv Q; & (2.1a) \\ w(0,x) \equiv w_t(0,x) \equiv 0 & \text{in } \Omega; & (2.1b) \\ w|_\Sigma \equiv 0 & \text{in } (0,T]\times\Gamma \equiv \Sigma, & (2.1c) \end{cases}$$

where $\delta(x)$ is the Dirac mass +1 at the point 0 (origin), assumed (without loss of generality) to be an interior point of the open bounded domain $\Omega \subset R^N$, N = 1, 2, 3. We define the positive, self-adjoint operator A

$$Ah = -\Delta h; \quad \mathcal{D}(A) = H^2(\Omega) \cap H_0^1(\Omega); \quad \mathcal{D}(A^{\frac{1}{2}}) = H_0^1(\Omega); \tag{2.2a}$$

$$\mathcal{D}(A^{\frac{1}{4}}) = [\mathcal{D}(A^{\frac{1}{2}}), L_2(\Omega)]_{\frac{1}{2}} = [H_0^1(\Omega), L_2(\Omega)]_{\frac{1}{2}} = H_{00}^{\frac{1}{2}}(\Omega) \tag{2.2b}$$

$$\mathcal{D}(A^{\frac{3}{4}}) = [\mathcal{D}(A), \mathcal{D}(A^{\frac{1}{2}})]_{\frac{1}{2}} \supset [H_0^2(\Omega), H_0^1(\Omega)]_{\frac{1}{2}} \equiv H_{00}^{\frac{3}{2}}(\Omega) \tag{2.2c}$$

(with equivalent norms) see [L-M.1, p. 66] for $H_{00}^{\frac{1}{2}}(\Omega)$. In all results below, the map $v \to \{w, w_t, w_{tt}\}$ is continuous between the respective function spaces, explicitly identified, by the closed graph theorem.

Theorem 2.1. With reference to problem (2.1), let

$$v \in L_2(0,T). \tag{2.3}$$

Then, continuously:

(a) for $N = \dim \Omega = 3$,

$$\begin{cases} w \in C([0,T]; L_2(\Omega)), & (2.4a) \\ w_t \in C([0,T]; H^{-1}(\Omega)), & (2.4b) \\ w_{tt} \in L_2(0,T; H^{-2}(\Omega)), & (2.4c) \end{cases}$$

(b) for $N = \dim \Omega = 2$

$$\begin{cases} w \in C([0,T]; H_{00}^{\frac{1}{2}}(\Omega) = \mathcal{D}(A^{\frac{1}{4}})), & (2.5a) \\ w_t \in C([0,T]; [H_{00}^{\frac{1}{2}}(\Omega)]' = [\mathcal{D}(A^{\frac{1}{4}})]'), & (2.5b) \\ w_{tt} \in L_2(0,T; [\mathcal{D}(A^{\frac{3}{4}})]') \subset L_2(0,T; [H_{00}^{\frac{3}{2}}(\Omega)]'), & (2.5c) \end{cases}$$

(c) for $N = \dim \Omega = 1$,

$$\begin{cases} w \in C([0,T];H_0^1(\Omega) = \mathcal{D}(A^{1/2})), & (2.6a) \\ w_t \in C([0,T];L_2(\Omega)), & (2.6b) \\ w_{tt} \in L_2(0,T;H^{-1}(\Omega) = [\mathcal{D}(A^{1/2})]') \quad \blacksquare & (2.6c) \end{cases}$$

Remark 2.1. If one studies the regularity of problem (2.1) by using only that, by Sobolev embedding, $\delta \in [H^{\alpha}(\Omega)]'$, where $\alpha = \frac{3}{2}+\epsilon$ for $N = 3$; $\alpha = 1+\epsilon$ for $N = 2$; $\alpha = \frac{1}{2}+\epsilon$ for $N = 1$, then one would obtain a regularity result for, say, w which is *lower* by "$\frac{1}{2}+\epsilon$" in space regularity, measured in Sobolev space order, than those of Theorem 2.1: e.g., for $N = 3$ one would get only $w \in H^{-\frac{1}{2}-\epsilon}(\Omega)$ rather than $L_2(\Omega)$ as in (2.4a); for $N = 2$, one would get only $w \in H^{-\epsilon}(\Omega)$ rather than $H_{00}^{\frac{1}{2}}(\Omega)$ as in (2.5a); for $N = 1$ one would get only $w \in H^{\frac{1}{2}-\epsilon}$ rather than $H_0^1(\Omega)$ as in (2.6a). To see this, we use $[H^{\alpha}(\Omega)]' \subset [\mathcal{D}(A^{\alpha/2})]'$, so that $A^{-\alpha/2}\delta \in L_2(\Omega)$ for the second-order operator A in (2.2). Then the solution w of (2.1) satisfies abstractly

$$A^{\frac{1-\alpha}{2}} w(t) = \int_0^t A^{\frac{1}{2}} S(t-\tau)A^{-\alpha/2}\delta v(\tau)d\tau \in C([0,T];L_2(\Omega))$$

by convolution properties between $A^{-\alpha/2}\delta v \in L_2(0,T;L_2(\Omega))$ and $t \to A^{\frac{1}{2}}S(t)$ strongly continuous on $L_2(\Omega)$, where $S(t) = \int_0^t C(\tau)d\tau$ and $C(t)$ is the cosine operator generated by $-A$ in (2.2a). $\blacksquare$

In order to state corresponding duality results, obtained by a duality argument on those of Theorem 2.1, we introduce the following homogeneous system in the unknown $z(t,x)$:

$$\begin{cases} z_{tt} = \Delta z & \text{in } (0,T]\times\Omega \equiv Q; & (2.7a) \\ z(T,\cdot) = z_0;\ z_t(T,\cdot) = z_1 & \text{in } \Omega; & (2.7b) \\ z|_{\Sigma} \equiv 0 & \text{in } (0,T]\times\Gamma \equiv \Sigma, & (2.7c) \end{cases}$$

Then, with reference to problems (2.1) and (2.7) we readily have that:

the map $v(t) \to \{w(T,\cdot),w_t(T,\cdot)\}$ is dual to the map

$$\{z_1,z_0\} \to z(t,0). \qquad (2.8)$$

Applying the duality relationship (2.8) on the results of Theorem 2.1 we obtain

Theorem 2.2. With reference to problem (2.7) we have that

$$z(t,0) \in L_2(0,T) \tag{2.9}$$

continuously on $\{z_0,z_1\}$, where

(a) for $N = \dim \Omega = 3$,

$$\{z_0,z_1\} \in H_0^1(\Omega)\times L_2(\Omega); \tag{2.10}$$

(b) for $N = \dim \Omega = 2$,

$$\{z_0,z_1\} \in H_{00}^{1/2}(\Omega)\times[H_{00}^{1/2}(\Omega)]' = \mathcal{D}(A^{1/4})\times[\mathcal{D}(A^{1/4})]'; \tag{2.11}$$

(c) for $N = \dim \Omega = 1$,

$$\{z_0,z_1\} \in L_2(\Omega)\times H^{-1}(\Omega) = L_2(\Omega)\times[\mathcal{D}(A^{1/2})]'. \quad \blacksquare \tag{2.12}$$

2.2. Proof of Theorem 2.1

Step 1 (free space problem). Following a standard strategy, used e.g., also in [D-Z.1], we consider first the free space problem, corresponding to problem (2.1), in the unknown $\phi(t,x)$:

$$\begin{cases} \phi_{tt} = \Delta\phi + \delta(x)v(t) & \text{in } R^1_{t^+}\times R^N_x, \quad (2.13a) \\ \phi(0,x) \equiv \phi_t(0,x) \equiv 0 & \text{in } R^N_x, \quad (2.13b) \end{cases}$$

after extending $v(t)$ by zero for $t > T$, where $R^1_{t^+} = (0,\infty)$ in t.

Proposition 2.3. With reference to problem (2.13), let v satisfy (2.3). Then, continuously

(a) for $N = \dim \Omega = 3$,

$$\begin{cases} \phi \in C([0,T];L_2(R^3)), & (2.14a) \\ \phi_t \in C([0,T];H^{-1}(R^3)), & (2.14b) \\ \phi_{tt} \in L_2(0,T;H^{-2}(R^3)), & (2.14c) \end{cases}$$

(b) for $N = \dim \Omega = 2$ and for any $\psi \in C_0^\infty(\Omega)$, $\psi(0) = 1$:

$$\begin{cases} \phi \in C([0,T];H^{\frac12}(R^2)); & \psi\phi \in C([0,T];H_{00}^{\frac12}(\Omega)= \mathcal{D}(A^{\frac14})) , & (2.15a) \\ \phi_t \in C([0,T];H^{-\frac12}(R^2)); & \psi\phi_t \in C([0,T];[H_{00}^{\frac12}(\Omega)]' = [\mathcal{D}(A^{\frac14})]') , & (2.15b) \\ \phi_{tt} \in L_2(0,T;H^{-\frac32}(R^2)); & \psi\phi_{tt} \in L_2(0,T;[\mathcal{D}(A^{\frac34})]') , & (2.15c) \end{cases}$$

(c) for $N = \dim \Omega = 1$,

$$\begin{cases} \phi \in C([0,T];H^1(R^1)) , & (2.16a) \\ \phi_t \in C([0,T];L_2(R^1)) , & (2.16b) \\ \phi_{tt} \in L_2(0,T;H^{-1}(R^1)) . \quad \blacksquare & (2.16c) \end{cases}$$

<u>Proof</u>. Let $\hat{\phi}(\lambda,\xi)$, $\lambda = \gamma+i\omega$, $\gamma > 0$ and $\omega \in R^1$, $\xi \in R^N$, be the Laplace (in t) - Fourier (in x) transform of $\phi(t,x)$:

$$\hat{\phi}(\lambda,\xi) = \frac{1}{(2\pi)^{N/2}} \int_0^\infty e^{-\lambda t} \int_{R^N} e^{-ix\cdot\xi}\phi(t,x)dx\, dt, \tag{2.17}$$

so that the transformed version of problem (2.13) is

$$\hat{\phi}(\lambda,\xi) = \frac{\hat{v}(\lambda)}{\lambda^2+|\xi|^2} . \tag{2.18}$$

It suffices to show the results for $\{\phi,\phi_t\}$ in (2.14)-(2.16) with $C[0,T]$ replaced by $L_2(0,T)$, and then appeal to a general result [L-T.2], [L-T.3] for time reversible dynamics (groups of operators) to lift $L_2(0,T)$ to $C[0,T]$, while preserving the space regularity.

<u>Proof of (2.14a)</u>. $N = 3$. Accordingly, it suffices to show that

$$e^{-\gamma t}\phi(t,x) \in L_2(R^1_{t^+}\times R^N_x), \tag{2.19}$$

or equivalently by the Parseval identity [D.1, p. 212] that

$$\hat{\phi}(\gamma+i\omega,\xi) \in L_2(R^1_\omega\times R^N_\xi). \tag{2.20}$$

To this end we compute via (2.18),

$$\|\hat{\phi}\|^2_{L_2(R^1_\omega \times R^N_\xi)} = \int_{R^1_\omega} \int_{R^N_\xi} |\hat{\phi}(\gamma+i\omega,\xi)|^2 d\xi \; d\omega$$

(by (2.18))

$$= \int_{R^1_\omega} |\hat{v}(\gamma+i\omega)|^2 \left(\int_{R^N_\xi} \frac{d\xi}{|\lambda^2+|\xi|^2|^2} \right) d\omega. \qquad (2.21)$$

Since $|\lambda^2+|\xi|^2|^2 = (|\xi|^2+\gamma^2-\omega^2)^2+4\gamma^2\omega^2$, we see that (2.21) yields

$$\|\hat{\phi}\|^2_{L_2(R^1_\omega \times R^N_\xi)} \le C \int_{R^1_\omega} |\hat{v}(\gamma+i\omega)|^2 d\omega = \frac{C}{2\pi} \int_0^\infty e^{-2\gamma t} |v(t)|^2 dt$$

$$\le \text{const } \|v\|^2_{L_2(0,T)} \qquad (2.22)$$

as desired, as soon as we prove that for N = 3,

$$\int_{R^N_\xi} \frac{d\xi}{|\lambda^2+|\xi|^2|^2} = \int_{R^N_\xi} \frac{d\xi}{(|\xi|^2+\gamma^2-\omega^2)^2+4\gamma^2\omega^2} \le C < \infty \qquad (2.23)$$

uniformly in $\omega \in R^1$. To this end, we use spherical coordinates (ρ,ϕ,θ) so that $d\xi = d\xi_1 d\xi_2 d\xi_3 = \rho^2 \sin\phi \; d\phi \; d\rho \; d\theta$, and then see that estimate (2.23) holds true provided

$$\int_0^\infty \frac{\rho^2 d\rho}{(\rho^2+\gamma^2-\omega^2)^2+4\gamma^2\omega^2} \le C < \infty, \text{ uniformly in } \omega \in R^1, \qquad (2.24)$$

which is precisely estimate (1.1) asserted by Lemma 1.1. Then (2.23), hence (2.22) and (2.20), (2.19) are all proved.

<u>Proof of (2.15a)</u>. N = 2. Here, it suffices to show that

$$e^{-\gamma t}\phi(t,x) \in L_2(R^1_{t^+}; H^{1/2}(R^N_x)), \text{ equivalently}$$

$$|\xi|^{1/2}\hat{\phi}(\gamma+i\omega,\xi) \in L_2(R^1_\omega \times R^N_\xi). \qquad (2.25)$$

To this end we compute via (2.18),

$$\| |\xi|^{1/2}\hat{\phi}(\gamma+i\omega,\xi)\|^2_{L_2(R^1_\omega \times R^N_\xi)} = \int_{R^1_\omega} |\hat{v}(\gamma+i\omega)|^2 \left(\int_{R^N_\xi} \frac{|\xi| \; d\xi}{|\lambda^2+|\xi|^2|^2} \right) d\omega \le C \int_{R^1_\omega} |\hat{v}(\gamma+i\omega)|^2 d\omega$$

$$= \frac{C}{2\pi} \int_0^\infty e^{-2\gamma t} |v(t)|^2 dt = C\|v\|^2_{L_2(0,T)} \tag{2.26}$$

as desired, as soon as we prove that for N = 2,

$$\int_{R^N_\xi} \frac{|\xi|\, d\xi}{|\lambda^2+|\xi|^2|^2} = \int_{R^N_\xi} \frac{|\xi|\, d\xi}{(|\xi|^2+\gamma^2-\omega^2)^2+4\gamma^2\omega^2} \leq C < \infty \tag{2.27}$$

uniformly in $\omega \in R^1$. Using now, for N = 2, polar coordinates (r,ω) so that $d\xi = d\xi_1 d\xi_2 = r\, dr\, d\theta$, we see that (2.27) leads to the same integral estimate (2.24) with ρ replaced by r, i.e., to estimate (1.1). The regularity of $\psi\phi$ in (2.15a) uses that $\psi \equiv 0$ near Γ.

<u>Proof of (2.16a)</u>. N = 1. Now, it suffices to show that

$$e^{-\gamma t}\phi(t,x) \in L_2(R^1_{t^+};H^1(R^N_x)), \text{ equivalently,}$$

$$|\xi|\hat{\phi}(\gamma+i\omega,\xi) \in L_2(R^1_\omega \times R^N_\xi), \tag{2.28}$$

which is seen to hold true precisely as in the two preceding cases, by virtue of the same estimate (1.1) of Lemma 1.1.

<u>Proof of (2.14c)</u>. It follows from (2.14a) already proved in the $L_2(0,T)$-sense, via [L-M.1, p. 85], that

$$\Delta\phi \in L_2(0,T;H^{-2}(R^N_x)). \tag{2.29}$$

Moreover, by Sobolev embedding with N = 3: $H^2_0(\Omega) \subset H^{3/2+\varepsilon}(\Omega) \subset C^0(\Omega)$, hence

$$\delta \in H^{-2}(\Omega) \text{ and } \delta v \in L_2(0,T;H^{-2}(\Omega)) \quad ; N = 3. \tag{2.30}$$

Hence (2.29) and (2.30), used on the right hand side of (2.13a), yield $\phi_{tt} \in L_2(0,T;H^{-2}(R^N_x))$ as desired.

<u>Proof of (2.14b)</u>. Applying the intermediate derivative theorem [L-M.1; p. 15] to (2.14a) and (2.14c) yields: $\phi_t \in L_2(0,T;H^{-1}(R^N_x))$. By virtue of the lifting property in time recalled below (2.18), the regularity of $\{\phi,\phi_t\}$ in the $L_2(0,T)$-sense is sufficient to establish the regularity (2.14a-b).

Proof of (2.16c). It follows from (2.16a) already proved in the $L_2(0,T)$-sense, via [L-M.1, p. 85] that

$$\Delta\phi \in L_2(0,T;H^{-1}(R_x^N)). \tag{2.31}$$

Moreover, by Sobolev embedding with $N = 1$: $H_0^1(\Omega) \subset H^{1/2+\epsilon}(\Omega) \subset C^0(\Omega)$, hence

$$\delta \in H^{-1}(\Omega) \text{ and } \delta v \in L_2(0,T;H^{-1}(\Omega)); \; N = 1. \tag{2.32}$$

Hence (2.31) and (2.32), used on the right hand side of (2.13a), yield $\phi_{tt} \in L_2(0,T;H^{-1}(R_x^N))$ as desired.

Proof of (2.16b). The intermediate derivative theorem [L-M.1, p. 15] applied to (2.16a) and (2.16c) yields $\phi_t \in L_2(0,T;L_2(R_x^N))$. The regularity of $\{\phi,\phi_t\}$ in the $L_2(0,T)$-sense is then lifted in time to the regularity (2.16a-b), as recalled below (2.18).

Proof of (2.15c). It follows from (2.15a) already proved in the $L_2(0,T)$-sense, via [L-M.1, p. 85] that

$$\Delta\phi \in L_2(0,T;H^{-3/2}(R_x^N)); \quad \Delta\phi \in L_2(0,T;[H_{00}^{3/2}(R_x^N)]'). \tag{2.33}$$

Moreover, by Sobolev embedding with $N = 2$: $H_{00}^{3/2}(\Omega) \subset H_0^{3/2}(\Omega) \subset H^{1+\epsilon}(\Omega) \subset C^0(\Omega)$, hence

$$\delta \in H^{-3/2}(\Omega) \subset [H_{00}^{3/2}(\Omega)]'; \quad \delta v \in L_2(0,T;[H_{00}^{3/2}(\Omega)]'); \; N = 2, \tag{2.34}$$

and (2.33), (2.34), used on the right hand side of (2.13a), yield $\phi_{tt} \in L_2(0,T;H^{-3/2}(R_x^N))$, $\phi_{tt} \in L_2(0,T;[H_{00}^{3/2}(\Omega)]')$ as desired.

Moreover, to see that $\psi\phi_{tt} \in L_2(0,T;[\mathcal{D}(A^{3/4})]')$, we note that multiplying (2.13a) by ψ, $\psi \in C_0^\infty(\Omega)$, $\psi(0) = 1$, we obtain $\psi\phi_{tt} = \psi\Delta\phi+\delta v$. Since $\mathcal{D}(A^{3/4}) \subset H^{3/2}(\Omega) \subset H^{1+\epsilon}(\Omega) \subset C^0(\Omega)$ for $N = 2$, we have $\delta \in [H^{1+\epsilon}(\Omega)]' \subset [\mathcal{D}(A^{3/4})]'$). It remains therefore to show that $\psi\Delta\phi \in L_2(0,T;[\mathcal{D}(A^{3/4})]')$. To this end we pick $g \in L_2(0,T;\mathcal{D}(A^{3/4})) \subset L_2(0,T;H^{3/2}(\Omega))$ so that $\psi g \in H_{00}^{3/2}(\Omega) \subset H_{00}^{1/2}(\Omega)$ a.e. in t, since $\psi \equiv 0$ near Γ. Moreover $\phi \in L_2(0,T;H^{1/2}(R_x^2))$ by (2.15a) already proved in the $L_2(0,T)$-sense, implies $\Delta\phi \in L_2(0,T;[H_{00}^{3/2}(\Omega)]')$ by [L-M.1, p. 85] and so the integral

$$\int_0^T \int_\Omega \Delta\phi \; g \; \psi \; d\Omega \; dt$$

is well defined, as desired.

Proof of (2.15b). The intermediate derivative theorem [L-M.1, p. 15] applied on (2.15a), (2.15c) yields (2.15b) with $L_2(0,T)$-time regularity. This--as mentioned below (2.18)--is all what suffices to show. The proof of Proposition 2.3 is complete. ∎

Step 2 (auxiliary problems). Following a general strategy, used also in [D-Z.1], we return from the free space ϕ-problem (2.13) to the original w-problem (2.1), via two intermediary problems. We introduce two new variables,

$$\phi_c(t,x) = \psi(x)\phi(t,x) \quad \text{and} \quad h(t,x) = \phi_c(t,x)-w(t,x), \tag{2.35}$$

where $\psi(x)$ is a C^∞-function with compact support in Ω, $\psi \in C_0^\infty(\Omega)$, so that ϕ_c is a 'cut off' of ϕ. Moreover, we impose $\psi(0) = 1$. Then, multiplying (2.13) by ψ yields ($' = \frac{d}{dt}$):

$$\begin{cases} \psi\phi'' = \psi\Delta\phi + \delta(x)v(t) & \text{in } (0,T]\times\Omega \equiv Q, & (2.36a)\\ \psi(x)\phi(0,x) \equiv \psi(x)\phi_t(0,x) \equiv 0 & \text{in } \Omega, & (2.36b)\\ \psi\phi|_\Sigma \equiv 0 & \text{in } (0,T]\times\Gamma \equiv \Sigma. & (2.36c)\end{cases}$$

Then, via (2.36), ϕ_c satisfies the problem

$$\begin{cases} \phi_c'' = \Delta\phi_c + \delta(x)v(t) + f(t,x) & \text{in } Q, & (2.37a)\\ \phi_c(0,x) \equiv \phi_c'(0,x) \equiv 0 & \text{in } \Omega, & (2.37b)\\ \phi_c|_\Sigma \equiv 0 & \text{in } \Sigma, & (2.37c)\end{cases}$$

$$f \equiv -(\Delta\psi)\phi-2\nabla\psi\cdot\nabla\phi. \tag{2.38}$$

Finally, via (2.37), h(t,x) satisfies the homogeneous (on Σ) problem

$$\begin{cases} h'' = \Delta h + f & \text{in } Q, & (2.39a)\\ h(0,x) = h'(0,x) \equiv 0 & \text{in } \Omega, & (2.39b)\\ h|_\Sigma \equiv 0 & \text{in } \Sigma. & (2.39c)\end{cases}$$

Since the regularity of ϕ (Proposition 2.3) determines that of ϕ_c by (2.35) (left), in order to obtain the regularity of w via (2.35) (right), it remains to establish the regularity of h(t,x). From the regularity of ϕ in (2.14a), (2.15a), (2.16a), it follows via [L-M.1, p. 85] and $\psi \in C_0^\infty(\Omega)$ that

$$\nabla\psi\cdot\nabla\phi, \text{ hence by (2.38), } f \in \begin{cases} C([0,T];H^{-1}(\Omega)) & N = 3 \quad (2.40a)\\ C([0,T];[H_{00}^{1/2}(\Omega)]') & N = 2 \quad (2.40b)\\ C([0,T];L_2(\Omega)) & N = 1. \quad (2.40c)\end{cases}$$

__Remark 2.1__. The results in (2.40a-b) are all that is needed in the present Dirichlet case. They will be strengthened in Section 3.2 to a form needed for the Neumann case, see (3.17)-(3.18) below. ∎

Then (2.40a-b) mean equivalently via (2.2a) and (2.2b),

$$A^{-1/2}f \in C([0,T];L_2(\Omega)), \quad N = 3; \tag{2.41a}$$

$$A^{-1/4}f \in C([0,T];L_2(\Omega)), \quad N = 2. \tag{2.41b}$$

Let now C(t) and $S(t) = \int_0^t C(\tau)d\tau$ denote the cosine and sine operators generated by the negative self-adjoint operator $-A$ on $L_2(\Omega)$ in (2.2a), $t \in R$. We recall that the maps: $t \to C(t)y$ and $t \to A^{1/2}S(t)y$, $y \in L_2(\Omega)$, are (strongly) continuous, *a fact which will be used freely below*. The solution of the h-problem (2.39) is (abstractly)

$$h(t) = \int_0^t S(t-\tau)f(\tau)d\tau; \tag{2.42}$$

$$h'(t) = \int_0^t C(t-\tau)f(\tau)d\tau; \quad h'' = -Ah+f. \tag{2.43}$$

__Proposition 2.4__. With reference to problem (2.39), equivalently (2.42), (2.43), we have

(a) for $N = \dim \Omega = 3$,

$$\begin{cases} h \in C([0,T];L_2(\Omega)), & (2.44a) \\ h_t \in C([0,T];[\mathcal{D}(A^{1/2})]' = H^{-1}(\Omega)), & (2.44b) \\ h_{tt} \in C([0,T];[\mathcal{D}(A)]'); & (2.44c) \end{cases}$$

(b) for $N = \dim \Omega = 2$,

$$\begin{cases} h \in C([0,T];\mathcal{D}(A^{1/4}) = H_{00}^{1/2}(\Omega)), & (2.45a) \\ h_t \in C([0,T];[\mathcal{D}(A^{1/4})]' = [H_{00}^{1/2}(\Omega)]'), & (2.45b) \\ h_{tt} \in C([0,T];[\mathcal{D}(A^{3/4})]') \subset C([0,T];[H_{00}^{3/2}(\Omega)]'); & (2.45c) \end{cases}$$

(c) for $N = \dim \Omega = 1$,

$$\begin{cases} h \in C([0,T];\mathcal{D}(A^{1/2}) = H_0^1(\Omega)), & (2.46a) \\ h_t \in C([0,T];L_2(\Omega)), & (2.46b) \\ h_{tt} \in C([0,T];[\mathcal{D}(A^{1/2})]' = H^{-1}(\Omega)). \quad \blacksquare & (2.46c) \end{cases}$$

__Proof__. We use convolution properties between C-spaces and L_2-spaces. For $N = 3$, via (2.41a), we obtain

$$h(t) = \int_0^t A^{1/2}S(t-\tau)A^{-1/2}f(\tau)d\tau \in C([0,T];L_2(\Omega)), \tag{2.47a}$$

$$A^{-1/2}h_t(t) = \int_0^t C(t-\tau)A^{-1/2}f(\tau)d\tau \in C([0,T];L_2(\Omega)), \tag{2.47b}$$

and equivalently (2.44a-b) follow. For $N = 2$, via (2.41b), we obtain

$$A^{1/4}h(t) = \int_0^t A^{1/2}S(t-\tau)A^{-1/4}f(\tau)d\tau \in C([0,T];L_2(\Omega)), \tag{2.48a}$$

$$A^{-1/4}h_t(t) = \int_0^t C(t-\tau)A^{-1/4}f(\tau)d\tau \in C([0,T];L_2(\Omega)), \tag{2.48b}$$

or equivalently (2.45a-b) by (2.2b). For $N = 1$, via (2.40c),

$$A^{1/2}h(t) = \int_0^t A^{1/2}S(t-\tau)f(\tau)d\tau \in C([0,T];L_2(\Omega)), \tag{2.49a}$$

$$h_t(t) = \int_0^t C(t-\tau)f(\tau)d\tau \in C([0,T];L_2(\Omega)), \tag{2.49b}$$

or equivalently (2.46a-b). Finally, the regularity (2.44c), (2.45c), (2.46c) for h_{tt} follows from that of h in (2.44a), (2.45a), (2.46a) via (2.43) (right) and that of f in (2.41a), (2.41b), (2.40c) respectively. The proof of Proposition 2.4 is complete. ∎

Step 3 (return to w-problem). We now return to the w-problem (2.1), via $w = \psi\phi - h$ (see (2.35)). Employing the regularity of ϕ in Proposition 2.3 and the regularity of h in Proposition 2.4, we obtain the regularity of w in Theorem 2.1. In analyzing w, note that for N = 2, $\phi \in C([0,T];H^{1/2}(R^2))$ by (2.15a) implies $\psi\phi \in C([0,T];H^{1/2}_{00}(\Omega))$ since $\psi \equiv 0$ near Γ, see [L-M.1, p. 66]; similarly for N = 1, $\phi \in C([0,T];H^1(R^1))$ by (2.16a) implies $\psi\phi \in C([0,T];H^1_0(\Omega))$. Finally, for w_{tt} in case N = 3, note that $H^2_0(\Omega) \subset \mathcal{D}(A)$ yields $[\mathcal{D}(A)]' \subset H^{-2}(\Omega)$. The proof of Theorem 2.1 is complete. ∎

3. Wave equation with homogeneous Neumann B.C. and interior point control

3.1. Statement of results

In this section we consider the following wave equation problem in the unknown w(t,x):

$$\begin{cases} w_{tt} = \Delta w + \delta(x)v(t) & \text{in } (0,T]\times\Omega \equiv Q, & (3.1a)\\ w(0,x) = w_t(0,x) \equiv 0 & \text{in } \Omega, & (3.1b)\\ w|_{\Sigma_0} \equiv 0 & \text{in } (0,T]\times\Gamma_0 = \Sigma_0, & (3.1c)\\ \frac{\partial w}{\partial \nu}|_{\Sigma_1} \equiv 0 & \text{in } (0,T]\times\Gamma_1 = \Sigma_1, & (3.1d)\end{cases}$$

where again the origin 0, the point at which the Dirac mass + 1 is applied, is an interior point of Ω. The boundary Γ is assumed divided into two parts $\Gamma = \Gamma_0 \cup \Gamma_1$, Γ_i open in Γ, where Γ_0 is possibly empty. We define the operator A by

$$Ah = -\Delta h; \quad \mathcal{D}(A) = \{h \in H^2(\Omega): h|_{\Gamma_0} = 0, \frac{\partial h}{\partial \nu}|_{\Gamma_1} = 0\}. \tag{3.2}$$

If Γ_0 is not empty, then A is positive self-adjoint on $L_2(\Omega)$ and (with equivalent norms)

$$\mathcal{D}(A^{1/2}) = H^1(\Omega); \quad \mathcal{D}(A^{1/4}) = H^{1/2}(\Omega) = H_0^{1/2}(\Omega); \tag{3.3a}$$

$$[\mathcal{D}(A^{1/2})]' = [H^1(\Omega)]'; \quad [\mathcal{D}(A^{1/4})]' = H^{-1/2}(\Omega). \tag{3.3b}$$

When Γ_0 is empty, then A is only non-negative self-adjoint on $L_2(\Omega)$, or positive self-adjoint on $L_2^0(\Omega) = L_2(\Omega)/\mathcal{N}(A)$, the quotient space of $L_2(\Omega)$ by the one-dimensional null space $\mathcal{N}(A)$ of A, given by constant functions. When Γ_0 is empty, we either consider (3.1) on spaces quotient with $\mathcal{N}(A)$, or else use

$$\mathcal{D}((A+I)^{1/2}) = H^1(\Omega); \quad \mathcal{D}((A+I)^{1/4}) = H^{1/2}(\Omega); \tag{3.4}$$

instead of (3.3a) and corresponding duality results.

<u>Theorem 3.1</u>. With reference to problem (3.1), let

$$v \in L_2(0,T). \tag{3.5}$$

Then, continuously,

(a) for $N = \dim \Omega = 3$,

$$\begin{cases} w \in C([0,T];L_2(\Omega)), & (3.6a) \\ w_t \in C([0,T];H^{-1}(\Omega)), & (3.6b) \\ w_{tt} \in L_2(0,T;H^{-2}(\Omega)); & (3.6c) \end{cases}$$

(b) for $N = \dim \Omega = 2$,

$$\begin{cases} w \in C([0,T];H^{1/2}(\Omega)), & (3.7a) \\ w_t \in C([0,T];[H_{00}^{1/2}(\Omega)]'), & (3.7b) \\ w_{tt} \in L_2(0,T;[H_{00}^{3/2}(\Omega)]'); & (3.7c) \end{cases}$$

(c) for $N = \dim \Omega = 1$,

$$\begin{cases} w \in C([0,T];H^1(\Omega)), & (3.8a) \\ w_t \in C([0,T];L_2(\Omega)), & (3.8b) \\ w_{tt} \in L_2(0,T;H^{-1}(\Omega)). \quad \blacksquare & (3.8c) \end{cases}$$

In order to state corresponding duality results, obtained by a transposition (duality) argument on those of Theorem 3.1, we introduce the following homogeneous system in the unknown $z(t,x)$,

$$\begin{cases} z_{tt} = \Delta z & \text{in } Q\,, & (3.9a)\\ z(T,\cdot) = z_0, \quad z_t(T,\cdot) = z_1 & \text{in } \Omega\,, & (3.9b)\\ z|_{\Sigma_0} \equiv 0 & \text{in } \Sigma_0\,, & (3.9c)\\ \frac{\partial z}{\partial \nu}\Big|_{\Sigma_1} \equiv 0 & \text{in } \Sigma_1\,. & (3.9d)\end{cases}$$

The duality between the w-problem (3.1) and the z-problem (3.9) is described by the correspondence of the maps

$$v \to \{w(T,\cdot), w_t(T,\cdot)\} \text{ is dual of } \{z_1, z_0\} \to z(t,0) \tag{3.10}$$

as in the Dirichlet case (2.8). Duality on Theorem 3.1 using (3.10) yields then

<u>Theorem 3.2</u>. With reference to problem (3.9) we have that

$$z(t,0) \in L_2(0,T) \tag{3.11}$$

continuously in $\{z_0, z_1\}$ where

(a) for $N = \dim \Omega = 3$,

$$\{z_0, z_1\} \in H^1(\Omega)\times L_2(\Omega); \tag{3.12}$$

(b) for $N = \dim \Omega = 2$,

$$\{z_0, z_1\} \in H^{1/2}_{00}(\Omega)\times H^{-1/2}(\Omega); \tag{3.13}$$

(c) for $N = \dim \Omega = 1$,

$$\{z_0, z_1\} \in L_2(\Omega)\times [H^1(\Omega)]'. \quad \blacksquare \tag{3.14}$$

3.2. <u>Proof of Theorem 3.1</u>

The proof is mostly the same as that in Section 2.2 for the Dirichlet case and is based on (Step 1) the regularity of the free problem ϕ in Proposition 2.3, and (Step 2) the regularity of the homogeneous problem

$$\begin{cases} h'' = \Delta h + f & \text{in } Q\,, & (3.15a)\\ h(0,x) = h'(0,x) \equiv 0 & \text{in } \Omega\,, & (3.15b)\\ h|_{\Sigma_0} \equiv 0 & \text{in } \Sigma_0\,, & (3.15c)\\ \frac{\partial h}{\partial \nu}\Big|_{\Sigma_1} \equiv 0 & \text{in } \Sigma_1\,, & (3.15d)\end{cases}$$

where again the changes of variables (2.35) are used, with, now, w solution of (3.1). The function f is the same as the one in (2.38), i.e.,

$$f = -(\Delta\psi)\phi - 2\nabla\psi\cdot\nabla\phi. \tag{3.16}$$

We now strengthen the regularity results for f, as given by (2.40a) for N = 3 and by (2.40b) for N = 2, to suit our present needs.

<u>Proposition 3.3</u>. With reference to (3.16), where ϕ solves (2.13) and $\psi \in C_0^\infty(\Omega)$, we have

(a) for $N = \dim \Omega = 3$,

$$f \in C([0,T];[H^1(\Omega)]'); \tag{3.17}$$

(b) for $N = \dim \Omega = 2$,

$$f \in C([0,T];H^{-\frac{1}{2}}(\Omega)) \tag{3.18}$$

(where, in fact, C[0,T] could replace $L_2(0,T)$ in (3.17), (3.18)).

<u>Proof</u>. (a) Let N = 3. We have $\phi \in L_2(\Omega)$ by (2.14a) so that $|\nabla\phi| \in H^{-1}(\Omega)$ by [L-M.1, p. 85]. Pick $g \in H^1(\Omega)$. Then $g|\nabla\psi| \in H_0^1(\Omega)$ since $\psi \equiv 0$ near Γ. Therefore the integral

$$\int_\Omega g\ \nabla\psi\cdot\nabla\phi\ d\Omega \tag{3.19}$$

is well defined, and then $\nabla\psi\cdot\nabla\phi \in [H^1(\Omega)]'$ as desired.

(b) Let N = 2. We have $\phi \in H^{\frac{1}{2}}(\Omega)$ by (2.15a) so that $|\nabla\phi| \in [H_{00}^{\frac{1}{2}}(\Omega)]'$ by [L-M.1, p. 85]. Pick $g \in H^{\frac{1}{2}}(\Omega)$. Then $g|\nabla\psi| \in H_{00}^{\frac{1}{2}}(\Omega)$ since $\psi \equiv 0$ near Γ. Therefore the integral in (3.19) is well defined also in this case and then $\nabla\psi\cdot\nabla\phi \in [H^{\frac{1}{2}}(\Omega)]' = H^{-\frac{1}{2}}(\Omega)$, as desired. ∎

Continuing with the proof of Theorem 3.1, we now recall (3.3) [or (3.4)] to obtain by (3.17) that

$$A^{-\frac{1}{2}}f \text{ or } (A+I)^{-\frac{1}{2}}f \in C([0,T];L_2(\Omega)),\quad N = 3, \tag{3.20}$$

according to $\Gamma_0 \neq \emptyset$ or $\Gamma_0 = \emptyset$. Similarly, by (3.18) and (3.3b) we obtain

$$A^{-\frac{1}{4}}f,\quad \text{or } (A+I)^{-\frac{1}{4}}f \in C([0,T];L_2(\Omega)),\quad N = 2, \tag{3.21}$$

respectively. Instead, for N = 1 we recall that

$$f \in C([0,T];L_2(\Omega)), \quad N = 1, \tag{3.22}$$

from (2.40c). Finally, (3.20)-(3.22) permit us to establish the regularity of the h-problem (3.15) by proceeding exactly as in the proof of Proposition 2.4, using (3.3). We obtain

Proposition 3.3. With reference to problem (3.15), equivalently to (2.42), (2.43) (where now C(t) is the cosine operator generated on $L_2(\Omega)$ by the operator -A defined in (3.2)), we have

(a) for $N = \dim \Omega = 3$,

$$\begin{cases} h \in C([0,T];L_2(\Omega)) \ , & (3.23a) \\ h_t \in C([0,T];[H^1(\Omega)]') \ , & (3.23b) \\ h_{tt} \in C([0,T];[\mathcal{D}(A)]') \ ; & (3.23c) \end{cases}$$

(b) for $N = \dim \Omega = 2$,

$$\begin{cases} h \in C([0,T];H^{1/2}(\Omega)) \ , & (3.24a) \\ h_t \in C([0,T];H^{-1/2}(\Omega)) \ , & (3.24b) \\ h_{tt} \in C([0,T];[\mathcal{D}(A^{3/4})]') \ ; & (3.24c) \end{cases}$$

(c) for $N = \dim \Omega = 1$,

$$\begin{cases} h \in C([0,T];H^1(\Omega)) \ , & (3.25a) \\ h_t \in C([0,T];L_2(\Omega)) \ , & (3.25b) \\ h_{tt} \in C([0,T];[H^1(\Omega)]'). \quad \blacksquare & (3.25c) \end{cases}$$

Step 3 (return to the w-problem). Recalling from (2.35) $w = \psi\phi - h$, $\psi \in C_0^\infty(\Omega)$, we see that the following results follow.

For N = 3, we use (2.14a-b) for $\{\phi, \phi_t\}$ and (3.23a-b) for $\{h, h_t\}$ to obtain the desired regularity (3.6a-b) for $\{w, w_t\}$ since $[H^1(\Omega)]' \subset H^{-1}(\Omega)$. Then (3.6c) follows for $w_{tt} = \Delta w + \delta(x)v(t)$ with $\Delta w \in C([0,T];H^{-2}(\Omega))$ by [L-M.1, p. 85] and (2.30) for δv; or else (3.6c) follows from (2.14c) and (3.23c) with $[\mathcal{D}(A)]' \subset H^{-2}(\Omega)$.

For N = 2, we use (2.15a-b) for $\{\phi,\phi_t\}$ and (3.24a-b) for $\{h,h_t\}$ to obtain the desired regularity (3.7a-b) for $\{w,w_t\}$ since $[H_{00}^{1/2}(\Omega)]' \subset H^{-1/2}(\Omega)$. Then (3.7c) follows from $w_{tt} = \Delta w + \delta v$, with $\Delta w \in C([0,T];[H_{00}^{3/2}(\Omega)]')$ by [L-M.1, p. 85] and (2.34) for δv; or else (3.7c) follows from (2.15c) (left) and (3.24c), with $\mathcal{D}(A^{3/4}) = [\mathcal{D}(A),L_2(\Omega)]_{1/4} \supset [H_0^2(\Omega),L_2(\Omega)]_{1/4} = H_{00}^{3/2}(\Omega)$, and hence $[\mathcal{D}(A^{3/4})]' \subset [H_{00}^{3/2}(\Omega)]'$.

Finally, for N = 1, we use (2.16a-b) for $\{\phi,\phi_t\}$ and (3.25a-b) for $\{h,h_t\}$, to obtain the desired regularity (3.8a-b) for $\{w,w_t\}$. Then (3.8c) follows for $w_{tt} = \Delta w + \delta v$ with $\Delta w \in C([0,T];H^{-1}(\Omega))$ and (2.32) for δv; or else (3.8c) follows from (2.16c) and (3.25c). The proof of Theorem 3.1 is complete. ∎

4. Euler-Bernoulli equation with homogeneous Dirichlet/Neumann B.C. ($w|_\Sigma = \frac{\partial w}{\partial \nu}|_\Sigma \equiv 0$) and interior point control

4.1. Statement of results

In this section we consider the following Euler-Bernoulli equation in the unknown $w(t,x)$:

$$\begin{cases} w_{tt} + \Delta^2 w = \delta(x)v(t) & \text{in } (0,T]\times\Omega \equiv Q\,, & (4.1a)\\ w(0,x) \equiv w_t(0,x) \equiv 0 & \text{in } \Omega\,, & (4.1b)\\ w|_\Sigma \equiv 0 & \text{in } (0,T]\times\Gamma \equiv \Sigma\,, & (4.1c)\\ \frac{\partial w}{\partial \nu}|_\Sigma \equiv 0 & \text{in } \Sigma, & (4.1d)\end{cases}$$

with the origin 0 as an interior point of Ω. We define the positive, self-adjoint operator A

$$Ah = \Delta^2; \quad \mathcal{D}(A) = \{h \in H^4(\Omega): h|_\Gamma = \frac{\partial h}{\partial \nu}|_\Gamma = 0\}. \tag{4.2}$$

We recall that (with equivalent norms)

$$\mathcal{D}(A^{3/4}) = \{h \in H^3(\Omega): h|_\Gamma = \frac{\partial h}{\partial \nu}|_\Gamma = 0\};\ \mathcal{D}(A^{1/2}) = H_0^2(\Omega);\ \mathcal{D}(A^{1/4}) = H_0^1(\Omega); \tag{4.3}$$

$$\mathcal{D}(A^{3/8}) = [\mathcal{D}(A^{1/2}),\mathcal{D}(A^{1/4})]_{1/2} = [H_0^2(\Omega),H_0^1(\Omega)]_{1/2} = H_{00}^{3/2}(\Omega); \tag{4.4}$$

$$\mathcal{D}(A^{1/8}) = [\mathcal{D}(A^{1/4}),L_2(\Omega)]_{1/2} = [H_0^1(\Omega),L_2(\Omega)]_{1/2} = H_{00}^{1/2}(\Omega). \tag{4.5}$$

<u>Theorem 4.1</u>. With reference to problem (4.1), let

$$v \in L_2(0,T). \tag{4.6}$$

Then, continuously:

(a) for $N = \dim \Omega = 3$,

$$\begin{cases} w \in C([0,T];H^{1/2}_{00}(\Omega) = \mathcal{D}(A^{1/8})) \ , & (4.7a) \\ w_t \in C([0,T];[H^{3/2}_{00}(\Omega)]' = [\mathcal{D}(A^{3/8})]') \ , & (4.7b) \\ w_{tt} \in L_2(0,T;[\mathcal{D}(A^{7/8})]') \ ; & (4.7c) \end{cases}$$

(b) for $N = \dim \Omega = 2$,

$$\begin{cases} w \in C([0,T];H^{1}_{0}(\Omega) = \mathcal{D}(A^{1/4})) \ , & (4.8a) \\ w_t \in C([0,T];H^{-1}(\Omega) = [\mathcal{D}(A^{1/4})]') \ , & (4.8b) \\ w_{tt} \in L_2(0,T;[\mathcal{D}(A^{3/4})]') \ ; & (4.8c) \end{cases}$$

(c) for $N = \dim \Omega = 1$,

$$\begin{cases} w \in C([0,T];H^{3/2}_{00}(\Omega) = \mathcal{D}(A^{3/8})) \ , & (4.9a) \\ w_t \in C([0,T];[H^{1/2}_{00}(\Omega)]' = [\mathcal{D}(A^{1/8})]') \ , & (4.9b) \\ w_{tt} \in L_2(0,T;[\mathcal{D}(A^{5/8})]') \ . \quad \blacksquare & (4.9c) \end{cases}$$

<u>Remark 4.1</u>. It was noted in Remark 2.1 that in the case of the wave problems (2.1) and (3.1) the present sharp approach produces regularity results which are "$\frac{1}{2}+\epsilon$" higher in space regularity (measured in Sobolev space order) than the one directly obtained by simply using that $\delta \in [H^{\alpha}(\Omega)]'$, $\alpha = \frac{3}{2}+\epsilon$, $1+\epsilon$, $\frac{1}{2}+\epsilon$, for $N = 3,2,1$, respectively. The same gain in regularity will be obtained in the case of Kirchhoff problems (also hyperbolic) see Part II of the present paper [T.1]. Instead, in the case of Euler-Bernoulli problems (both problem (4.1) as well as the subsequent problem (5.1)) the present sharp approach produces only an "ϵ-improvement" over Theorem 4.1 (or Theorem 5.1). To see this, we use $\delta \in [H^{\alpha}(\Omega)]' \subset [\mathcal{D}(A^{\alpha/4})]'$, equivalently that $A^{-\alpha/4}\delta \in L_2(\Omega)$ for the fourth-order operator A in (4.2) (or in (5.2)). Then the solution w to, say, problem (4.1) satisfies

$$A^{\frac{1}{2}-\alpha/4}w(t) = \int_0^t A^{\frac{1}{2}}S(t-\tau)A^{-\alpha/4}\delta v(\tau)d\tau \in C([0,T];L_2(\Omega))$$

by the usual convolution properties. This yields results which are "ε-worse," i.e., $w \in \mathcal{D}(A^{\frac{1}{8}-\varepsilon})$, $\mathcal{D}(A^{\frac{1}{4}-\varepsilon})$, $\mathcal{D}(A^{\frac{3}{8}-\varepsilon})$ in space regularity over those in (4.7a), (4.8a), (4.9a) respectively. ∎

In order to state corresponding duality results, obtained by a duality argument on those of Theorem 4.1, we introduce the following homogeneous system in the unknown $z(t,x)$:

$$\begin{cases} z_{tt} + \Delta^2 z = 0 & \text{in } Q\,, & (4.10a)\\ z(T,\cdot) = z_0; \quad z_t(T,\cdot) = z_1 & \text{in } \Omega\,, & (4.10b)\\ z|_\Sigma \equiv 0 & \text{in } \Sigma\,, & (4.10c)\\ \frac{\partial z}{\partial \nu}|_\Sigma \equiv 0 & \text{in } \Sigma\,. & (4.10d)\end{cases}$$

Then, with reference to the w-problem (4.1) and the z-problem (4.10) we have readily

the map $v(t) \to \{w(T,\cdot), w_t(T,\cdot)\}$ is dual to the map

$$\{z_1, z_0\} \to z(t,0) \qquad (4.11)$$

(as in the wave case, see (2.8)). By applying the duality relationship (4.11) on the results of Theorem 4.1 we obtain

<u>**Theorem 4.2**</u>. With reference to problem (4.10) we have that

$$z(t,0) \in L_2(0,T) \qquad (4.12)$$

continuously in $\{z_0, z_1\}$, where

(a) for $N = \dim \Omega = 3$,

$$\{z_0, z_1\} \in H_{00}^{\frac{3}{2}}(\Omega)\times[H_{00}^{\frac{1}{2}}(\Omega)]' = \mathcal{D}(A^{\frac{3}{8}})\times[\mathcal{D}(A^{\frac{1}{8}})]'; \qquad (4.13)$$

(b) for $N = \dim \Omega = 2$,

$$\{z_0, z_1\} \in H_0^1(\Omega)\times H^{-1}(\Omega) = \mathcal{D}(A^{\frac{1}{4}})\times[\mathcal{D}(A^{\frac{1}{4}})]'; \qquad (4.14)$$

(c) for $N = \dim \Omega = 1$,

$$\{z_0, z_1\} \in H_{00}^{1/2}(\Omega) \times [H_{00}^{3/2}(\Omega)]' = \mathcal{D}(A^{1/8}) \times [\mathcal{D}(A^{3/8})]'. \quad \blacksquare \tag{4.15}$$

4.2. <u>Proof of Theorem 4.1</u>

<u>Step 1</u> (free space problem). As in Section 2.2, we consider first the free space problem corresponding to problem (4.1), in the unknown $\phi(t,x)$:

$$\begin{cases} \phi_{tt} + \Delta^2 \phi = \delta(x) v(t) & \text{in } R^1_{t^+} \times R^N_x & (4.16a) \\ \phi(0,x) \equiv \phi_t(0,x) \equiv 0 & \text{in } R^N_x, & (4.16b) \end{cases}$$

after extending $v(t)$ by zero for $t > T$.

<u>Proposition 4.3</u>. With reference to problem (4.16), let v satisfy (4.6). Then, continuously,

(a) for $N = \dim \Omega = 3$, and $\psi \in C_0^\infty(\Omega)$, $\psi(0) = 1$,

$$\begin{cases} \phi \in C([0,T]; H^{3/2}(R^3)) & \psi\phi \in C([0,T]; H_{00}^{3/2}(\Omega) = \mathcal{D}(A^{3/8})), & (4.17a) \\ \phi_t \in C([0,T]; H^{-1/2}(R^3)) & \psi\phi_t \in C([0,T]; [H_{00}^{1/2}(\Omega)]' = [\mathcal{D}(A^{1/8})]'), & (4.17b) \\ \phi_{tt} \in L_2(0,T; H^{-5/2}(R^3)) & \psi\phi_{tt} \in L_2(0,T; [H_{00}^{5/2}(\Omega)]') & (4.17c) \\ & \qquad \subset L_2(0,T; [\mathcal{D}(A^{5/8})]'); & \end{cases}$$

(b) for $N = \dim \Omega = 2$,

$$\begin{cases} \phi \in C([0,T]; H^2(R^2)), & (4.18a) \\ \phi_t \in C([0,T]; L_2(R^2)), & (4.18b) \\ \phi_{tt} \in L_2(0,T; H^{-2}(R^2)); & (4.18c) \end{cases}$$

(c) for $N = \dim \Omega = 1$,

$$\begin{cases} \phi \in C([0,T]; H^{5/2}(R^1)), & (4.19a) \\ \phi_t \in C([0,T]; H^{1/2}(R^1)), & (4.19b) \\ \phi_{tt} \in L_2(0,T; H^{-3/2}(R^1)), \ \phi_{tt} \in L_2(0,T; [H_{00}^{3/2}(\Omega)]'). \quad \blacksquare & (4.19c) \end{cases}$$

<u>Proof</u>. Let $\hat{\phi}(\lambda,\xi)$, $\lambda = \gamma+i\omega$, $\gamma > 0$, $\omega \in R^1$, $\xi \in R^N$, be the Laplace (in t) - Fourier (in x) transform of $\phi(t,x)$ so that the transformed version of problem (4.16) is

$$\hat{\phi}(\lambda,\xi) = \frac{\hat{v}(\lambda)}{\lambda^2+|\xi|^4} . \tag{4.20}$$

As in the wave equation case, it suffices to show the results for $\{\phi,\phi_t\}$ in (4.17)-(4.19) with $C[0,T]$ replaced by $L_2(0,T)$ (see paragraph below (2.18)).

<u>Proof of (4.17a)</u>. $N = 3$. Accordingly, it suffices to show that

$$e^{-\gamma t}\phi(t,x) \in L_2(R^1_{t^+};H^{3/2}(R^N_x)) \text{ equivalently}$$

$$|\xi|^{3/2}\hat{\phi}(\gamma+i\omega,\xi) \in L_2(R^1_\omega\times R^N_\xi). \tag{4.21}$$

To this end we compute via (4.20) and Parseval identity [D.1, p. 212]

$$\left\| |\xi|^{3/2}\hat{\phi}\right\|^2_{L_2(R^1_\omega\times R^N_\xi)} = \int_{R^1_\omega} \|\hat{v}(\gamma+i\omega)|^2\left[\int_{R^N_\xi} \frac{|\xi|^3\, d\xi}{|\lambda^2+|\xi|^4|^2}\right]d\omega$$

$$\leq C\int_{R^1_\omega} |\hat{v}(\gamma+i\omega)|^2 d\omega = \frac{C}{2\pi}\int_0^\infty e^{-2\gamma t}|v(t)|^2 dt$$

$$\leq \text{Const } \|v\|^2_{L_2(0,T)} \tag{4.22}$$

as desired, as soon as we prove that for $N = 3$,

$$\int_{R^N_\xi} \frac{|\xi|^3 d\xi}{|\lambda^2+|\xi|^4|^2} = \int_{R^N_\xi} \frac{|\xi|^3 d\xi}{(|\xi|^4+\gamma^2-\omega^2)+4\gamma^2\omega^2} \leq C < \infty \tag{4.23}$$

uniformly in $\omega \in R^1$. Using spherical coordinates $d\xi = \rho^2\sin\phi\, d\phi\, d\rho\, d\theta$, we see that estimate (4.23) holds true provided that (set $\rho^2 = y$)

$$\int_0^\infty \frac{\rho^5 d\rho}{(\rho^4+\gamma^2-\omega^2)^2+4\gamma^2\omega^2} = \tfrac{1}{2}\int_0^\infty \frac{y^2 dy}{(y^2+\gamma^2-\omega^2)^2+4\gamma^2\omega^2} \leq C < \infty \tag{4.24}$$

uniformly in $\omega \in R^1$, where the uniform bound in (4.24) attains by (1.1) of Lemma 1.1. Then (4.17a) for $\psi\phi$ follows since $\psi \equiv 0$ near Γ.

Proof of (4.18a). N = 2. Now we need to show that

$$e^{-\gamma t}\phi(t,x) \in L_2(R^1_{t^+};H^2(R^N_x)) \text{ equivalently}$$

$$|\xi|^2\hat{\phi}(\gamma+i\omega,\xi) \in L_2(R^1_\omega \times R^N_\xi). \tag{4.25}$$

Proceeding as above, via (4.20), we see that (4.25) holds indeed true provided that

$$\int_{R^N_\xi} \frac{|\xi|^4 d\xi}{|\lambda^2+|\xi|^4|^2} = \int_{R^N_\xi} \frac{|\xi|^4 d\xi}{(|\xi|^4+\gamma^2-\omega^2)^2 + 4\gamma^2\omega^2} \leq C < \infty, \tag{4.26}$$

or, after introducing polar coordinates, so that $d\xi = rdrd\theta$ provided that

$$\int_0^\infty \frac{r^5 dr}{(r^4+\gamma^2-\omega^2)^2+4\gamma^2\omega^2} = \tfrac{1}{2}\int_0^\infty \frac{y^2 dy}{(y^2+\gamma^2-\omega^2)^2+4\gamma^2\omega^2} \leq C < \infty \tag{4.27}$$

(using $r^2 = y$), which is in fact true by (1.1) of Lemma 1.1.

Proof of (4.19a). N = 1. Here we need to show that

$$e^{-\gamma t}\phi(t,x) \in L_2(R^1_{t^+};H^{5/2}(R^N_x)) \text{ equivalently}$$

$$|\xi|^{5/2}\hat{\phi}(\gamma+i\omega,\xi) \in L_2(R^1_\omega \times R^N_\xi). \tag{4.28}$$

Then, as before, it is readily seen that (4.28) holds true provided

$$\int_{R^N_\xi} \frac{|\xi|^5 d\xi}{|\lambda^2+|\xi|^4|^2} = \int_{R^N_\xi} \frac{|\xi|^5 d\xi}{(|\xi|^4+\gamma^2-\omega^2)^2 + 4\gamma^2\omega^2}$$

$$= \int_0^\infty \frac{y^2 dy}{(y^2+\gamma^2-\omega^2)^2+4\gamma^2\omega^2} \leq C < \infty \tag{4.29}$$

(using $|\xi|^2 = \xi^2 = y$) which is in fact true by (1.1) of Lemma 1.1.

<u>Proof of (4.17c), (4.18c), (4.19c)</u>. We use (4.16a). It follows from (4.17a), (4.18a), (4.19a) via [L-M.1, p. 85] that

$$\Delta^2\phi \in \begin{cases} L_2(0,T;[H^{5/2}_{00}(\Omega)]'); & \Delta^2\phi \in L_2(0,T;H^{-5/2}(R^N)); \quad N = 3; \quad (4.30a) \\ L_2(0,T;H^{-2}(\Omega)); & \Delta^2\phi \in L_2(0,T;H^{-2}(R^N)); \quad N = 2; \quad (4.30b) \\ L_2(0,T;[H^{3/2}_{00}(\Omega)]'); & \Delta^2\phi \in L_2(0,T;H^{-3/2}(R^N)); \quad N = 1. \quad (4.30c) \end{cases}$$

Then by (4.16a) we see that (4.30a-b-c), together with

$$\text{for } N = 3\text{: } H^{3/2+\varepsilon}(\Omega) \subset C^0(\Omega), \text{ hence } \delta v \in L_2(0,T;[H^{3/2+\varepsilon}(\Omega)]'), \tag{4.31a}$$

$$\text{for } N = 2\text{: } H^{1+\varepsilon}(\Omega) \subset C^0(\Omega), \text{ hence } \delta v \in L_2(0,T;[H^{1+\varepsilon}(\Omega)]'), \tag{4.31b}$$

$$\text{for } N = 1\text{: } H^{1/2+\varepsilon}(\Omega) \subset C^0(\Omega), \text{ hence } \delta v \in L_2(0,T;[H^{1/2+\varepsilon}(\Omega)]'), \tag{4.31c}$$

yield (4.17c), (4.18c), (4.19c), respectively, since

$$H^{5/2}_{00}(\Omega) \subset H^{5/2}_0(\Omega) \subset H^{3/2+\varepsilon}_0(\Omega) \subset H^{3/2+\varepsilon}(\Omega), \tag{4.32a}$$

$$H^{3/2}_{00}(\Omega) \subset H^{3/2}_0(\Omega) \subset H^{1/2}_0(\Omega) \subset H^{1/2+\varepsilon}(\Omega). \tag{4.32b}$$

Finally, for $\psi\phi_{tt} = -\psi\Delta^2\phi+\delta v$, we note that by (4.3)

$$H^{5/2}_{00}(\Omega) = [H^3_0(\Omega),H^2_0(\Omega)]_{1/2} \supset [\mathcal{D}(A^{3/4}),\mathcal{D}(A^{1/2})]_{1/2} = \mathcal{D}(A^{5/8}).$$

<u>Proof of (4.17b), (4.18b), (4.19b)</u>. Finally the regularity of ϕ_t at least in the $L_2(0,T)$-sense follows by the intermediate derivative theorem [L-T.1, p. 15]. After this, the $L_2(0,T)$ regularity of $\{\phi,\phi_t\}$ is lifted to $C[0,T]$, as noted below (4.20), (or (2.18)). The proof of Proposition 4.3 is complete. ∎

<u>Step 2</u> (auxiliary problems). As in Section 2, 3, we introduce two new variables

$$\phi_c(t,x) = \psi(x)\phi(t,x) \text{ and } h(t,x) = \phi_c(t,x)-w(t,x) \tag{4.33}$$

with $\psi(x) \in C^\infty_0(\Omega)$, $\psi(0) = 1$, ϕ solution of (4.16) and w solution of (4.1). Then multiplying (4.16) by ψ yields

$$\begin{cases} \psi\phi'' + \psi\Delta^2\phi = \delta(x)v(t) & \text{in } (0,T]\times\Omega = Q, \quad (4.34a) \\ \psi(x)\phi(0,x) \equiv \psi(x)\phi'(0,x) \equiv 0 & \text{in } \Omega, \quad (4.34b) \\ \psi\phi|_\Sigma \equiv 0 & \text{in } (0,T]\times\Gamma = \Sigma, \quad (4.34c) \\ \psi \frac{\partial\phi}{\partial\nu}|_\Sigma \equiv 0 & \text{in } \Sigma. \quad (4.34d) \end{cases}$$

Then, via (4.34), we see that $\phi_c = \psi\phi$ satisfies the problem

$$\begin{cases} \phi_c'' + \Delta^2\phi_c = \delta(x)v(t) + f(t,x) & \text{in } Q, \quad (4.35a) \\ \phi_c(0,x) = \phi_c'(0,x) \equiv 0 & \text{in } \Omega, \quad (4.35b) \\ \phi_c|_\Sigma \equiv 0 & \text{in } \Sigma, \quad (4.35c) \\ \frac{\partial\phi_c}{\partial\nu}|_\Sigma \equiv 0 & \text{in } \Sigma; \quad (4.35d) \end{cases}$$

$$f = (\Delta^2\psi)\phi + 2(\Delta\psi)\Delta\phi + 2\nabla(\Delta\psi)\cdot\nabla\phi + 2\nabla\psi\cdot\nabla(\Delta\phi) + 2\Delta(\nabla\psi\cdot\nabla\phi); \quad (4.36a)$$

$$\Delta(\nabla\psi\cdot\nabla\phi) = \nabla\psi\cdot\nabla(\Delta\phi) + \nabla(\Delta\psi)\cdot\nabla\phi + 2\sum_{i=1}^{N} \nabla(\psi_{x_i})\cdot\nabla(\phi_{x_i}). \quad (4.36b)$$

Finally, via (4.35), $h(t,x)$ in (4.33) satisfies the homogeneous (on Σ) problem

$$\begin{cases} h'' + \Delta^2 h = f & \text{in } Q, \quad (4.37a) \\ h(0,x) = h'(0,x) \equiv 0 & \text{in } \Omega, \quad (4.37b) \\ h|_\Sigma \equiv 0 & \text{in } \Sigma, \quad (4.37c) \\ \frac{\partial h}{\partial\nu}|_\Sigma \equiv 0 & \text{in } \Sigma. \quad (4.37d) \end{cases}$$

Hence, the regularity of ϕ (Proposition 4.3) determines that of ϕ_c by (4.33) (left). Thus, in order to determine the regularity of w via (4.33) (right), it remains to establish the regularity of h. The solution h of problem (4.37) is (abstractly):

$$h(t) = \int_0^t S(t-\tau)f(\tau)d\tau; \quad h'(t) = \int_0^t C(t-\tau)f(\tau)d\tau; \quad h'' = -Ah+f. \quad (4.38)$$

Lemma 4.4. With reference to f in (4.36) we have

$$\nabla\psi\cdot\nabla(\Delta\phi),\ \text{hence } f \in \begin{cases} C([0,T];[H_{00}^{3/2}(\Omega)]' = [\mathcal{D}(A^{3/4})]'), & N = 3, \quad (4.39a) \\ C([0,T];H^{-1}(\Omega) = [\mathcal{D}(A^{1/2})]'), & N = 2, \quad (4.39b) \\ C([0,T];[H_{00}^{1/2}(\Omega)]' = [\mathcal{D}(A^{1/4})]'), & N = 1. \quad (4.39c) \end{cases}$$

Proof. The regularity (4.17a), (4.18a), (4.19a) for ϕ implies by [L-M.1, p. 85]

$$\nabla(\Delta\phi) \in \begin{cases} C([0,T];[H_{00}^{3/2}(\Omega)]'), & N = 3, \quad (4.40a) \\ C([0,T];H^{-1}(\Omega)), & N = 2, \quad (4.40b) \\ C([0,T];[H_{00}^{1/2}(\Omega)]'), & N = 1. \quad (4.40c) \end{cases}$$

from which (4.39a-b-c) follow for $\nabla\psi\cdot\nabla(\Delta\phi)$ respectively, using (4.4) and (4.5). Finally, one readily sees that the lower order terms (in ϕ) in the definition (4.36) for f make f preserve the regularity of $\nabla\psi\cdot\nabla(\Delta\phi)$. For instance

$$\nabla(\Delta\psi)\cdot\Delta\phi \in \begin{cases} C([0,T];[H_{00}^{1/2}(\Omega)]' = [\mathcal{D}(A^{1/4})]'), & N = 3, \quad (4.41a) \\ C([0,T];L_2(\Omega)), & N = 2, \quad (4.41b) \\ C([0,T];H_{00}^{1/2}(\Omega) = \mathcal{D}(A^{1/4})), & N = 1. \quad \blacksquare \quad (4.41c) \end{cases}$$

Rewriting Lemma 4.4, equivalently, as

$$\left.\begin{matrix} A^{-3/4}f \\ A^{-1/2}f \\ A^{-1/4}f \end{matrix}\right\} \in C([0,T];L_2(\Omega)) \quad \begin{matrix} N = 3, & (4.42a) \\ N = 2, & (4.42b) \\ N = 1, & (4.42c) \end{matrix}$$

and using (4.42) in (4.38) yields in the usual way, by convolution properties, the following regularity of h.

Proposition 4.5. With reference to problem (4.37), equivalently to (4.38) we have

(a) for $N = \dim \Omega = 3$,

$$\begin{cases} h \in C([0,T];H_{00}^{1/2}(\Omega) = \mathcal{D}(A^{1/4})), & (4.43a) \\ h_t \in C([0,T];[H_{00}^{1/2}(\Omega)]' = [\mathcal{D}(A^{1/4})]'), & (4.43b) \\ h_{tt} \in C([0,T];[\mathcal{D}(A^{3/4})]'); & (4.43c) \end{cases}$$

(b) for $N = \dim \Omega = 2$,

$$\begin{cases} h \in C([0,T];H_0^1(\Omega) = \mathcal{D}(A^{1/2})), & (4.44a) \\ h_t \in C([0,T];[H^{-1}(\Omega) = [\mathcal{D}(A^{1/2})]'), & (4.44b) \\ h_{tt} \in C([0,T];[\mathcal{D}(A^{3/4})]'); & (4.44c) \end{cases}$$

(c) for $N = \dim \Omega = 1$,

$$\begin{cases} h \in C([0,T];H_{00}^{3/2}(\Omega) = \mathcal{D}(A^{3/4})), & (4.45a) \\ h_t \in C([0,T];[H_{00}^{1/2}(\Omega)]' = [\mathcal{D}(A^{1/4})]'), & (4.45b) \\ h_{tt} \in C([0,T];[\mathcal{D}(A^{3/4})]'). \quad \blacksquare & (4.45c) \end{cases}$$

<u>Proof</u>. For instance for $N = 3$ we have by (4.24a),

$$A^{-1/4}h(t) = \int_0^t A^{1/2}S(t-\tau)A^{-3/4}f(\tau)d\tau \in C([0,T];L_2(\Omega)), \tag{4.46}$$

and (4.43a) follows. The other cases are similar. $\blacksquare$

<u>Step 3</u> (return to w-problem). We now return to the w-problem (4.1), via $w = \psi\phi - h$, see (4.33).

<u>Proof of (4.7a), (4.8a), (4.9a)</u>. Recalling that $\psi \equiv 0$ near Γ, and invoking the regularity (4.17a), (4.18a), (4.19a) for ϕ, we obtain

$$\psi\phi \in \begin{cases} C([0,T];H_{00}^{3/2}(\Omega)), \quad N = 3, & (4.47a) \\ C([0,T];H_{00}^{2}(\Omega)), \quad N = 2, & (4.47b) \\ C([0,T];H_{00}^{5/2}(\Omega)), \quad N = 1. & (4.47c) \end{cases}$$

Pairing (4.47a) with (4.43a); (4.47b) with (4.44a); (4.47c) with (4.45a), we obtain the desired regularity (4.7a), (4.8a), (4.9a) for w, respectively.

Proof of (4.7c), (4.8c), (4.9c). We use $w_{tt} = \psi\phi_{tt} - h_{tt}$ along with (4.17a), (4.18c), (4.19c) for $\psi\phi_{tt}$ and (4.43c), (4.44c), (4.45c) for h_{tt} to obtain, respectively (4.7c), (4.8c), (4.9c). (The same conclusion is obtained by looking at $w_{tt} = -Aw+\delta v$ instead.)

Proof of (4.7b), (4.8b), (4.9b). The regularity of w_t, first in $L_2(0,T;\cdot)$, follows via the intermediate derivative theorem [L-M.1, p. 15] from that of w and w_{tt}. Finally, the regularity of w_t in $L_2(0,T;\cdot)$ is then lifted to $C([0,T];\cdot)$ while preserving the space regularity, as observed below (4.20) (or below (2.18) in the wave-case). The proof of Theorem 2.1 is complete. ∎

5. Euler-Bernoulli equation with homogeneous boundary conditions for $w|_\Sigma$ and $\Delta w|_\Sigma$ and interior point control

5.1. Statement of results

In this section we consider the following Euler-Bernoulli equation in the unknown $w(t,x)$:

$$\begin{cases} w_{tt} + \Delta^2 w = \delta(x)v(t) & \text{in } (0,T]\times\Omega \equiv Q, & (5.1a)\\ w(0,x) = w_t(0,x) \equiv 0 & \text{in } \Omega, & (5.1b)\\ w|_\Sigma \equiv 0 & \text{in } (0,T]\times\Gamma \equiv \Sigma, & (5.1c)\\ \Delta w|_\Sigma \equiv 0 & \text{in } \Sigma, & (5.1d)\end{cases}$$

with the origin 0 as an interior point of Ω. We define the positive, self-adjoint operator A

$$Ah = \Delta^2 h;\quad \mathcal{D}(A) = \{h \in H^4(\Omega): h|_\Gamma = \Delta h|_\Gamma = 0\}. \tag{5.2}$$

We recall that (with equivalent norms)

$$\mathcal{D}(A^{3/4}) = \{h \in H^3(\Omega): h|_\Gamma = \Delta h|_\Gamma = 0\};\quad \mathcal{D}(A^{1/4}) = H_0^1(\Omega); \tag{5.3}$$

$$A^{1/2}h = -\Delta h;\quad \mathcal{D}(A^{1/2}) = H^2(\Omega) \cap H_0^1(\Omega); \tag{5.4}$$

$$\mathcal{D}(A^{1/8}) = [\mathcal{D}(A^{1/4}), L_2(\Omega)]_{1/2} = [H_0^1(\Omega), L_2(\Omega)]_{1/2} = H_{00}^{1/2}(\Omega); \tag{5.5}$$

$$\mathcal{D}(A^{3/8}) = [\mathcal{D}(A^{1/2}), \mathcal{D}(A^{1/4})]_{1/2} = [H^2(\Omega) \cap H_0^1(\Omega), H_0^1(\Omega)]_{1/2}$$

$$\supset [H_0^2(\Omega), H_0^1(\Omega)]_{1/2} = H_{00}^{3/2}(\Omega). \tag{5.6}$$

<u>Theorem 5.1</u>. With reference to problem (5.1), let

$$v \in L_2(0,T). \tag{5.7}$$

Then, continuously,

(a) for $N = \dim \Omega = 3$,

$$\begin{cases} w \in C([0,T]; H^{1/2}_{00}(\Omega) = \mathcal{D}(A^{1/8})), & (5.8a) \\ w_t \in C([0,T]; [\mathcal{D}(A^{3/8})]') \subset C([0,T]; [H^{3/2}_{00}(\Omega)]'), & (5.8b) \\ w_{tt} \in L_2(0,T; [\mathcal{D}(A^{7/8})]'); & (5.8c) \end{cases}$$

(b) for $N = \dim \Omega = 2$,

$$\begin{cases} w \in C([0,T]; H^1_0(\Omega) = \mathcal{D}(A^{1/4})), & (5.9a) \\ w_t \in C([0,T]; H^{-1}(\Omega) = [\mathcal{D}(A^{1/4})]'), & (5.9b) \\ w_{tt} \in L_2(0,T; [\mathcal{D}(A^{3/4})]'); & (5.9c) \end{cases}$$

(c) for $N = \dim \Omega = 1$,

$$\begin{cases} w \in C([0,T]; \mathcal{D}(A^{3/8})), & (5.10a) \\ w_t \in C([0,T]; [H^{1/2}_{00}(\Omega)]' = [\mathcal{D}(A^{1/8})]'), & (5.10b) \\ w_{tt} \in L_2(0,T; [\mathcal{D}(A^{5/8})]'); \quad \blacksquare & (5.10c) \end{cases}$$

In order to state corresponding duality results, obtained by a duality argument on those of Theorem 5.1, we introduce the following homogeneous system in the unknown $z(t,x)$:

$$\begin{cases} z_{tt} + \Delta^2 z = 0 & \text{in } Q, & (5.11a) \\ z(T,\cdot) = z_0;\ z_t(T,\cdot) = z_1 & \text{in } \Omega, & (5.11b) \\ z|_\Sigma \equiv 0 & \text{in } \Sigma, & (5.11c) \\ \Delta z|_\Sigma \equiv 0 & \text{in } \Sigma. & (5.11d) \end{cases}$$

Then, with reference to the w-problem (5.1) and the z-problem (5.11), we have

$$\text{the map } v(t) \to \{w(T,\cdot), w_t(T,\cdot)\} \text{ is dual to the map}$$

$$\{z_1, z_0\} \to z(t,0) \tag{5.12}$$

counterpart of (4.11). By applying the duality relationship (5.12) on the results of Theorem 5.1, we obtain

<u>Theorem 5.2</u>. With reference to problem (5.11) we have that

$$z(t,0) \in L_2(0,T) \tag{5.13}$$

continuously in $\{z_0, z_1\}$, where

(a) for $N = \dim \Omega = 3$,

$$\{z_0, z_1\} \in \mathcal{D}(A^{3/8}) \times [\mathcal{D}(A^{1/8})]'; \tag{5.14}$$

(b) for $N = \dim \Omega = 2$,

$$\{z_0, z_1\} \in H_0^1(\Omega) \times H^{-1}(\Omega) = \mathcal{D}(A^{1/4}) \times [\mathcal{D}(A^{1/4})]'; \tag{5.15}$$

(c) for $N = \dim \Omega = 1$,

$$\{z_0, z_1\} \in \mathcal{D}(A^{1/8}) \times [\mathcal{D}(A^{3/8})]'. \quad \blacksquare \tag{5.16}$$

5.2. <u>Proof of Theorem 5.1</u>

The proof is mostly the same as that in Section 4.2 (for the Euler-Bernoulli equation with B.C. $w|_\Sigma \equiv \frac{\partial w}{\partial \nu}|_\Sigma \equiv 0$) and is based on (Step 1) the regularity of the free problem ϕ in Proposition 4.3 and (Step 2) the regularity of the homogeneous problem

$$\begin{cases} h'' + \Delta^2 h = f & \text{in } Q, & (5.17a) \\ h(0,x) = h'(0,x) \equiv 0 & \text{in } \Omega, & (5.17b) \\ h|_\Sigma \equiv 0 & \text{in } \Sigma, & (5.17c) \\ \Delta h|_\Sigma \equiv 0 & \text{in } \Sigma, & (5.17d) \end{cases}$$

where again the changes of variables (4.33) are used, with, now, w solution of (5.1). The function f is the same as that in (4.36), i.e.,

$$f = 4\nabla\psi \cdot \nabla(\Delta\phi) + 2(\Delta\psi)\Delta\phi + 4\nabla(\Delta\psi)\cdot\nabla\phi + 4\sum_i^N \nabla(\psi_{x_i})\cdot\nabla(\phi_{x_i}) + (\Delta^2\psi)\phi. \tag{5.18}$$

We now strengthen the regularity results for f, as given by (4.39a) in the case $N = 3$.

<u>Proposition 5.3</u>. With reference to (5.18), where ϕ solves the free problem (4.16) and $\psi \in C_0^\infty(\Omega)$, we have

(a) for $N = \dim \Omega = 3$,

$$\nabla\psi \cdot \nabla(\Delta\phi), \text{ hence } f \in L_2(0,T;[H^{3/2}(\Omega)]') \subset L_2(0,T;[\mathcal{D}(A^{3/8}]') \tag{5.19}$$

where, in fact, $L_2(0,T)$ could be replaced by $C[0,T]$). ∎

Remark 5.1. Since, by (5.6),

$$H_{00}^{3/2}(\Omega) \subset \mathcal{D}(A^{3/8}) \subset H^{3/2}(\Omega); \quad \text{hence } [H^{3/2}(\Omega)]' \subset [\mathcal{D}(A^{3/8}]' \subset [H_{00}^{3/2}(\Omega)]', \tag{5.20}$$

then (5.19) is sharper than (4.39a). ∎

Proof of Proposition 5.3. We have $\phi \in H^{3/2}(\Omega)$ by (4.17a), hence $|\nabla(\Delta\phi)| \in [H_{00}^{3/2}(\Omega)]'$ by [L-M.1, p. 85]. Thus, if we pick $g \in H^{3/2}(\Omega)$, then $g|\nabla\psi| \in H_{00}^{3/2}(\Omega)$ since $\psi \equiv 0$ near Γ. Therefore $\int_\Omega g\nabla\psi \cdot \nabla(\Delta\phi)d\Omega$ is well defined, and (5.19) follows, at least for $\nabla\psi \cdot \nabla(\Delta\phi)$. Finally, one similarly sees that the lower order terms (in ϕ) in the definition (5.18) of f make f preserve the regularity of $\nabla\psi \cdot \nabla(\Delta\phi)$. ∎

Proceeding with the proof of Theorem 5.1, we see that for $N = 2,1$ the regularity results (4.39b-c) for f continue to hold true also in the present case, where A is defined by (5.2), while in (4.39b-c) A is the operator defined by (4.2): compare (4.3), (4.5) with, respectively, (5.3), (5.5), showing that the domain of fractional powers $\mathcal{D}(A^\Theta)$, $\Theta = 1/4, 1/8$, coincide in both cases with $H_0^1(\Omega)$ and $H_{00}^{1/2}(\Omega)$, respectively. These remarks along with (5.19) then yield

$$\left.\begin{array}{l} A^{-3/8}f \\ A^{-1/4}f \\ A^{-1/8}f \end{array}\right\} \in L_2(0,T;L_2(\Omega)) \qquad \begin{array}{l} N = 3, \\ N = 2, \\ N = 1 \end{array} \tag{5.21}$$

(in fact, in $C[0,T];L_2(\Omega)$), also in the present case. The conclusion of the proof of Theorem 5.1 proceeds then as in Section 4, from Proposition 4.5 on.

References

[D.1] G. Doetsch, Introduction to the theory and applications of the Laplace transformation, Springer-Verlag, 1974.

[D-Z.1] F. Dexing and D. Zhonghai, The approximation problem for hyperbolic pointwise control systems, Systems Science and Mathematical Sciences, vol. 1 (1988), 47-56.

[F-L-T.1] F. Flandoli, I. Lasiecka, and R. Triggiani, Algebraic Riccati equations with non-smoothing observation arising in hyperbolic and Euler-Bernoulli equations, Annali di Matematica Pura e Applicata, (iv) vol. CLIII (1988), 307-382.

[G.1] P. Grisvard, A characterization de quelques espaces d'interpolation, Arch. Rat. Mech. and Anal., vol. 25 (1967), 40-63.

[L.1] J. L. Lions, Exact controllability, stabilization and perturbations for distributed systems, SIAM Review, vol. 30 (1988), 1-68.

[L.2] J. L. Lions, Pointwise control for distributed systems, SIAM Publication, to appear (based on colloquium given at Workshop held in Tampa, Florida, February 1985).

[L.3] J. L. Lions, Control des systemes distribues singuliers, Gauthier Villars, 1983.

[L-L-T.1] I. Lasiecka, J. L. Lions, and R. Triggiani, Nonhomogeneous boundary value problems for second order hyperbolic operators, J. Math. Pures et Appliques 65 (1986), 149-192.

[L-M.1] J. L. Lions and E. Magenes, Nonhomogeneous Boundary Value Problems, I, II, Springer-Verlag (1972).

[L-T.1] I. Lasiecka and R. Triggiani, A cosine operator approach to modeling $L_2(0,T;L_2(\Gamma))$-boundary input hyperbolic equations, Appl. Math. and Optimiz. 7 (1981), 35-83.

[L-T.2] I. Lasiecka and R. Triggiani, Regularity of hyperbolic equations under $L_2(0,T;L_2(\Gamma))$-Dirichlet boundary terms, Appl. Math. and Optimiz. 10 (1983), 275-286.

[L-T.3] I. Lasiecka and R. Triggiani, A lifting theorem for the time regularity of solutions to abstract equations with unbounded operators and applications to hyperbolic equations, Proc. Amer. Math. Soc., Vol. 10 (1988), 745-755.

[L-T.4] I. Lasiecka and R. Triggiani, Differential and Algebraic Riccati equations with applications to boundary/point control problems: Continuous theory and approximation theory, 190-page manuscript, August 1990, to appear.

[M.1] Y. Meyer, Etude d'um modéle mathématique issu du contrôle des structures spatiales déformables, in Nonlinear Partial Differential Equations and their Applications, College de France Seminar, vol II (H. Brezis and J. L. Lions, eds), Research Notes in Mathematics, Pitman, Boston 1985, pp. 234-242.

[P-S.1] A. Pritchard and D. Salamon, Pointwise estimates for the three-dimensional wave equation, Control Theory Centre Report, n. 154, 1st January, 1989, University of Warwick.

[T.1] R. Triggiani, Regularity with point control. Part II: Kirchhoff equations, February 1991.

NEW RESULTS in SHAPE OPTIMIZATION

Piero Villaggio (Pisa) and Jean Paul Zolésio (Nice)

One of most delicate questions which must be faced in optimizing the shape of a thin elastic plate, loaded by forces acting in its mid plane, is that of predicting the possible losses of connection occurring in the optimal solutions. To render the problem more precise, let us consider a system of coplanar forces, applied along the edge CD of a rectangular elastic plate (Fig.1(a)) which is free of stresses along the vertical sides, and clamped along the basis AB. This gives a well-defined problem, and, if the customary assumptions of the thery of classical elasticity are admitted, the solution can be found explicitly. However, a different problem is often posed ; leaving the base AB and the top CD fixed, it is possible to change the shape of the vertical sides BC and AD, while leaving the area fixed, in such a way as to minimize the strain energy stored in the plate ? Problems of thie kind are called "problem of shape optimization", because a part of the boundary is not prescribed and must be determined by solving a problem of the calculus of variations. The most obvious idea to put the problem into simple terms is that of deforming the vertical sides, which are variable, so that, while still preserving the area, they describe two regular curves joining the vertices B,C and A,D, respectively. But there may be solutions that are less regular then

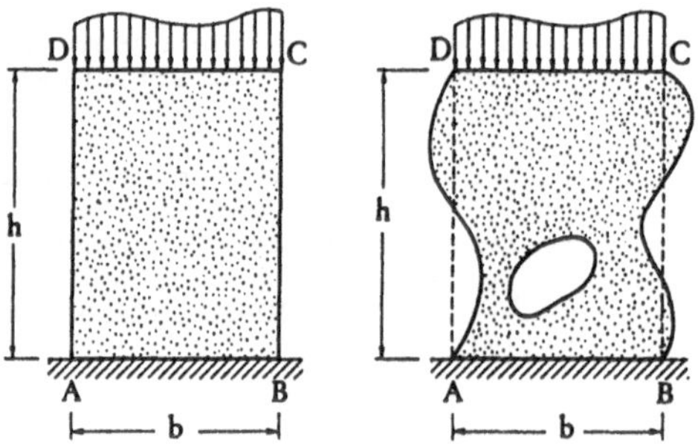

(a) (b)

Fig. 1

those just described, for example solutions that admit corners or even jumps. Such solutions could not be obtained by solving differential equations derived from a minimum principle. In addition, other difficulties arise if it is conceivable that the domain of least strain energy contains holes in its interior in order to remove material from parts of the plate which are potentially less loaded and place it instead in the regions of greater stress. The improvement of the resistence of an elastic body by carving cavities or notches inside it has been known for more than a hundred years by engineers, as Neuber showed in his book on notch effects [1937,ch VII]. On the other hand, similar methods of displacing superfluous material are even observable in nature, in the configuration of the branches of some trees.

Consequently, in order to treat the problem of shape optimization of a body permitting the presence of cavities, Villaggio and Zolésio [1990] have considered the following typical problem in plane elasticity.

The portion of the body above the x-axis, hereafter called E is prescribed (Fig.2). The part of the boundary of E with $x > 0$ is called Γ, and a system of forces acting on Γ is given. The portion G of the plate below the x-axis is left free, except that the total area of $\Omega = E \cup G$ is equal to a constant α. The boundary of Ω is thus composed of Γ, Σ, and Σ_0 (Fig.2), where Σ contains the points with $y < 0$ and Σ_0 those with $y = 0$ of the boundary. Since $\Sigma \cup \Sigma_0$ is free of forces, it is necessary to require that the forces

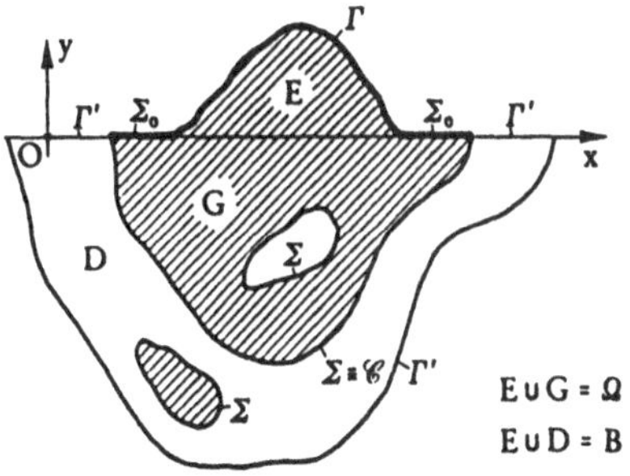

Fig. 2

on Γ have zero resultant and zero resultant moment with respect to any point of the x,y-plane. The part of the boundary of Ω constituting the prolongation of Γ is denoted by $\mathscr{C}$, and the boundaries of interior holes or of possible detached pieces of G are generically indicated by Σ.

Since the problem involves planar stresses, the average stresses, taken over the thickness of the plate, are denoted by $\bar{\sigma}_x, \bar{\sigma}_y, \bar{\tau}_{xy}$ and these latter derive from an Airy stress function F(x,y) such that

$$\bar{\sigma}_x = F_{yy} \quad , \quad \bar{\sigma}_y = F_{xx} \quad , \quad \bar{\tau}_{xy} = -F_{xy} \tag{1}$$

The function F(x,y) is biharmonic in Ω; it satisfies boundary conditions of the type $F = g$, $\frac{\partial F}{\partial n} = h$ along Γ, where h and g are two known functions depending on the boundary data; it satisfies the conditions $F = 0$, $\frac{\partial F}{\partial n} = 0$ on the part $\mathscr{C}$ of the boundary; in addition, along the remaining part of Σ, F and $\frac{\partial F}{\partial n}$ take the values (cf. Love [1924,Art 144])

$$F = a_1 x + b_1 y + c_1 \quad , \quad \frac{\partial F}{\partial n} = a_1 n_x + b_1 n_y \ , \tag{2}$$

where n_x, n_y are the direction cosione of the outer unit normal and a_1, b_1, c_1 are constants determined by the requirement that the displacement must be single valued along any loop surrounding each hole in G or each separated component of G.

In terms of F(x,y), the strain energy stored in the plate is a quadratic functional of the form

$$\mathscr{E} = \frac{1}{2\bar{E}} \iint_\Omega \left\{ (F_{xx} + F_{yy})^2 + 2(1-\sigma)(F_{xy}^2 - F_{xx}F_{yy}) \right\} dxdy \ , \tag{3}$$

$\bar{E}$ being the Young modulus and σ the Poisson ratio of the material. $\mathcal{E}$ obviously depends on the shape of Ω and it is necessary to find G so that $\mathcal{E}$ is a minimum under the condition that the area of Ω is a constant α.

But, before considering the problem abstractly, it is convenient to derive those properties of solutions that can be obtained by the discussion of the Euler equations of the variational problem, whenever solutions are smooth enough to render these equations meaningful. Under this assumptions one can say that extrema must satisfy the following conditions :

(a) Along the free boundary Σ the normal derivative of the stress function must satisfy

$$\frac{\partial F}{\partial n} = 0 \quad , \quad \left(\frac{\partial^2 F}{\partial n^2} \right)^2 = \lambda = \text{constant.} \tag{4}$$

(b) Along each separated boundary Σ_i $(i=1,\ldots,n)$ we have

$$F = c_i = \text{constant} . \tag{5}$$

(c) The optimal domain cannot have detached pieces, like those represented in Fig.2, because the stresses vanish in them, and so does the strain energy stored in their interior.

(d) The optimal domains cannot have cracks, that is holes enclosing sets of zero measure.

Of these properties, derived from the Euler equations, the second turns out to be essential for the formulation of the existence problem.

In fact the natural space where solutions may be sought would seem to be the Sobolev space $H^2_0(\Omega)$ or better, the space$H^2_{\mathscr{C}}(\Omega)$ consisting of those functions in $H_2(\Omega)$ with null trace on that part $\mathscr{C}$ of the boundary which is the prolongation of Γ. Since the non-homogeneous data h and g are prescribed on Γ, the solution belongs to the convex subset $K(\Omega)$ of $H^2_{\mathscr{C}}(\Omega)$ that consists of functions such that $F = g$, $\frac{\partial F}{\partial n} = h$ on Γ. However, it is not known in advance how regular Ω is and therefore $K(\Omega)$ is not exactly defined. It is thus necessary to consider a larger sufficiently smooth region D containing G in its interior. Denoting by Γ' the part of boundary of D that prolongs Γ (Fig.2), the Sobolev space $H^2_{\Gamma'}(B)$ $(B = E \cup D)$ of functions with null trace on Γ' is well defined, and so is the convex subset K(B) of functions satisfying the non-homogeneous conditions $F = g$, $\frac{\partial F}{\partial n} = h$ on Γ. Moreover, since F is constant along the boundary Σ_i of each conceivable hole within G, it can be extended as a constant on eachconnected component of D\G. Thus the convex set $K(\Omega)$ is defined as the subset of K(B) consisting of functions with constant values on each connected component of D\G. But, even in this form, the definition of $K(\Omega)$ is not general enough since the connected components of D\G might be not regular domains in D. It is therefore convenient to redefine $K(\Omega)$ so as to consist of functions ϕ whose gradient $\nabla\phi$ is equal to zero "quasi everywhere" in D\G, that is within sets with zero capacity with respect B.

Having so defined $K(\Omega)$, the existence theorem follows by considering a sequence of domains Ω_n and of functions F_n, characterizing the functionals

$$\frac{1}{2\overline{E}} \iint_{\Omega_n} \left\{ \left[\left(F_n\right)_{xx} + \left(F_n\right)_{yy} \right]^2 + 2(1-\sigma) \left[\left(F_n\right)^2_{xy} - \left(F_n\right)_{xx} \left(F_n\right)_{yy} \right] \right\} dxdy, \quad (5)$$

which give rise to a minimizing sequence for the originary minimization problem. Then, by introducing a sequence of characteristic functions χ_{Ω_n} the functionals become

$$\frac{1}{2\bar{E}}\iint_B \chi_{\Omega_n}\left\{\left[\left(F_n\right)_{xx}+\left(F_n\right)_{yy}\right]^2+2(1-\sigma)\left[\left(F_n\right)_{xy}^2-\left(F_n\right)_{xx}\left(F_n\right)_{yy}\right]\right\}dxdy, \quad (6)$$

and these converge to a minimum in the convex set $K(\Omega)$. There is, however, also another consequence of the fact that the functionals (6) converge to a minimum : namely, that the characteristic functions χ_{Ω_n} weakly converge in $L^2(B)$ to a function λ, with $0 \le \lambda \le 1$.

R E F E R E N C E S

[1927] LOVE,A.H.E. : A Treatise on the Mathematical Theory of Elasticity . Cambridge : University Press.

[1937] NEUBER,H. : Kerbspannungslehre. Berlin : Julius Springer.

[1990] VILLAGGIO,P. and ZOLÉSIO,J.P. : "The Optimal Shape of a Plate Stretched by Forces in its Plane". Submitted to Comm. Pure Appl. Math.

SHAPE FORMULATION OF FREE BOUNDARY PROBLEMS

WITH NON LINEARIZED BERNOULLI CONDITION

Jean-Paul Zolésio

Institut Non Lineaire de Nice

URM CNRS 129 . Parc Valrose, 06034 Nice . France

Abstract.

We consider shape optimization problems of the type Min $\{E(\Omega) \mid \Omega$ measurable set in $D, meas(\Omega)=\alpha,...\}$ associated to energy functionals of the form $E(\Omega)=Inf\{J(\Omega,u) \mid u$ belongs to $H(\Omega)\}$ where $H(\Omega)$ is a relaxed version of some classical closed subspace of $H^1(\Omega)$ when Ω is just a measurable subset in a smooth given domain D. For several situations connected to the Bernoulli free boundary condition we give existence results and first necessary conditions. Assuming smoothness of the solution we give second order necessary conditions.

INTRODUCTION

Several classical free boundary value problems can be formulated in the following way: D being a bounded domain in $\mathbf{R}^n$, H a Hilbert space of functions defined on D, K a closed convex set in H and J a functional on K, minimize J over K. If y is a minimizer to that problem then, roughly speaking, the free boundary problem is solved by setting

$$\Omega = \{ x \text{ in } D \mid \text{the constraints associated to K are not active} \}$$

assuming here obviously K to be a pointwise constraint in H . When J is weakly lower semicontinuous on H (w.l.s.c.), the minimization problem can be understood as a variational inequality , v.i. (or more generally an implicit variational inequality) . In others situation , for example in Alt & Caffarelli [1980], J takes the form :

$$J(y) = \int_D \{ (\text{grad } y)^2 + \chi_{\{y>o\}} G \} \, dx$$

it is not w.l.s.c. on $H = H^1(D)$ nevertheless some minimization can be performed on an appropriate K (assuming $G \geq 0$) by the vertu of lemma 2.1 bellow. We shall refer to that situation as being a weak variational formulation , w.v.f. . Using extra regularity results the v.i. leads to the solution of a free boundary problem in strong version . The same kind of conclusion can be derived for the w.v.f. , at least in some very specific situations for which we refer to A. Fasano [1990].

In order to explain briefly what is the weak shape formulation w. s.f., introduced in Zolésio[1979],[1981], we shall specify to the convex $K=\{ \phi \geq 0$ a.e. in D $\}$ of $H^1(D)$. Both in v.i. and w.v.f , the domain Ω, whose boundary is the solution to the free boundary problem, is given by $\Omega = \{ x \text{ in } D \mid y(x) > 0 \}$. The minimization problem is performed on the variable ϕ lying in K while the domain Ω is obtained from the minimizing term y . The fact that the domain Ω does not appears as an explicit variable in the minimization may be an advantage with respect to several view points.

In many examples arising from structural mechanics, fluid dynamics, electrostatic..., the free boundary is searched as being the boundary (or in fact a part of the boundary) of a domain Ω which is assigned to several constraints . The simplest one is the prescribed measure : $\text{meas}(\Omega) = \alpha$, α given . If the domain Ω is identified to its characteristic function X_Ω ,that constraint is linear , but expressed on ϕ , $\text{meas}\{\phi>0\}=\alpha$, it is a very severe constraint . The Weak Shape Formulation ,w.s.f. , introduces Ω as the variable , Ω belongs to { measurables subsets E of D | meas(E) = α } , and on that set we minimize the functional

$$J(\Omega) = \text{Min} \{ \int_\Omega |\text{grad}\, \phi|^2 + G + \ldots \mid \phi \text{ in } H^1_0(\Omega) \}$$

The positivness of y will eventually derives from the maximum principle , but the minimization of $J(\Omega)$ will be worked out in section 2 without any hypothesis on the sign of G . That w.s.f. is related to the homogeneous dirichlet boundary condition so will be denoted by $(\mathcal{D}_\alpha)$, with the convention that $\alpha = 0$ means the situation without constraint on the volume of Ω . The first section deals with an illustrative example with a v.i. and its w.s. f. that we denote $(\mathcal{P}_\alpha)$.

The problems $(\mathcal{P}_\alpha)$ and $(\mathcal{D}_\alpha)$ in the two first sections are then introduced as relaxed formulations respectively of a v.i. and a w.v.f. , the v.i. under consideration being not the well known obstacle problem (for the membrane) to which we refered but a very simplified version of a free boundary problem arising from plasmas physics .

In the next sections we shall investigate other boundary conditions , $\mathcal{N}$ standing for Neumann condition ,$\mathcal{T}$ for transmission condition , σ being associated to a constraint on the perimeter (with the same convention , when σ=0, as for α).

We shall be concerned by five classical free boundary situations related to scalar elliptic problems and the Bernoulli free boundary condition. Without loss of generality we restrict our study to the Laplace equation and to basic optimization problems $(\mathcal{D}_0)$,$(\mathcal{D}_\sigma{}^\alpha)$,$(\mathcal{D}\Sigma\mathcal{N}^\alpha{}_\sigma)$,$(\mathcal{N}^\alpha{}_\sigma)$,$(\mathcal{T}^\sigma{}_\alpha)$ for which we give existence results.The two first deal with the Dirichlet condition on the boundary of the domain , they correspond for example to the free boundary condition associated to a perfect 2-D fluid (using a stream function representation) . The fourth one corresponds to the neuman condition , which can be related to the 3-D perfect fluid (using a potential formulation) . The last one corresponds to transmission conditions through the boundary while the third one is a mixed situation. The free boundary in these weak shape formulations is the boundary of a measurable subset Ω of D . Following the previous results of Zolésio J.P.[1980b] we explicit the extra boundary condition in (7.6),(7.9) in an unified expression for these problems and finely at the last section, assuming the optimal solutions smooth enough we identify (7.6),(7.9) with a strong boundary condition in each problem. For that purpose we assume that the boundary $\partial\Omega$ has a generalized mean curvature $\mathfrak{H}$, it is by definition the shape gradient of the mapping $\Omega \dashrightarrow P_D(\Omega)$, a vectorial distribution on D (i.e. an element of $\mathcal{D}'(D;R^n)$) having its support in $\partial\Omega = cl(\Omega) \cap cl(cl(D) \setminus \Omega)$.

The main difficulty is the existence question.The first step is to define the Hilbert Spaces $H^1_0(\Omega), H^1(\Omega)$,...when Ω is just a measurable subset in D. The two first problems ($\mathcal{D}_0$) and ($\mathcal{D}_\sigma^\alpha$) are related to Dirichlet condition and are associated to two different relaxations of the Hilbert space $H^1_0(\Omega)$.In the first one, as we explain at the second section the idea of that relaxation is closed to Caffarelli's work but here we chose to have Ω explicitely as a control parameter, so even in that first simple problem we face to capacity questions in the relaxation. We chose to consider Ω as a control , that is an explicit variable in the problems, to be able to impose some constraints on it, for example on its volume (α refer to the constraint meas(Ω)=α), or on its perimeter and σ refer to the mean curvature H of the boundary $\partial\Omega$. The term σ physically (i.e.in the Bernoulli condition) is the surface tension . In the Dirichlet problems ($\mathcal{D}_0$) and ($\mathcal{D}_\sigma^\alpha$) σ can be taken equal to zero, that is to say that the surface tension is not necessary to get existence results for the relaxed problem. Now it turns out that when $\sigma>0$ we can relax the Dirichlet condition in a different way which is more convenient for modelling of potential problems, for example hydrodynamical problems. The basic idea is that when $\partial\Omega$ is smooth but having several connected componants the potential y should be constant on each of these componants but equal to zero only on one of thems. When $\sigma>0$ the existence question is helped for Ω has a bounded perimeter; and for any limit of sequence of such measurable sets Ω_n we show (Lemma 4.6) that the Hausdorf limit is easily related to the BPS(D) limit.

The problems we consider have the following form :

INF { E(Ω) | Ω is a measurable subset in D }

In order to derive existence results we use the three compactness results concerning the family of measurable sets in D, D being smooth enough and bounded:

{ λ | $0 \le \lambda \le 1$ a.e. in D } is weakly compact in $L^2(D)$.

{ C | C is a compact set in cl(D) } is compact for the Hausdorf metric .

Any bounded family in BPS(D) is compact in $L^2(D)$, see section IV concerning the bounded perimeter sets in D .

If $\lambda_n = X_{\Omega_n}$ is a sequence of characteristic functions weakly converging to λ in $L^2(D)$ and such that cl(Ω_n) converges for the Hausdorf metric to C (so that Ω_n itself also converges to C) we have the following inclusions : up to sets with zero measures in D , C contains the support of the measure λ , which itselfwhich contains the set Ω = { x in D | $\lambda(x)=1$ } .

I. Free Interface With Continuity Condition.

we coinsider a very simple free boundary problem which is not related to the Bernoulli condition but which permit easily to introduce the weak shape formulation for a free boundary problem and to underline that , even when exists a variational principle , say for example a variational inequality , the shape formulation is not equivalent and permit to handle more general free boundary conditions. In fact the problem developed in that section is a simplified version of a free boudary value problem arising in plasmas physics .

I. Free Interface With Continuity Condition.

We consider a free boundary problem solved by a variational inequality, at the section 1.2 we give the associated "shape variational formulation " and then we show the shape extension of that problem as a shape optimization problem which cannot be reformulated as a variational inequality . That extension is obtained by introducing the constraint on the volume of the domain

1.1 Variational inequality .

We consider here the classical solution of the free boundary problem obtain by the minimization of a coercive functional leading to variational inequality . The most famous such problem is the well known obstacle problem for the membrane in which the functional to be minimized is quadratic while the convex set on which the minimization is performed is bounded. In view to avoid that example we chose here an example in which the cost to be minimized has no gradient while the convex set is the hole space .That example is in fact a very simple version of a free boundary problem arising from plasma physics and studied by the author in after 78 ,in particular see Zolésio [79]. The free boundary appears as a level curve of the solution of the variational inequality and the difficulties are related to the possible existence of level sets with non zero measure.The main point in the following variational formulation is that the non linear term in the variational inequality will force the the level set under consideration to have a zero measure.

D is a bounded smooth domain of $\mathbf{R}^n$ and the unknown domain is a measurable set Ω in D whose boundary $\Gamma=cl(\Omega) \cap cl(D\backslash\Omega)$ is considered as an interface for a BVP posed in the domain D. In that section we are concerned with interface Γ which is a level curve of the solution u to the boundary value problem.

We consider the following problem: D is a bounded smooth domain in $\mathbf{R}^n$, $f \geq o$ in $L^2(D)$, find a measurable set Ω in D and u in $H^1{}_o(D) \cap H^2(D)$ such that

$$- \Delta u(x) = \begin{cases} f(x) \text{ a.e. } x \text{ in } \Omega \\ 0 \text{ a.e. } x \text{ in } D\backslash\Omega \end{cases} \tag{1.1}$$

with $\quad \text{meas}(x \mid u(x)=1) = 0 \quad (1.2)$

$u(x) > 1 \text{ in } \Omega \ , \ u(x) < 1 \text{ in } D\backslash\Omega \quad (1.3)$

in other words $\Omega = \{ x \text{ in } D \mid u(x)>1 \}$ and X_Ω being the characteristic function of Ω we can write the problem (1.1)-(1.3) equivalently as follows:

$$-\Delta u = X_\Omega f, \Omega=\{x \mid u(x)>1)\}, \qquad \text{meas}(\{x \mid u(x)=1\})=0 \qquad (1.4)$$

We consider the energy functional W: $H^1_0(D) \longrightarrow \mathbf{R}$ defined by

$$W(\phi) = \int_D (1/2 \mid \text{grad}\phi \mid^2 - f (\phi-1)^+)dx \qquad (1.5)$$

<u>lemma 1.1</u>

The Hadamard semi-derivative $W'(\phi;\gamma)$ exists for each ϕ and γ in $H^1_0(D)$ and is given by:

$$W'(\phi;\gamma) = \int_D \text{grad}\phi . \text{grad}\gamma \, dx - \int_D f X_{\phi>1} \gamma \, dx - \int_D f X_{\phi=1} \gamma^+ \, dx$$

Proof: directely using Lebesgue dominated convergence theorem .

Obviously W is weakly lower semi continuous and coercive on $H^1_0(D)$ so that it reaches its minimum on $H^1_0(D)$. Let Φ be a local minima of W , the first order optimality necessary condition can be written as follows : for any γ in $H^1_0(D)$ we have

$$\int_D \text{grad}\phi . \text{grad}\gamma \, dx - \int_D f X_{\phi>1} \gamma \, dx \geq \int_D f X_{\phi=1} \gamma^+ \, dx \qquad (1.6)$$

Chosing $+,- \gamma$ in that variational inequality and adding the two equalities we obtain at any local minima Φ of W :

$$\int_D f X_{\phi=1} |\gamma| \, dx = 0 \quad \text{for any } \gamma \text{ in } H^1_0(D) \qquad (1.7)$$

from (1.7) it derives easily that , assuming that $f>0$ a.e. in D , $\text{meas}(\{x \mid \phi(x)=1\})=0$, i.e. (1.2) holds.Then from (1.6) we get
that u is solution of the problem (1.1)-(1.3),we get the following results.

<u>Proposition 1.2</u>

Let f be given in $L^2(D)$ verifying $f>0$ a.e. in D , then W reaches its minimum on $H^1_0(D)$ .Let u be a local minima of W then $\text{meas}(\{x \mid u(x)=1\})=0$ and u is a solution to the problem (1.1)-(1.3).

<u>Proposition 1.3</u>

Assume that f is given in $L^p(\Omega)$, with $p>N/2$ and verifying $f>o$ a.e. in D ,then W reaches its minimum on $H^1_0(D)$ .Let u be a local minima of W then $\text{meas}(\{x \mid u(x)=1\})=0$ and u is a solution to the problem (1.1)-(1.3), u being in $W^{2,p}(D)$ the set $\Omega = \{ x \text{ in } D \mid u(x) > 1\}$ is open in D.

Remark 1.4

In the particular situation $f \geq 0$ a.e. over D then W can be written as follows:

$$W(\phi) = \text{MIN} \{ \int_D (1/2 \mid \text{grad}\phi \mid^2 - f\mu(\phi-1)) \, dx \mid \mu \text{ in } M \} \tag{1.8}$$

$$\text{where} \quad M = \{ \mu \mid 0 \leq \mu(x) \leq 1 \text{ ,a.e. } x \text{ in } D \} \tag{1.9}$$

So that the minimization of W over $H^1_0(D)$ is equivalent to the following problem:

$$\text{MIN} \ \{ \int_D (1/2 \mid \text{grad}\phi \mid^2 - f\mu(\phi-1)) \, dx \mid \phi \text{ in } H^1_0(D), \mu \text{ in } M \} \tag{1.10}$$

Remark 1.5

In the problem (1.10) let us assume f given in $W^{s,\infty}(D)$, $0< s < 1/2$. As $(\phi-1)$ is an element of $H^s(D)$, the multiplier μ can be taken in $H^{-s}(D)$, $0< s < 1/2$, and as the unit ball of that Hilbert space is strictly convex it turns out that in fact for each ϕ in $H^1(D)$ there exists a unique minimizer μ_ϕ in the unit ball of $H^{-s}(D)$: $\int_D f \mu_\phi (\phi-1)) \, dx = \int_D f \, (\phi-1)^+ \, dx$.

1.2 The Weak Shape Formulation .

The minimization problem

$$\text{MIN} \ \{ \int_D (1/2 \mid \text{grad}\phi \mid^2 - f(\phi-1)^+) \, dx \mid \phi \text{ in } H^1_0(D) \}$$

posesses solutions for any f in $H^{-1}(D)$. That problem can be written as a shape optimization one in the following way . For any measurable set Ω in D let us consider the functionals defined by :

$$J_1(\Omega) = \text{MIN}\{ \int_{D\setminus\Omega} 1/2 \mid \text{grad}\phi \mid^2 \, dx \mid \phi \text{ in } H^1_0(D)_+ , \phi=1 \text{ a.e in } \Omega \}$$

where

$$H^1_0(D)_+ = \{ \phi \text{ in } H^1_0(D) \ / \ \phi \geq 0 \text{ a.e. in } D \}$$

and

$$J_2(\Omega) = \text{MIN}\{ \int_\Omega (1/2 \mid \text{grad}\phi \mid^2 - f\phi) \, dx \mid \phi \text{ in } H^1_0(D)_+ \}$$

then the shape optimization problem is : given α, $0 < \alpha < \text{meas}(D)$,

$$(\mathcal{P}_\alpha) \quad \text{MIN} \ \{ J_1(\Omega) + J_2(\Omega) \ / \ \Omega \text{ measurable subset in D, meas}(\Omega) = \alpha \ \}$$

If v and w are the minimizers of the problems $J_1(\Omega)$ and $J_2(\Omega)$, then we consider $U = (v+1) - w$, element of $H^1_0(D)$. u being the solution of $(\mathcal{P})$ with $f > 0$ a.e., i.e. u is solution of (1.1)-(1.4), let $\alpha_0 = \text{meas}\{x \text{ in } D / u(x) > 1\}$. Then

$w = (u-l)^-$ and $v = (u-l)^+$ are solutions of J_1 and J_2 in $\Omega_o=\{x \text{ in } D / u(x)>l\}$ and that Ω_o is solution of $(\mathcal{P}_{\alpha o})$. So the problem $(\mathcal{P}_{\alpha o})$, for that value of α_o , has at least one solution. Conversely if Ω is a smooth solution to $(\mathcal{P}_{\alpha o})$, U is a solution for (1.1)-(1.4).That fact derives from the necessary conditions for optimality of Ω minimizing $J(\Omega) = J_1(\Omega) + J_2(\Omega)$ which will force the normal derivatives of v and w on $\partial\Omega$ to be equals so that "is in $H^1_0(D) \cap H^2(D)$. In general for given α the problems $(\mathcal{P})$ and $(\mathcal{P}_\alpha)$ are differents.

At the next section we shall be concerned with a classical free boundary problem posed in Ω (i.e. without equation in the complement $D\setminus\Omega$) . That problem was studied by Alt and Caffarelli [1980] , after recalling the weak vcariational formulation in an appropriate setting ,th we shall formulate the weak shape formulation associated and give existence results for that new problem.

II. Bernoulli Condition Associated to Homogeneous Dirichlet Condition.

We turn now to the situation of the free boundary problem of finding Ω in D and a function y on Ω such that, on $\partial\Omega$, we have y=0 and the Neumann condition $\partial_n y= Q^2$ where Q is given over D. Alt H.W and Caffarelli L.A.[1980] introduced the following functional:

$$J(\phi)= \int_D (1/2\, |\mathrm{grad}\phi|^2 - f\phi)\, dx + \int_D Q^2 X_{\{\phi>0\}}\, dx \qquad (2.1)$$

to be minimized on

$$K = \{ u \text{ in } H^1_0(D) \text{ such that } u(x) \geq 0 \text{ for a.e. } x \text{ in } D\} \qquad (2.2)$$

The existence results is based on the following result:

Lemma 2.1

Let u_n and X_n be two converging sequences; u_n ---->u in $L^2(D)$ and X_n a characteristic function,$X_n(1-X_n) = 0$,weakly converging in $L^2(D)$ to an element λ .Then we have :

$$(1 - X_n)\, u_n = 0 \text{ for all } n \quad \text{implies} \quad \lambda \geq X_{\{u\neq 0\}} \qquad (2.3)$$

proof:

we have $(1 - X_n)u_n = 0$ then in the limit we get $(1 - \lambda)u = 0$,then on the set $\{x \mid u(x)\neq 0\}$ we have $\lambda=1$;on the other hand, as a weak limit of characteristic functions, λ lies beetwen 0 and 1 .

<u>Proposition 2.2</u>

Let f and Q be two element of $L^2(D)$ then there exists u in K which minimizes the functional J over the positive cone K of $H^1_0(D)$.

Proof:

Let u_n be a minimizing sequence of the functional J over the convex set K, u_n in K.We denote X_n the characteristic function of the set $\{x \text{ in } D \mid u_n(x)>0\}$, which is in fact the same subset, then $\{x \text{ in } D \mid u_n(x) \neq 0\}$ .It is immediate to verify that the sequence u_n remains bounded in $H^1_0(D)$ so that we shall now denote by u_n a subsequence which weakly converges in $H^1_0(D)$ to an element u of K.That convergence holds in $L^2(D)$ so that the Lemma 2.1 applies and we get $\lambda \geq X_{u\neq 0}$ for any weak limiting element of the sequence X_n (which is bounded in $L^2(D)$).

Let j denote the minimum of J over K,then $J(u_n)$ converges to j but in the weak limit we get

$$\int_D (1/2\, |\text{grad } u|^2 - fu)\, dx \leq \liminf \int_D (1/2\, |\text{grad } u_n|^2 - fu_n)\, dx \quad \text{and}$$

$$\int_D X_{u\neq 0}\, Q^2\, dx \leq \int_D \lambda Q^2\, dx = \lim \int_D X_n\, Q^2\, dx$$

finely by adding these two majorations we get $J(u) \leq j$.

Remark 2.3

If we assume f to be non negative , $f \geq 0$ a.e. in D and u to be smoothly defined in D we shall see that the set $\Omega = \{x \text{ in } D \mid u(x) > 0\}$ and the element $u|_\Omega$ are a weak solution to the free boundary problem

$$-\Delta u = f \quad \text{in } \Omega, \qquad u = 0, \quad \partial_n u = Q^2 \text{ on } \partial\Omega \tag{2.4}$$

So the idea is now to consider Ω as an independent variable.

The minimization problem (2.1),(2.2) can be written as a shape optimization problem as follows:

for any measurable subset Ω of D define the Sobolev space $H^1_0(\Omega)$ in the following way:

$$H^1_0(\Omega) = \{u \text{ in } H^1_0(D) \mid u(x) = 0 \text{ q.e. } x \text{ in } D\setminus\Omega\} \tag{2.5}$$

and the positive cone :

$$H^1_0(\Omega)_+ = \{u \text{ in } H^1_0(\Omega) \mid u(x) \geq 0 \text{ a.e. } x \text{ in } D\}.$$

Then we consider the shape optimization problem:

$$\text{INF}\ \{ E(\Omega) \mid \Omega \text{ is a measurable subset in } D \} \tag{2.6}$$

where the energy functional E is given by :

$$E(\Omega) = \text{MIN}\ \{ \int_\Omega (1/2\, |\text{grad}\phi|^2 - f\phi)\, dx + \int_\Omega Q^2\, dx \mid \phi \text{ in } H^1_0(\Omega)_+ \} \tag{2.7}$$

From the definition we have

$$H^1_0(\Omega) = H^1_0(\Omega \cup E) \quad \text{for any E such that cap(E)=0}$$

we recall here that the capacity of E in D is classiquely defined as

$$\text{cap}(E) = \text{MIN}\ \{ (\int_D |\text{grad}\ \phi|^2\, dx)^{1/2} \mid \phi \text{ in } H^1_0(D),\ \phi \geq X_E \text{ a.e. in a neighborhood of D} \}.$$

We know , Deny J., Lions J.L.[1951], Federer H., Ziemer W.P.[1972] , that any element u of $H^1_0(D)$ can be defined quasi everywhere and that if u_n is a bounded sequence in $H^1_0(D)$ we can extract a subsequence which converges quasi everywhere to an element u of $H^1_0(D)$. We say that u(x) = 0 q.e. x in $D\setminus\Omega$ if there exists E in D with zero capacity in D such that the equality u(x) = 0 holds for any x in $(D\setminus\Omega)\setminus E$. Let us also recall that if E is a measurable in D with cap(E)=0 then meas(E)=0 but the converse is false .Equipped with the norm of $H^1_0(D)$,

$H^1_0(\Omega)$ is a Hilbert space so that for any measurable subset Ω in D the problem (2.7) do has a unique solution y in the closed convex $H^1_0(\Omega)_+$ and we have the following equivalence between problems (2.6)-(2.7) and the minimization of J over K:

Proposition 2.4

Let u be a minimizing element of J over K.Then $\Omega:=\{x \text{ in } D \mid u(x)>0\}$ is a solution of problem (2.6) while $y=u_{|\Omega}$ is a solution of (2.7). Conversely,if Ω and y are solution of (2.6)-(2.7) , Ω being a measurable set in D and y in $H^1_0(\Omega)_+$, the element u defined by

$u(x) = y(x)$ in Ω and $u(x) = 0$ in $D\backslash\Omega$, lies in K and minimizes J over K.

The sets $\{x \text{ in } D \mid u(x)>0\}$, $\{x \text{ in } D \mid u(x)=0\}$ and $\{x \text{ in } D \mid u(x)\neq 0\}$ are defined as measurable subsets of D up to a set E with cap(E)=0, this fact derives from the quasi everywhere definition of u, element of $H^1(D)$. It is also interesting to built these sets as follows: we recall that any element u in $H^1_0(D)$ posseses a quasi-continuous representant; for any positive ε there exists a set E_ε with capacity less then ε sucht that u is continous in $D\backslash E_\varepsilon$.

Lemma 2.5

Any u in $H^1_0(D)$ belongs to $H^1_0(\Omega_u)$.

Proof.

By construction of Ω_u we have u=0 q.e. in $D\backslash\Omega$.

We turn now to general situations for which the sign of f is not prescribed and the measure of Ω can be given (or bounded) but the control problem Min $E(\Omega)$ is well defined and do posesses solutions while it does not correspond to the minimization of functional $J(\phi)$ as in the two first sections.

III. Shape Existence of Weak Solutions.

3.1 Dirichlet Problem without Constraint.

The problem (2.6)-(2.7) can be relaxed as follows : given any f in $L^2(D)$, G in $L^1(D)$

$$(\mathcal{D}_0) \qquad \text{INF} \; \{ \; E(\Omega) \mid \Omega \text{ measurable in } D \; \} \qquad (3.1)$$

where the energy functional is defined by

$$E(\Omega) = \text{MIN} \; \{ \int_\Omega (1/2 \,|\text{grad}\phi|^2 - f\phi \,)\, dx + \int_\Omega G\, dx \mid \phi \text{ in } H^1_0(\Omega) \; \} \qquad (3.2)$$

The Hilbert space $H^1_0(\Omega)$ being defined at (2.5) for any measurable subset Ω in D.From a classical result by G.Stampacchia [1960] we know that for any element u in $H^1_0(\Omega)$ we have gradu(x)=0 a.e. x in $D\backslash\Omega$ so that the functional E can be rewritten as follows:

$$E(\Omega)= \text{MIN} \; \{ \; E(\Omega, \phi\,)+ \int_\Omega G\, dx \mid \phi \text{ in } H^1_0(\Omega) \; \} \qquad (3.3)$$

where $E(\Omega, \phi) = \int_\Omega (1/2\,|\,\mathrm{grad}\phi|^2 - f\phi)\,dx = \int_D (1/2\,|\,\mathrm{grad}\phi|^2 - f\phi)\,dx$ for any ϕ in $H^1_0(\Omega)$

We have the following existence result for the problem $(\mathfrak{D}_0)$:

Theorem 3.1

For any f in $L^2(D)$, $G = Q^2$ in $L^1(D)$, there exists a solution (at least one) to the problem $(\mathfrak{D}_0)$.

Proof.

Let Ω_n be a minimizing sequence for the problem $(\mathfrak{D}_0)$,and for each n let u_n be the (unique) solution to the problem (3.3).If X_n is the characteristic function of the measurable set Ω_n we have u_n in $H^1_0(\Omega_n)$ that implies $(1-X_n)\,u_n = 0$.On the other hand the sequence u_n remains bounded in $H^1_0(D)$ (taking $\phi = 0$ in (3.3) we get $\int_D (1/2\,|\,\mathrm{grad}u_n|^2 - fu_n)\,dx \le 0$ and the conclusion derives from the equivalence of $H^1(D)$ and $H^1_0(D)$ norms) .We can assume that X_n weakly converges in $L^2(D)$ to an element λ and u_n weakly converges in $H^1_0(D)$ to an element u. From Lemma 2.1 we get $\lambda \ge X_{u\neq 0}$ a.e. in D .

Let us define $\Omega = \Omega(u) := \{x \text{ in } D \mid u(x) \neq 0\}$. This measurable set Ω is defined up to a set with zero capacity. Then u belongs to $H^1_0(\Omega(u))$ and we have $\mathrm{meas}(\Omega) = \mathrm{meas}(\{x \text{ in } D \mid \lambda(x) = 1\}) \le \alpha$ (as we have $\alpha = \mathrm{meas}(\{x \mid \lambda(x)=1\}) + \mathrm{meas}(\{x \mid 0 \le \lambda(x) < 1\})$).In the limit we get

$$\int_\Omega (1/2\,|\,\mathrm{grad}u|^2 - fu)\,dx = \int_D (1/2\,|\mathrm{grad}u|^2 - fu)\,dx \le \liminf \int_D (1/2\,|\,\mathrm{grad}u_n|^2 - fu_n)\,dx \quad (3.4)$$

and

$$\int_\Omega G\,dx \le \int_D \lambda\,G\,dx = \lim \int_{\Omega_n} G\,dx \quad (3.5)$$

by adding (3.4) and (3.5) we get that Ω minimises E and u minimizes $E(\Omega,.)$

3.2 Dirichlet Problem with Constraint on the measure of the domain Ω.

The problem (2.6)-(2.7) can also be relaxed as follows : given any f in $L^2(D)$, G in $L^1(D)$ and a real number α ,$0<\alpha<\mathrm{meas}(D)$

$$(\mathfrak{D}^\alpha_0) \qquad \mathrm{INF}\ \{\ E(\Omega)\ \mid\ \mathrm{meas}(\Omega)=\alpha,\ \Omega \text{ in } D\ \} \quad (3.6)$$

We have the following existence result for the problem $(\mathfrak{D}^\alpha_0)$:

Theorem 3.2

For any f in $L^2(D)$, $G = 0$ and any real number α, $0<\alpha<\mathrm{meas}(D)$,there exists a solution (at least one) to the problem $(\mathfrak{D}^\alpha_0)$.

Proof.

Let Ω_n be a minimizing sequence for the problem $(\mathfrak{D}^\alpha_0)$,and for each n let u_n be the (unique) solution to the problem (3.3).If X_n is the characteristic function of the measurable set

Ω_n we have u_n in $H^1_0(\Omega_n)$ that implies $(1-X_n)\, u_n = 0$.On the other hand the sequence u_n remains bounded in $H^1_0(D)$ (taking $\phi = 0$ in (3.3) we get $\int_D (1/2\, |\mathrm{grad} u_n|^2 - f u_n)\, dx \leq 0$ and the conclusion derives from the equivalence of $H^1(D)$ and $H^1_0(D)$ norms) .We can assume that X_n weakly converges in $L^2(D)$ to an element λ and u_n weakly converges in $H^1_0(D)$ to an element u.In the limit we get $\int_D \lambda(x)\, dx = \alpha$ .From Lemma 2.1 we get $\lambda \geq X_{u\neq 0}$ a.e. in D . Let us define $\Omega = \Omega(u) := \{x \text{ in } D \mid u(x) \neq 0\}$ then u belongs to $H^1_0(\Omega(u))$ and we have $\mathrm{meas}(\Omega) = \mathrm{meas}(\{x \text{ in } D \mid \lambda(x) = 1\}) \leq \alpha$ (as we have $\alpha = \mathrm{meas}(\{x \mid \lambda(x)=1\}) + \mathrm{meas}(\{x \mid 0 \leq \lambda(x) < 1\})$).In the limit we get

$$\int_\Omega (1/2\, |\mathrm{grad} u|^2 - fu)\, dx = \int_D (1/2\, |\mathrm{grad} u|^2 - fu)\, dx \leq \liminf \int_D (1/2\, |\mathrm{grad} u_n|^2 - f u_n)\, dx \tag{3.7}$$

and

$$\int_D \lambda\, G\, dx = \lim \int_{\Omega_n} G\, dx \tag{3.8}$$

so that

$$\int_\Omega (1/2\, |\mathrm{grad} u|^2 - fu)\, dx + \int_D \lambda\, G\, dx \leq \mathrm{INF}\ \{E(\Omega) \mid \mathrm{meas}(\Omega) = \alpha\}$$

as $G \geq 0$ we get $E(\Omega) \leq \mathrm{INF}\ \{E(\Omega) \mid \mathrm{meas}(\Omega) = \alpha\}$; but Ω does not verify the constraint on the measure.

Let us note that for any measurable set Ω' such that Ω' is between Ω and D we have $\int_{\Omega'} (1/2\, |\mathrm{grad} u|^2 - fu)\, dx = \int_\Omega (1/2\, |\mathrm{grad} u|^2 - fu)\, dx$ so that in (3.7) Ω can be increased to any such Ω' .The inclusion of Ω in Ω' implies the inclusion of $H^1_0(\Omega)$ in $H^1_0(\Omega')$ so that in (3.3) it is immediat that ,G being non positive over D ($G \leq 0$ a.e. x in D),we have $E(\Omega') \leq E(\Omega)$ .To conclude the proof we just have to select Ω' with $\mathrm{meas}(\Omega') = \alpha$.That measurable set Ω' is admissible and minimizes the coast in (3.6) and we have

$$E(\Omega') = E(\Omega) = \mathrm{INF}\{E(\Omega) \mid \mathrm{meas}(\Omega) = \alpha\}$$

Corollary 3.3.

Assume f in $L^2(D)$ and $f = Q^2$ in $L^1(D)$, then the following problem do have an optimal solution

$$(\mathcal{D}^{\alpha-}_{0}) \qquad \mathrm{INF}\ \{E(\Omega) \mid \mathrm{meas}(\Omega) \leq \alpha,\ \Omega \text{ in } D\} \tag{3.9}$$

proof: it is similar to the proof of theorem 3.2,the minimizing sequence is chosen such that $\mathrm{meas}(\Omega_n) \leq \alpha$, so that in the weak limit we get

$$\mathrm{meas}(\Omega) \leq \int_D \lambda(x)\, dx \leq \alpha$$

Remark 3.4.

In these problems f has been supposed given in $L^2(D)$.This was necessary in the proofs to get in the limit in the terms $\int_{\Omega_n} f u_n\, dx$ which is $\int_D f u_n\, dx$. Neverveless in many applications f turns to be given in $H^{-1}(D)$ but with a compact support in D. We briefly show now that the previous existence results are easily extended to that situation.

3.3 The situation in which f is Given in $H^{-1}(D)$.

3.3.1. The case without constraint.

Let G be given in $L^1(D)$ and f in $H^{-1}(D)$ such that

$$\mathbf{C} = \text{support of } f \text{ is a compact set in } D \tag{3.10}$$

Then we consider the following problem:

$$(\mathfrak{D}\mathbf{C}_0) \qquad \mathrm{INF}\ \{\ E(\Omega)\ |\ \Omega \text{ measurable in } D, \mathbf{C} \text{ included in } \Omega\} \tag{3.11}$$

where the energy functional is defined by

$$E(\Omega) = \mathrm{MIN}\ \{\ \int_\Omega 1/2\,|\mathrm{grad}\phi|^2\,dx + \langle f, \phi^o\rangle + \int_\Omega G\,dx\ |\ \phi \text{ in } H^1_0(\Omega)\ \} \tag{3.12}$$

Where <,> is the bilinear form pairing between $H^{-1}(D)$ and its dual space $H^1_0(D)$,the Hilbert space $H^1_0(\Omega)$ being defined in Remark 2.3 for any measurable subset Ω in D and ϕ^o being the extension of ϕ by 0,element of $H^1_0(D)$. For any element u in $H^1_0(\Omega)$ we have gradu(x)=0 a.e. x in $D\backslash\Omega$ so that the functional E can be rewritten as follows:

$$E(\Omega) = \mathrm{MIN}\ \{\ E(D, \phi) + \int_\Omega G\,dx\ |\ \phi \text{ in } H^1_0(\Omega)\ \} \tag{3.13}$$

where $E(\Omega, \phi) = \int_D 1/2\,|\mathrm{grad}\phi|^2\,dx + \langle f, \phi^o\rangle\!\int)$ for any ϕ in $H^1_0(\Omega)$

We have the following existence result for the problem $(\mathfrak{D}\mathbf{C}_0)$:

Theorem 3.5

For any f in $H^{-1}(D)$ such that $\mathbf{C}$ = support of f is a compact set in D , $G = Q^2$ in $L^1(D)$,there exists a solution (at least one) to the problem $(\mathfrak{D}\mathbf{C}_0)$.

Proof.

Let Ω_n be a minimizing sequence for the problem $(\mathfrak{D}\mathbf{C}_0)$,and for each n let u_n be the (unique) solution to the problem (3.12).If X_n is the characteristic function of the measurable set Ω_n we have $(1\text{-}X_n)\,u_n = 0$.On the other hand the sequence u_n remains bounded in $H^1_0(D)$ (taking $\phi = 0$ in (3.3) we get $\int_D (1/2\,|\mathrm{grad}u_n|^2 dx - \langle f,u_n\rangle \le 0$ and the conclusion derives from the equivalence of $H^1(D)$ and $H^1_0(D)$ norms) .We can assume that X_n weakly converges in $L^2(D)$ to an element λ and u_n weakly converges in $H^1_0(D)$ to an element u. From Lemma 2.1 we get $\lambda \ge X_{u\neq 0}$ a.e. in D .

Let us define $\Omega=\Omega(u):=\{x \text{ in } D\ |\ u(x)\neq 0\ \}$. This measurable set Ω is defined up to a set with zero capacity. Then u belongs to $H^1_0(\Omega(u))$ and we have $\mathrm{meas}(\Omega) = \mathrm{meas}(\{x \text{ in } D|\lambda(x) = 1\}) \le \alpha$ (as we have $\alpha=\mathrm{meas}(\{x|\lambda(x)=1\})+\mathrm{meas}(\{x|0\le\lambda(x)<1\})$).In the limit we get

$$\int_\Omega 1/2\,|\mathrm{grad}u|^2\,dx - \langle f,u\rangle = \int_D (1/2\,|\mathrm{grad}u|^2 dx - \langle f,u\rangle \le \liminf \int_D (1/2\,|\mathrm{grad}u_n|^2\,dx - \langle f,u_n\rangle$$

and $\int_\Omega G\,dx \le \int_D \lambda\,G\,dx = \lim \int_{\Omega_n} G\,dx$

3.3.2. Constraint on the measure of the domain Ω.

Let us define the problem

$$(\mathfrak{D}C^{\alpha}{}_{0}) \quad \mathrm{INF}\{\ E(\Omega)\ \mid\ \Omega \text{ measurable in } D,\ C \text{ included in } \Omega, \mathrm{meas}(\Omega)=\alpha\} \qquad (3.14)$$

where $E(\Omega)$ is defined at (3.12).Let G=0 and (Ω_n,u_n) be a minimizing sequence as previously,then exactely as for problem $(\mathfrak{D}^{\alpha}{}_{0})$ we get the following existence result.

Theorem 3.6

For any f in $H^{-1}(D)$ having its support in C .Taking G = 0 and any real number α , $0<\alpha<\mathrm{meas}(D)$,there exists a solution (at least one) to the problem $(\mathfrak{D}C^{\alpha}{}_{0})$.

IV. Shape Weak Existence with Bounded Perimeter Sets. Dirichlet Condition

The problems $(\mathfrak{D}_0)$, $(\mathfrak{D}^{\alpha}{}_{0})$ and $(\mathfrak{D}^{\alpha-}{}_{0})$ do have optimal solutions but as they are associated to the homogeneous Dirichlet condition, u in $H^1{}_0(\Omega)$,the optimal domain Ω is not permitted in general to posesses holes; that is to say, roughly speaking ,that the topology of Ω is a priori given.In many examples it turns out that the solution u is physically interpreted as a potential so that the homogeneous Dirichlet condition turns to be non adequate.The physical condition is that the potential u should be constant on each connected componant of the boundary $\partial\Omega$ in D.When Ω is a simply connected domain then $\partial\Omega$ has a single connected component so the constant can be taken as zero,but in general this constant can be fixed only on one connected component,on the others the constant should be an unknown of the problem.To illustate the reason for which holes cannot occur in the previous problems let us consider a simple example.D is the square $)0,1(^2$ and f=1,G=0. For any smooth domain Ω in D we have $E(\Omega) = -1/2 \int_\Omega u(x)\,dx$ and it can easily be verified that

$$\mathrm{INF}\ \{\ E(\Omega)\ \mid\ \Omega \text{ measurable in } D\ \}$$

is achieved at $\Omega = D$ .In particular if one modifies this optimal domain by substracting a closed subset E such that $|E|_{\mathcal{H}^{n-1}} = 0$ but with positive capacity cap(E) >0,for example E is a line, then it is possible to construct a sequence Ω_n which converges to D but such that the optimal solutions $u_n = u(\Omega_n)$ does not converges to u(D) but to $u(D\setminus E)$ for the homogeneous Dirichlet condition in $H^1{}_0(D\setminus E)$ implies u to be zero on E.In other words the mapping X_Ω ----> $u(\Omega)$ is not continuous from $L^2(D)$ in $H^1{}_0(D)$; neverveless the infimum in the problem is achieved but ,at least when f is positive,no hole is allowed in the optimal solutions.These considerations justify the introduction of the following Hilbert space:

For any measurable set Ω in D we consider

$H^1{}_0(\Omega) = \{\phi \text{ in } H^1{}_0(D) \mid (1-X_\Omega)\ \text{grad}\ \phi(x) = 0 \text{ a.e. } x \text{ in } D\}$ (4.1)

When Ω is an open subset in D such that D\cl(D) is not simply connected from classical results of Distribution's theory we know that f is constant on each connected component of D\cl(Ω), this constant being zero in the component whose ∂D is a part of the boundary.

The minimization problems ($\mathfrak{D}_0$), ($\mathfrak{D}^\alpha{}_0$) and ($\mathfrak{D}^{\alpha-}{}_0$) associated to that Hilbert space fail (in the sens that the previous technics for existence of optimal Ω fail).The main reason is that the Lemma 2.1 is false when u_n is replaced by grad u_n weakly converging in $L^2(D)^n$. The tool is to recover the equivalent of Lemma 2.1 by imposing the strong $L^2(D)$ convergence of the sequence u_n.In practice u_n stands for the sequence of characteristic functions X_{Ω_n} of a minimizing sequence. To obtain the strong $L^2(D)$ convergence of a subsequence we add a constraint on the

perimeters.We consider the family of Bounded Perimeter Sets in D defined as follows:

$$BPS(D) = \{ \text{ subsets } \Omega \text{ of } D \mid \sup \{\textstyle\int_\Omega \text{div}(g)\, dx \mid g \text{ in } C^\infty{}_{comp}(D;\mathbf{R}^n) ;\ \|g(x)\| \leq 1, x \text{ in } D\} < \infty \}.$$

It is immediat to verify that $\{ X_\Omega \mid \Omega \text{ is in BPS(D)} \}$ is included in BV(D) .The norm of X_Ω is given by

$$\| X_\Omega \|_{BV(D)} = \text{meas}\Omega + \| \text{grad } X_\Omega \|_{M^0(D)}$$

Where the norm of grad X_Ω in the Banach space $M^0(D)$ is given by

$$\| \text{grad } X_\Omega \|_{M^0(D)} = \sup \{< \text{grad } X_\Omega , g > \mid g \text{ in } C_{comp}(D;\mathbf{R}^n) ;\ \|g(x)\| \leq 1, x \text{ in } D\}$$

where $< \text{grad } X_\Omega , g > = - \lim \int_\Omega \text{div}(g_n)\, dx$, g_n being a sequence in $C^\infty{}_{comp}(D;\mathbf{R}^n)$ which converges to g in $C_{comp}(D;\mathbf{R}^n)$.(it can easily be verified that that limit is independent on the choice of such sequence g_n).

The perimeter of Ω in D is given by

$$P_D(\Omega) = \sup \{\textstyle\int_\Omega \text{div}(g)\, dx \mid g \text{ in } C^\infty{}_{comp}(D;\mathbf{R}^n) ;\ \|g(x)\| \leq 1, x \text{ in } D\}$$

When Ω is a smooth subdomain of D the n-1 dimensional measure of its boundary is given by $|\partial\Omega|_{\mathcal{H}^{n-1}} = P_D(\Omega) + |\partial(\Omega \cap D)|_{\mathcal{H}^{n-1}}$. The inclusion of BPS(D) in BV(D) permit us to obtain the following compactness result concerning the family BPS(D).

<u>Lemma 4.1.</u>

Let Ω_n a sequence in BPS(D) such that $P_D(\Omega) \leq M$, then there exists Ω in BPS(D) and a subsequence, still denoted Ω_n such that the characteristic functions converge in $L^1(D)$:

$X_n \dashrightarrow X$ in $L^1(D)$ as $n \dashrightarrow \infty$; for any g in $C_{comp}(D;\mathbf{R}^n)$,

$$< \text{grad } X_{\Omega_n} , g > \dashrightarrow < \text{grad } X_\Omega , g >$$

and $P_D(\Omega) \leq \liminf P_D(\Omega_n)$.

Proof. This is a simple translation of the classical 'compact embedding' of BV(D) in $L^1(D)$,see for example R.Temam [1983] .

We define the perimeter for all measurable subsets Ω in D as follows:

$$P_D(\Omega) = \sup \{ \int_\Omega \operatorname{div}(g)\, dx \mid g \text{ in } C^\infty_{comp}(D;\mathbf{R}^n) ;\ \|g(x)\| \leq 1, x \text{ in } D \}$$

belonging to $\mathbf{R} \cup \{+\infty\}$

We introduce the following problem,for any $\sigma > 0$

$$(\mathfrak{D}^\alpha_\sigma) \qquad \text{INF} \ \{ E_\sigma(\Omega) \mid \operatorname{meas}(\Omega)=\alpha, \Omega \text{ in } D \} \tag{4.2}$$

where

$$E_\sigma(\Omega) = E(\Omega) + \sigma P_D(\Omega) \tag{4.3}$$

$$E(\Omega) = \text{MIN} \ \{ \int_\Omega (1/2 \,|\operatorname{grad}\phi|^2 - f\phi)\, dx + \int_\Omega G\, dx \mid \phi \text{ in } H^1_0(\Omega) \} \tag{4.4}$$

Theorem 4.2.

Let f in $L^2(D)$, G in $L^1(D)$, $\sigma > 0$, $0 \leq \alpha <$ meas(D) , then the problem $(\mathfrak{D}^\alpha_\sigma)$ posesses (at least) one optimal solution Ω in BPS(D).

Proof.

Let Ω_n be a minimizing sequence for $(\mathfrak{D}^\alpha_\sigma)$; taking $\phi = 0$ in (4.4) we get $P_D(\Omega_n) \leq \sigma^{-1} (E(D) + \int_\Omega |G|\, dx)$,then we consider the subsequence $X_n = X_{\Omega_n}$ described by Lemma 4.1. For each n , let u_n in $H^1_0(\Omega_n)$ the unique minimizer of (4.4) .That sequence remains bounded in $H^1_0(D)$ so that we assume that it is weakly converging to an element u in $H^1_0(D)$.

From $(1 - X_n)$ grad $u_n = 0$ a.e. in D we get in the limit $(1-X_\Omega)$ grad $u = 0$ a.e. in D , so that the limiting element u belongs to $H^1_0(\Omega)$ .We have:

$$\int_\Omega (1/2|\operatorname{grad}u|^2 - fu)dx = \int_D (1/2|\operatorname{grad}u|^2)dx - \int_\Omega fu\, dx \leq$$

$$\leq \quad \liminf \int_D (1/2 \,|\operatorname{grad}u_n|^2)dx - \int_{\Omega_n} fu_n\, dx \tag{4.5}$$

and

$$\int_\Omega fu\, dx = \lim \int_{\Omega_n} fu_n\, dx, \quad \int_\Omega G dx = \lim \int_{\Omega_n} G dx \tag{4.6}$$

so that

$$\int_\Omega (1/2 \,|\operatorname{grad}u|^2 - fu)\, dx + \int_\Omega G\, dx + \sigma P_D(\Omega) \leq \text{INF} \ \{E_\sigma(\Omega) \mid \operatorname{meas}(\Omega)=\alpha \} \tag{4.7}$$

As in section 3 we could consider the situation when f is given in $H^{-1}(D)$ with a compact support $\mathbf{C}$ in D. We introduce the following problem,for any $\sigma > 0$

$$(\mathfrak{D}\mathbf{C}^\sigma_\alpha) \qquad \text{INF} \ \{ E_\sigma(\Omega) \mid \Omega \text{ in } D, \Omega \text{ contains } \mathbf{C} \}$$

where

$$E_\sigma(\Omega) = E(\Omega) + \sigma P_D(\Omega)$$

$$E(\Omega) = \text{MIN} \ \{ \int_\Omega 1/2 \,|\operatorname{grad}\phi|^2\, dx - \langle f,\phi \rangle + \int_\Omega G\, dx \mid \phi \text{ in } H^1_0(\Omega) \}$$

Theorem 4.3.

Let f in $H^{-1}(D)$ with support included in the compact subset $\mathbf{C}$ of D , G in $L^1(D)$, $\sigma > 0$, meas($\mathbf{C}$)$< \alpha <$ meas(D) Then the problem $(\mathfrak{D}^\sigma_\alpha)$ posesses (at least) one optimal solution Ω in BPS(D).

Proof. It is similar to that of theorem 4.2 .

An important situation is when the 'forcing term' f is not a distributed function but a boundary term.When working with smooth domains Ω it is easy to includ a Neumann condition $\partial_n u = g$ on a part $\Gamma_0$of $\partial\Omega$ but now Ω is just a measurable subset in D.We shall relax in the weak formulation that Neuman condition as follows : we chose f =0 and Σ_0 being a fixed given subset of ∂D we consier $\Gamma_0 = \Sigma_0 \cap \mathrm{cl}(\Omega)$. Of course Γ_0 can be empty but the minimization with respect to the set Ω will force Γ_0 to have a strictely positive n-l dimensional measure.

Let us define

$$(\mathfrak{D}\Sigma\mathcal{N}^{\alpha}_{\sigma}) \qquad \mathrm{INF}\ \{\ E\Sigma_{\sigma}(\Omega)\ \mid\ \mathrm{meas}(\Omega)=\alpha,\ \Omega \text{ in } D\ \} \tag{4.8}$$

where

$$E\Sigma_{\sigma}(\Omega)=E\Sigma(\Omega) + \sigma\, P_D(\Omega) \tag{4.9}$$

and

$$E\Sigma(\Omega) = \mathrm{MIN}\{\ \int_{\Omega} (1/2\ |\ \mathrm{grad}\phi|^2\ )\, dx + \int_{\Sigma_0 \cap \mathrm{cl}(\Omega)} g\phi\, d\mu + \int_{\Omega} G\, dx \mid \phi \text{ in } H^1\Sigma_0(\Omega)\ \} \tag{4.10}$$

where

$$H^1\Sigma_0(\Omega) =\{\ \phi \text{ in } H^1(D) \mid \phi|_{\partial D}=0 \text{ a.e.in } \partial D\setminus(\Sigma_0 \cap \mathrm{cl}(\Omega)),\ (1-X_{\Omega}\ )\ \mathrm{grad}\phi =0 \text{ a.e.in } D\} \tag{4.11}$$

The subset Σ_0 of ∂D is chosen in such a way that on $H^1\Sigma_0(\Omega)$ the norm of $H^1_0(D)$ is equivalent to the norm of $H^1(D)$.

<u>Theorem 4.4.</u>

Let G and g be given , respectively in $L^1(D)$ and $L^2(\Sigma_0)$ .Then there exist a solution (at least one) Ω in BPS(D) to the problem $(\mathfrak{D}\Sigma\mathcal{N}^{\alpha}_{\sigma})$.

Proof.

Let Ω_n be a minimizing sequence.For each n let u_n be the unique solution to the problem (4.10).

u_n belongs to $H^1\Sigma_0(\Omega)$ so that $\int_{\Omega} (1/2\ |\ \mathrm{grad}u_n|^2\ )\, dx= \int_D (1/2\ |\ \mathrm{grad}u_n|^2\ )\, dx$

and

$$\int_{\Sigma_0 \cap \mathrm{cl}(\Omega_n)} g u_n\, d\mu = \int_{\partial D} g u_n\, d\mu \tag{4.12}$$

taking ϕ =0 in (4.10) we get the following boundness

$$(\ \int_D\ |\ \mathrm{grad}u_n|^2\ dx\ )^{1/2}\ \leq\ c\ \|g\|_{L^2(\partial D)}$$

We can then substract a subsequence which converges weakly in $H^1(D)$ to an element u. As in the proof of theorem 1.2 we can consider that Ω_n converges to Ω , Ω in BPS.

Lemma 4.5.
The element u is in the space $H^1\Sigma_0(\Omega)$.

The main point is to note that

$$j(\Omega_n):= 2(E\Sigma(\Omega_n) - \int_{\Omega_n} G\,dx) = \int_{\Sigma_0\cap cl(\Omega_n)} gu_n \; d\mu = - \int_D |gradu_n|^2 \,dx \qquad (4.13)$$

Obviously if the set $w_n := \Sigma_0 \cap cl(\Omega_n)$ is empty then in the minimum j =lim j_n will be zero,then ,by one more extraction of subsequence we can assume that

for all n ,

$$|\Sigma_0 \cap cl(\Omega_n)|_{\mathcal{H}^{n-1}} > 0 . \qquad (4.14)$$

From the two equalities (4.12) we get in the limit:

$$\int_\Omega (1/2\,|gradu|^2)\,dx = \int_D (1/2\,|gradu|^2)\,dx \le \liminf \int_D (1/2\,|gradu_n|^2)\,dx$$

$$= \liminf \int_{\Omega_n} (1/2\,|gradu_n|^2)\,dx \qquad (4.15)$$

$$\int_{\Sigma_0\cap cl(\Omega)} gu \; d\mu = \int_{\partial D} gu\, d\mu = \lim \int_{\partial D} gu_n\, d\mu = \lim \int_{\Sigma_0\cap cl(\Omega_n)} gu_n \; d\mu \qquad (4.16)$$

So that in the limit we get

$$\int_\Omega (1/2\,|gradu|^2)\,dx + \int_{\Sigma_0\cap cl(\Omega)} gu \; d\mu + \sigma P_D(\Omega) \le \text{INF}\,\{ E\Sigma_\sigma(\Omega) \mid meas(\Omega)=\alpha,\ \Omega \text{ in } D \}$$

and as u belongs to $H^1\Sigma_0(\Omega)$ we get that Ω is a solution to $(\mathbf{P}\Sigma^\alpha{}_\sigma)$ while u is the solution of(4.10).

Proof of Lemma 4.5.
We have to prove that the trace of u on ∂D is zero a.e in $\Sigma_0 \cap cl(\Omega)$ .Let us chose a point x_o in $\Sigma_o \cap cl(\Omega)$ such that the sequence $u_n(x_o)$ ---> $u(x_o)$ .Such point exists for,after substracting a subsequence still denoted u_n, we have u_n---> u a.e. on ∂D.We shall prove that,for n large enough, x_o does not belong to $cl(\Omega_n)$.Assume that for any N one could find n $\ge$ N such that x_o belongs to $cl(\Omega_n)$.Then one could built a diagonal sequence y_k, y_k in Ω_{n_k} and y_k --->x_o. This implies that x_o belongs to the Hausdorf limit C of the sequence Ω_{n_k} (It si classical that a subsequence $cl(\Omega_{n_k})$ converges in Hausdorf measure to some closed subset C of cl(D) , then also Ω_{n_k} converges to C).But now Σ_0 being open in ∂D there exists a ball B centered in x_o such that B is included in $\Sigma_o \cap cl(\Omega)$ .From lemma 15 we have C=cl(Ω) so that

x_o would belong to cl(Ω) which is contrary to our hypothesis.Then x_o does not belong to any cl(Ω_n) so that $u_n(x_o)=0$ and in the limit $u(x_o)=0$.

Lemma 4.6 .

Let Ω_n converges to Ω in BPS(D) . Then Ω_n converges to cl(Ω) in the Hausdorf topology.

proof.

Let Ω_n be a bounded sequence in BPS(D) which converges to Ω in BPS(D) . We know that in BPS(D) each Ω_n can be approached by a sequence Ω^p_n of polyhedral open set in D :

$$\Omega^p_n \text{ ---> } \Omega_n \quad \text{as } p\text{-->}\infty, \text{ in BPS(D)}.$$

Let $\mathcal{H}$ be the Hausdorf limit of the sequence Ω_n we always have the inclusion :

there exists a set E with zero measure in D such that

$$\Omega \backslash E \text{ is included in } \mathcal{H}$$

(effectively, as one can find a subsequence which converges a.e. in D to X_Ω,that is that a subset E with zero measure in D can be found such that $X_{\Omega_n}(x)$ --->$X_\Omega(x)$ for any x in D\E, for any x in Ω \E the sequence $x_n = x$ verifies:

$$x_n \text{ belongs to } \Omega_n \quad , \quad x_n \text{ --->} x$$

which are the two conditions characterizing the elements x of $\mathcal{H}$)

One can built a diagonal subsequence $O_n = \Omega^{p(n)}_n$ which converges in BPS(D) to Ω , O_n being a polyhedral open subset in D.There exist subsets E and E_n in D with zero measres such that : for any x in Ω\E, x belongs to Ω_n for $n\geq N(x)$

for any y_n in$\Omega_n \backslash E_n$, y_n belongs to Ω^p_n for $p\geq P(n,y_n)$

So that with $\mathcal{E}=E \cup_{n\geq o} E_n$ we have ,meas($\mathcal{E}$)=0,

and for any x in $\Omega \backslash \mathcal{E}$, x belongs to O_n for $n\geq \mathcal{N}(x)$.

Assume now that $G=\mathcal{H}\backslash\Omega$ do have a strictely positive measure in D.Let x_o in G.As x_o is in $\mathcal{H}$ there exists a sequence x_n ,x_n in cl(O_n),such that x_n --->x_o, but as x_o does not belongs to Ω , by the previous considerations, x_o does not belong to O_n for $n \geq \mathcal{N}(x_o)$.As O_n is a polyhedral open set in D we have that x_n belongs to the boundary S_n of O_n , S_n=cl(O_n)$\cap$ cl(D\$O_n$).The idea is now to show tat the perimeter of O_n,that is $|S_n|_{N-1}$, is arbitrary large which will be a contradiction,then G will have a zero measure in D.For any set $M=\{x_1,...x_m\}$ of points in G the N-1 dimensional piecewise smooth surface S_n should be arbitrary closed to M,this clearly implies that $|S_n|_{N-1}$ is arbitrarely large .

V . Shape weak Existence with Bounded Perimeter Sets. Neuman Condition .

5. 1 Min Min formulation.

We turn to the minimization of energy functional associated to the Neumann condition.The main question is to relax the definition of the Sobolev space $H^1(\Omega)$ when Ω is a measurable subset of D with finite perimeter, i.e. Ω in BPS(D).

$$H^1(\Omega) = \{ (f,h) \text{ in } L^2(\Omega)xL^2(\Omega)^n \mid \text{exists polyhedral open sets } \Omega_n, X_{\Omega_n} \dashrightarrow X_\Omega \text{ in } L^2(D),$$
$$f_n \text{ in } H^1(\Omega_n) \text{ with } X_{\Omega_n} f_n \dashrightarrow X_\Omega f \text{ weakly in } L^2(D), \tag{5.1}$$
$$X_{\Omega_n} \text{grad } f_n \dashrightarrow X_\Omega h \text{ weakly in } L^2(D)^n \}.$$

Note that in that definition the sequence of polyhedral open sets may depend on the element (f,h).

For shotness in the sequel the situation (5.1) will be denoted by:

$$(\Omega_n, f_n) \text{------>>} (\Omega, f, h)$$

Proposition 5.1.

Equipped with the norm $\| (f,h) \|_{H^1(\Omega)} = ((\|f\|_{L^2(\Omega)})^2 + (\|h\|_{L^2(\Omega)^n})^2)^{1/2}$, $H^1(\Omega)$ is a Hilbert space.

proof.

Let (f^k,h^k) be a Cauchy sequence for the previous norm.Then there exist f and h in $L^2(\Omega)$ and $L^2(\Omega)^n$ such that f^k --->f and h^k--->h strongly in $L^2(\Omega)$ and $L^2(\Omega)^n$ .From (5.1) ,for any k there exist a sequence $\Omega^k{}_n$ of polyhedral open subsets of D such that $\Omega^k{}_n$ converges in $L^2(D)$ to Ω^k and elements $f^k{}_n$ in $H^1(\Omega^k{}_n)$ such that , as n--->∞

$X_{\Omega k_n}$ $f^k{}_n$ ----> $X_\Omega f^k$ weakly in $L^2(\Omega)$, $X_{\Omega k_n}$ grad $f^k{}_n$ ---> $X_\Omega h^k$ weakly in $L^2(\Omega)^n$

by a 'diagonal'subsequence (selecting n=n(k)) we get the existence of $\Omega_k = \Omega^k{}_{n(k)}$ converging to Ω in $L^2(D)$ and elements $f_k = f^k{}_{n(k)}$ in $L^2(\Omega_k)$ $h_k = h^k{}_{n(k)}$ in $L^2(\Omega_k)^n$ such that

f_k in $H^1(\Omega_k)$ with $X_{\Omega_k} f_k$ --->$X_\Omega f$ weakly in $L^2(D)$,

X_{Ω_k}grad f_k ----> $X_\Omega h$ weakly in $L^2(D)^n$ } .So that f belongs to $H^1(\Omega)$.

We consider the fllowing problem

$$(N^\alpha{}_\sigma) \qquad \text{I N F } \{ EN_\sigma(\Omega) \mid \text{meas}(\Omega)=\alpha, \Omega \text{ in D} \} \tag{5.2}$$

where

$$EN_\sigma(\Omega) = EN(\Omega) + \sigma P_D(\Omega) \tag{5.3}$$

$$EN(\Omega) = \text{M I N} \{ \int_\Omega 1/2(| h |^2 + f^2) \, dx + \int_\Omega F f \, d\mu + \int_\Omega G \, dx \mid (f,h) \text{ in } H^1(\Omega) \} \tag{5.4}$$

Lemma 5.2.

For any measurable subset Ω in D the minimum in (5.4) is achieved.

Proof.

Let (f^k,h^k) be a minimzing sequence in $H^1(\Omega)$.chosing the first element of that sequence to be zero we get

$$1/2 \int_\Omega (|h^k|^2 + (f^k)^2) \, dx - \int_\Omega F f^k \, dx \leq 0 \tag{5.5}$$

From which we get

$$\| f^k \|_{L^2(\Omega)} \leq \| F \|_{L^2(\Omega)}, \| h^k \|_{L^2(\Omega)^n} \leq 2 \| F \|_{L^2(\Omega)}$$

We can then substract subsequences,still denoted by f^k and h^k, which weakly converge to elements f and h respectively in $L^2(\Omega)$ and $L^2(\Omega)^n$ .One can verify that the element (f,h) belongs to $H^l(\Omega)$,this proof is done by builting a 'diagonal' subsequence exactly as in the proof of lemma 16 but the strong $L^2(\Omega)$ convergence of the sequence being replace by the weak one(using the fact that on the ball $\{f \text{ in } L^2(\Omega) \mid \|f\| \leq 2\|F\| \}$ the weak topology of $L^2(\Omega)$ is metrisable ,so that to substract the diagonal subsequence one can write the triangular inequality for that distance and then classiquely chose first k and then n=n(k) large enough in the other term)).Then the fact that (f,h) is a minimizing element to problem (5.4) classiquely derives from weak lower semi continuity of the functional under consideration .

<u>Proposition 5.3.</u>

The problem $(\mathcal{N}^{\alpha}_{\sigma})$ posesses solutions Ω (at least one) in BPS(D) with meas $(\Omega)=\alpha$.

Proof.

Let Ω_k be a minimizing sequence for the problem $(\mathcal{N}^{\alpha}_{\sigma})$ and for any k (f^k,h^k) be a minimizing element to (5.4) in $H^l(\Omega_k)$ .We do have

$\int_{\Omega_k} 1/2\, (|h^k|^2 + (f^k)^2)\, dx - \int_{\Omega_k} Ff^k dx + \sigma P_D(\Omega_k) \leq E_\sigma(\Omega_o) := E(\Omega_o) + \sigma P_D(\Omega_o)$ for any admissible set Ω_o .But for any Ω_o (5.5) implies $E(\Omega_o) \leq 0$ so that we get

$$\int_{\Omega_k} 1/2\, (|h^k|^2 + (f^k)^2)\, dx - \int_{\Omega_k} Ff^k dx + \sigma P_D(\Omega_k) \leq \sigma\alpha \qquad (5.6)$$

in particular it derives:

$$\| f^k \|_{L^2(\Omega_k)} \leq \| F \|_{L^2(D)} + (\| F \|_{L^2(D)})^2 + 2\sigma)^{1/2} = b$$

$$\|h^k\|_{L^2(\Omega_k)^n} \leq 2(\sigma\alpha + b \| F \|_{L^2(D)})$$

$$\sigma P_D(\Omega_k) \leq \sigma\alpha + b \| F \|_{L^2(D)}$$

then after substracting a subsequence there exists two elements λ and μ in $L^2(D)$ and $L^2(D)^n$ such that the following weak convergences hold:

$X_{\Omega_k} f^k \dashrightarrow \lambda$; $X_{\Omega_k} h^k \dashrightarrow \mu$, $\Omega_k \longrightarrow \Omega$, Ω in BPS(D) and then we prove now that (λ,μ) belongs to $H^l(\Omega)$ .For each k ,(f^k,h^k) is in $H^l(\Omega_k)$ then by definition there exists a sequence $\Omega_k{}^n$ of polyhedral open sets in D and $f_k{}^n$ in $H^l(\Omega_k{}^n)$ such that $X_{\Omega_k n} \dashrightarrow X_{\Omega_k}$ in $L^2(D)$,

$X_{\Omega_k n} f_k{}^n \dashrightarrow X_{\Omega_k} f_k$ weakly in $L^2(D)$,

$X_{\Omega_k n}\, \mathrm{grad}\, f_k{}^n \dashrightarrow X_{\Omega_k} h^k$,weakly in $L^2(D)^n$, when $n \dashrightarrow \infty$.

The weak topology of $L^2(D)$ on the ball of radius $2(\sigma\alpha+b \| F \|_{L^2(D)})$ is metrisable.Let $d(.,.)$ be that distance,then we write

$$d(X_{\Omega_k n} f_k{}^n , \lambda) \leq d(X_{\Omega_k n} f_k{}^n, X_{\Omega_k} f_k) + d(X_{\Omega_k} f_k, \lambda) \tag{5.7}$$

$$d(X_{\Omega_k n} \text{grad } f_k{}^n , \mu) \leq d(X_{\Omega_k n} \text{grad } f_k{}^n, X_{\Omega_k} h_k) + d(X_{\Omega_k} h_k, \mu) \tag{5.8}$$

given some arbitrary positive ε we chose k large enough, $k=k(\varepsilon)$,so that the two last terms in the right hand sides of (5.7) and (5.8) are less then ε ;in a second time we select n large enough $n=n(k(\varepsilon))$ so that also the other terms in the right hand sides of (5.7),(5.8) are less then ε.The 'diagonal' subsequence defined by

$(X_{\Omega_{k(n(\varepsilon))}n(e)} f_{k(n(\varepsilon))}{}^{n(\varepsilon)} , X_{\Omega_{k(n(\varepsilon))}n(e)} \text{grad } f_{k(n(\varepsilon))}{}^{n(\varepsilon)})$ weakly converges in $L^2(D)^{n+1}$ to

(μ,λ) and the domain $\Omega_{k(n(\varepsilon))}$ converges ,as ε goes to zero, to Ω in BPS(D).Obviously (μ,λ) is in fact in $L^2(\Omega)^{n+1}$, to see it it is enough to get in the limit in the following equalities $0=(1-X_{\Omega_k})X_{\Omega_k} f_k$ and $0=(1-X_{\Omega_k})X_{\Omega_k} h_k$ so that $0=(1-X_\Omega)\mu$ and $0=(1-X_\Omega)\lambda$ .We conclud that (μ,λ) belongs to $H^1(\Omega)$.

Also in the limit we do have

$$\int_D(\lambda^2+\mu^2)dx - \int_D F\lambda \, dx +\sigma P_D(\Omega) \leq \liminf \int_{\Omega_k} 1/2 \, (|h^k|^2 +(f^k)^2) \, dx -\int_{\Omega_k} Ff^k dx +\sigma P_D(\Omega_k)$$

as (λ,μ) belongs to $H^1(\Omega)$ we have

$\int_D(\lambda^2+\mu^2)dx - \int_D F\lambda \, dx=\int_\Omega(\lambda^2+\mu^2)dx - \int_\Omega F\lambda \, dx$, so that Ω is a solution to $(\mathcal{P}N^\alpha{}_\sigma)$ and (λ,μ) is the associated solution to problem (5.4).

<u>Remark 5.4.</u>

In the definition of $H^1(\Omega)$ the sequence Ω_n of polyhedral open sets in D can be replaced by a sequence of smooth C^∞ open domains in D.We know from De Giorgi E.Colombini F.,Piccinini [1972] that any set in BPS(D) can be approached in $L^2(D)$ (in the sens of $L^2(D)$ convergence of characteristic functions and weak convergence of the gradients as bounded measures on D) by a sequence of polyhedral open sets in D.

5. 2 Max Inf Formulation.

We turn now to the easiest situation :

$$\underset{\Omega}{\text{Max}} \ \underset{\phi}{\text{Inf}} \ (\ L(\Omega,\phi) \ / \ \Omega \text{ in D, mes}(\Omega)=\alpha, \ \phi \text{ in } H^1{}_0(D) \ ) \tag{5.9}$$

where

$$L(\Omega,\phi) = \int_\Omega (1/2 \, |\text{grad}\phi|^2 -f\phi \,) \, dx + \int_\Omega G \, dx - \sigma P_D(\Omega) \tag{5.10}$$

The mapping $\Omega \dashrightarrow \mathrm{Inf}\ (\ L(\Omega,\phi)\ /\ \phi \text{ in } H^1_o(D)\ )$ is upper semi continuous on BPS(D) so that the maximum is reached in (5.9) .

VI. Free Interface with Transmission Conditions.

The optimization problem associated to the transmission condition that we shall define in an analogoud way is the easiest one . It could be compare to the problem (P) we studied at the begining for $\partial\Omega$ will an interface on which we shall have first order differential conditions solved by the solution y. Let be given two positive real number a and b,a<b and F in $H^1(D)$.We consider the following problem:

$(\mathcal{T}_\sigma^\alpha)$ MIN $\{ E_\sigma(\Omega)\ |\ \Omega$ measurable set in D, measure of Ω equal to $\alpha \}$

where

$$E_\sigma(\Omega) = \mathrm{MIN}\ \{ \int_D (1/2\chi(a,b,\Omega)\ |\mathrm{grad}\phi|^2 - F\phi)\ dx\ |\phi \text{ belongs to } H^1_o(D)\} + \sigma P_D(\Omega) \quad (6.1)$$

where

$$\chi(a,b,\Omega) = bX_\Omega + a\ X_{D\setminus\Omega} = a + (b-a)X_\Omega \quad (6.2)$$

for short we shall write $\chi(\Omega)$ or simply χ.
That problem can be written

$$\mathrm{Min}\{\int_D (1/2\chi(\Omega)\ |\mathrm{grad}\phi|^2 - F\phi)\ dx\ |\phi \text{ belongs to } H^1_o(D),\Omega \text{ measurable in } D\} + \sigma P_D(\Omega) \quad (6.3)$$

Let (ϕ_n,Ω_n) be a minimizing sequence it is immediat to verify that it remains bounded in $H^1_o(D)\mathrm{x}BPS(D)$ as we assume $\sigma>0$).So that ,after substracting a subsequence,we can assume that that sequence converges in ($H^1_o(D)\mathrm{x}BPS(D)$)-weak to (y,Ω).The element ϕ_n can be chosen as a solution to $E_\sigma(\Omega_n)$ so we have:

$$\int_D (\chi(\Omega_n)\ \mathrm{grad}\phi_n.\mathrm{grad}\psi - F\psi)\ dx = 0 \text{ for all y in } H^1_o(D) \quad (6.4)$$

As $\chi(\Omega_n)$ converges in $L^2(D)$ strongly to $\chi(\Omega)$,takinf ψ in $C^1(cl(D))$, also we get $\chi(\Omega_n)\psi$ converges in $L^2(D)$ strongly to $\chi(\Omega)\psi$.On the other hand gradϕ_n converges weakly in $L^2(D)$ to grady.Then in the limit we get that (Ω,y) verify (6.4) so that y is the solution to (6.1).

In fact we can show that ϕ_n converges strongly to ϕ in $H^1_o(D)$ .Taking $\psi=\phi_n$ in (6.4) we get $E_\sigma(\Omega_n) = -1/2 \int_D F\,\phi_n\ dx$ which converges to $-1/2 \int_D F\,y\ dx = E_\sigma(\Omega)$,so that $E_\sigma(\Omega)$ realises the minimum in $(\mathcal{T}_\sigma^\alpha)$ .The strong $H^1_o(D)$ convergence of ϕ_n derives as follows: $\mathrm{grad}\phi_n = (\chi(\Omega_n))^{-1}(\chi(\Omega_n)\ \mathrm{grad}\phi_n)$, but $(\chi(\Omega_n))^{-1}$ converges strongly in $L^2(D)$ to $(\chi(\Omega))^{-1}$ while $\chi(\Omega_n)\ \mathrm{grad}\phi_n$ converges strongly in $L^2(D)$ to $\chi(\Omega)$ grady as it weakly converges and its norm ,whose square is $-\int_D F\phi_n\ dx$, converges.We can state that result as follows.

Proposition 6.1.

For any s>0 and a,0<a<meas(D) there exist solutions Ω to the problem $(\mathcal{T}_\sigma^\alpha)$ .If Ωn converges to Ω in the sens that the characteristic functions of those measurable sets converges in the $L^2(D)$ norms, then we have the strong $H^1{}_0(D)$ convergence of the solutions to $E_\sigma(\Omega_n$ and $E_\sigma(\Omega)$:$y(\Omega_n)$ --->$y(\Omega)$ in $H^1{}_0(D)$.

VII. First Order Necessary Conditions.

7.1 The Flow Mapping T_t of a Speed Vector field $\mathcal{V}$

We shall show that the solutions to problems $(\mathcal{D}_0)$,$(\mathcal{D}_0{}^\alpha)$,$(\mathcal{D}_\sigma{}^\alpha)$,$(\mathcal{D}\Sigma\mathcal{N}_\sigma{}^\alpha)$,$(\mathcal{N}_\sigma{}^\alpha)$, $(\mathcal{T}_\sigma{}^\alpha)$ furnish weak solutions to the Free Boundary Problems under consideration.

We have to introduce perturbations Ω_t for t little of an optimal solution to one of the previous problems.The simpliest idea is to use perturbations of the identity mapping as follows.Let V belongs to $C^\infty(cl(D),R^n)$ with $V(x).n(x)=0$ on ∂D at each point x of ∂D at which the normal field n exists and V(x)=0 in the other points of ∂D.For t little enough,$0\le t\le\tau(V)$, the mapping $T_t = I_d+tV$ is one to one from D onto D and from ∂D onto ∂D.We define the perturbed domain as $\Omega_t = T_t(\Omega)$ with $\Omega_0=\Omega$. T_t and $T_t{}^{-1}$ being elements on $C^\infty(cl(D),R^n)$ the boundary $\partial\Omega_t$ is given by $\partial\Omega_t=T_t(\partial\Omega)$.The jacobian matrix will be denoted $DT_t = I_d+ t\,DV$ and its determinant $j(t)= \det(DT_t) = \det(I_d+ t\,DV)$ verifies $j'(t)= j(t)\,\mathrm{div}(\mathcal{V}(t))$ where $\mathcal{V}(t)=VoT_t{}^{-1}$ is the velocity field associated to the transformation T_t . V is an autonomeous vector field while $\mathcal{V}(t)$ is not . Then we get $j(t)= \exp(\int_0^t \mathrm{div}(\mathcal{V}(s))\,ds)$ .To handle the constraint on the measure of the fith problems indexed by α we need transformations preserving the measure of Ω .This condition is achieved when j(t)=1 and to reach it we shall consider the transformation T_t as being built by a free divergence vectot field $\mathcal{V}$: for any t , $\mathrm{div}(\mathcal{V}(t)=0$ implies j(t)=1 . Let be given $\mathcal{V}$ in $C^\infty([0,t[,C^\infty(cl(D),R^n)$ with $\mathcal{V}(t,x).n(x)=0$ on ∂D at each point x of ∂D at which the normal field n exists and $\mathcal{V}(t,x)=0$ in the other points of ∂D.Then the transformation T_t is the flow of the field $\mathcal{V}$, given by

$$T_t(x)=x+\int_0^t \mathcal{V}(s,T_s(x))\,ds \quad \text{or} \quad \partial_t(T_t(x)) = \mathcal{V}(t,T_t(x)) , (T_t(x))_{t=0} = x \qquad (7.1)$$

while the field V is given by $V=\mathcal{V}(0)$. In that definition the field $\mathcal{V}$ can be chosen autonomeous, i.e $\mathcal{V}$ independant on time t ,what we shall always do in the sequel, then the transformation T_t cannot be written in the form I_d+tV but $T_t = I_d+t\,V(t,.)$ .That is to say that when $\mathcal{V}$ is autonomeous V is not (and conversely as we have seen) .When there is no constraint on the measure of the optimal domain Ω the transformation $T_t=I_d+tV$,with V autonomeous, is adequate to obtain the first order necessary condition but will be completely unadequate concerning the first order necessary conditions of problems indexed with α. On the other hand the transformations

$T_t(x)=x+\int_0^t \mathcal{V}(s,T_s(x))\,ds$ are adequate in all situations ,including second order necessary conditions.For this purpose it will be enough to consider such transformations associated to autonomeous vector field V. Of course when both $T_t(x)=x+\int_0^t \mathcal{V}(s,T_s(x))\,ds$ and $T_t=I_d+tV$ are adequate to derive the first order necessary conditions ,this occurs only for problem $(\mathfrak{D}_o)$, these two families of transformations lead to the same first order conditions,i.e. the same free boundary problem solved by the optimal domain Ω.This is false concerning the second order conditions.

For constraints such that Ω should contain a given measurable subset $\mathfrak{E}$ of D it is enough to take $\mathcal{V}(t,.)=0$ on $\mathfrak{E}$.

7.2 Free Boundary Conditions In Weak Form.

In the problems c the optimality of the measurable set Ω can be written as follows .

There exist a measurable set Ω in D and $y=y(\Omega)$ in $H(\Omega)$,y being uniquely associated to Ω such that :

for any admissible field $\mathcal{V}$,

for any t , $0\le t\le \tau(\mathcal{V})$,

for any ϕ_t in $H(\Omega_t)$,$\Omega_t=T_t(\Omega)$, T_t being the flow mapping of $\mathcal{V}$

$$\int_\Omega \{ 1/2\, |\, \text{grad } y\, |^2 - fy + G \}\, dx + \delta \int_{\Sigma_0 \cap\, cl(\Omega)} gy\, d\mu + P_D(\Omega)$$

$$\le \int_{\Omega_t} \{ 1/2\, |\, \text{grad } \phi_t\, |^2 - f\,\phi_t + G \}\, dx + \delta \int_{\Sigma_0 \cap\, cl(\Omega_t)} g\,\phi_t\, d\mu + P_D(\Omega_t) \tag{7.2}$$

where $\delta \ge 0$,f and G could be zero and $H(\Omega)$ is the Hilbert space under consideration in each problem: respectively $H^1_0(\Omega)$,$H^l_0(\Omega)$ .In the two problems the 'natural' boundary condition attached to $H(\Omega)$ is y=0 on $\partial\Omega$; The additive optimality condition (7.2) implies an extra boundary condition solved by $y(\Omega)$ on $\partial\Omega$. This extra condition is given by (7.6) at lemma 22 bellow.In the next section we shall explicit the boundary expression for (7.6) when we assume Ω and y to be smooth enough.

Proposition 7.1.

Let Ω be a measurable subset in D.

i) When $H(\Omega)$ ranges among the following hilbert spaces $H^1_0(\Omega)$,$H^l_0(\Omega)$, we do have ϕ belongs to $H(\Omega)$ iff $\phi_t = \phi o T_t^{-1}$ belongs to $H(\Omega_t)$.

Proof.

ii) We use the following lemma.

Lemma 7.2.

Let f_n be a sequence in $L^2(D)$ which converges weakly to f in $L^2(D)$.Then the sequence $f_n o T_t^{-1}$ converges weakly to $f o T_t^{-1}$ in $L^2(D)$.

In view of Proposition 7.1 we can take $\phi_t = y o T_t^{-1}$ in (7.2) .We get :

for any admissible field $\mathcal{V}$,

for any t , $0 \leq t \leq \tau(\mathcal{V})$,

T_t being the flow mapping of $\mathcal{V}$

$$\int_\Omega \{ 1/2 \,|\, \text{grad } y \,|^2 - fy + G \} \, dx$$
$$\leq \int_{\Omega_t} \{ 1/2 \,|\, \text{grad } y o T_t^{-1} |^2 - f y o T_t^{-1} + G \} \, dx$$

Using the change of variable $X=T_t(x)$ in the right hand side of (7.3) we get :

$$\int_\Omega \{ 1/2 \,|\, \text{grad } y \,|^2 - fy + G \} \, dx \qquad \leq$$
$$\int_\Omega \{ 1/2 \langle A(t).\, \text{grad } y, \text{grad} y\rangle - j(t)\, f o T_t\, y + j(t) G o T_t \} \, dx \qquad (7.4)$$

where n is the unitary normal field on ∂D (which is assumed to exist on Σ_o) and A(t) is the symetric matrix $j(t)\, DT_t^{-1.*}(DT_t)^{-1}$

Lemma 7.4. (J.P.Zolésio[1979],[1980],J.Sokolowski,J.P. Zolésio[1990])

Let $2\varepsilon(\mathcal{V}) = D\mathcal{V} + {}^*D\mathcal{V}$, then for all integer k we have

$$\| t^{-1}(A(t)-I_d) - (\text{div}(\mathcal{V})\, I_d - 2\varepsilon(\mathcal{V}) \|_{W^{k,\infty}(cl(D),R^n)} \text{--->}0, \text{ as } t\text{--->}0$$

Let us define the 'Eulerian semi-derivative ' of the perimeter $P_D(\Omega)$ at Ω in the direction V by :

$$d\_\, P_D(\Omega; \mathcal{V}) \;=\; \liminf_{t>0,\, t\text{-->}0} \; t^{-1} \left(P_D(\Omega_t) - P_D(\Omega) \right) \qquad (7.5)$$

where $\Omega_t = T_t(\Omega)$,T_t being the flow mapping of the field $\mathcal{V}$ .That lim inf can be expressed as follows:

$$d\_\, P_D(\Omega; \mathcal{V}) = \inf \{ \liminf_{n\text{-->}\infty} \; t_n^{-1} \left(P_D(\Omega_{t_n}) - P_D(\Omega) \right) \mid \{t_n\}, t_n>0, t_n\text{-->}0 \}$$

In view of lemm 21 and (7.4) we get the

Proposition 7.5.

Let f and G be an elements of $H^1(D)$ and Ω be an optimal solution to one of the problems $(\mathfrak{D}_o)$,$(\mathfrak{D}_o^{\alpha})$,$(\mathfrak{DC}_o)$,$(\mathfrak{DC}_o^{\alpha})$,$(\mathfrak{D}_\sigma^{\alpha})$). Then for any admissible field $\mathcal{V}$ we have

$$\int_{\Omega} \{ 1/2 <[\mathrm{div}(V)\, I_d - 2\varepsilon(V)].\, \mathrm{grad}\, y, \mathrm{grad} y> - \mathrm{div}(f V)\, y + \mathrm{div}(G V) \}\, dx + \\ + \ \sigma\ d\_P_D(\Omega; V) \ \geq 0.$$

and then $\quad d\_P_D(\Omega; V) > -\infty$.

7.3 Weak Condition at Ω having a curvature $\mathfrak{H}$ in $\mathcal{E}'(R^n,R^n)$

The two variational inequalities (7.6),(7.9) become equalities as soon that $\sigma = 0$ or $\sigma > 0$ but Ω has a curvature $\mathfrak{H}$.

Definition 7.8.

We say that the measurable set Ω has a curvature if Ω belongs to BPS(D) and if the mapping $V \dashrightarrow d\_P_D(\Omega; V)$ is linear and continuous on $C^\infty(\mathrm{cl}(D),R^n)$.

Then there exists an element $\mathfrak{H}$ in $\mathcal{E}'(R^n,R^n)$, i.e. a vectorial distribution over R^n having its support contained in $\partial\Omega$, such that

$$d\_P_D(\Omega; V) = < \mathfrak{H} , V >_{\mathcal{D}'(R^n,R^n)\times\mathcal{D}(R^n,R^n)} \tag{7.10}$$

Proposition 7.9.

Assume that Ω is an optimal solution to problem (Q),when Q ranges in $(\mathcal{D}_0)$,$(\mathcal{D}_0{}^\alpha)$,$(\mathcal{D}_\sigma{}^\alpha)$. Then for any admissible field V we have :

$$< -1/2\ \mathrm{grad}(X_\Omega\ |\mathrm{grad} y|^2) + \mathrm{div}(\ X_\Omega\ (\mathrm{grad} y * \mathrm{grad} y)\ ) + (fy - G)\ \ \mathrm{grad} X_\Omega, V> \ = 0$$

where the brackets <,> stand for the duality pairing between $\mathcal{D}'(R^n,R^n)$ and $\mathcal{D}(R^n,R^n)$, ${}^*\gamma_{\partial D}$ is the transposed of the trace operator $\gamma_{\partial D}$ on ∂D and grady*grady is the matrix $\partial_i y \partial_j y$.

7.4 Optimality condition when Ω is a Smooth Open Set.

We assume now that $\partial\Omega$ is smoth enough manifold ,Ω being located on one side of $\partial\Omega$.S_0 that,when s>0 the mean curvature H exists a.e. on $\partial\Omega$ and $\mathfrak{H} = {}^*\gamma_{\partial\Omega}(\ Hn\ )$,n being the normal field on $\partial\Omega$.From the smoothness of Ω we obtain that $H^1(\Omega)$,$H^1{}_0(\Omega)$... are classiquely defined and by the classical regularity results for the solution of elliptic boundary value problems we obtain that for the problems $(\mathcal{D}_0)$,$(\mathcal{D}_0{}^\alpha)$,$(\mathcal{D}_\sigma{}^\alpha)$ the solution y=y(Ω) belongs to $H^2(\Omega)$ .Using the Stoke's formula we obtain easily from (7.11) the free boundary condition verified by y in each problems $(\mathcal{D}_0)$,$(\mathcal{D}_0{}^\alpha)$,$(\mathcal{D}_\sigma{}^\alpha)$,For any admissible fiels V ,y verifies

$$\int_\Omega (\ \Delta y + F\ ) < \mathrm{grad} y, V > \ dx + \int_{\partial\Omega} \{ [1/2 |\mathrm{grad} y|^2 - fy + G]\ V.n - \partial_n y\ <\mathrm{grad} y, V> \quad \}\ d\mu \\ + \ \sigma \int_{\partial\Omega} H\, V.n\ \ d\mu = 0 \tag{7.12}$$

for the Dirichlet situations y is zero on $\partial\Omega$ or constant on each connected componant of $\partial\Omega$ so that in both cases we get $\text{grad}y=\partial_n y\, n$ and (7.12) reduces to

$$\int_{\partial\Omega} [\, -1/2\, |\text{grad}y|^2 - fy + G]\ \mathcal{V}.n\, d\mu \ + \ \sigma \int_{\partial\Omega} H\, \mathcal{V}.n\ d\mu = 0 \qquad (7.13)$$

In the fist problem there is no constraint on the measure of the domain so there is no constraint on the divergence of the admissible field $\mathcal{V}$, we only have $\mathcal{V}.n = 0$ a.e. on ∂D ($\mathcal{V}=0$ at exceptional points at which the normal n does not exist on ∂D) , so from (7.13) we get:

$$-1/2\, |\, \partial_n y\, |^2 + G + \sigma H = 0 \quad \text{on } \partial\Omega \cap D \qquad (7.14)$$

and when $\sigma=0$,i.e. for $(\mathfrak{D}_0)$

$$\partial_n y = 0 \quad \text{on } \partial\Omega \cap D \qquad (7.15)$$

In the problems $(\mathfrak{D}_0{}^{\alpha})$,$(\mathfrak{D}_\sigma{}^{\alpha})$ the admissible field is free divergence which implies that the normal componant $\mathcal{V}.n$ on $\partial\Omega$ verifies the following constraint:

$$\int_{\partial\Omega} \mathcal{V}.n\ d\mu = 0 \qquad (7.16)$$

So that from (7.13) we get :
There exists a constant c such that,

$$-1/2\, |\, \partial_n y\, |^2 - fy + G + \sigma H = c \quad \text{on } \partial\Omega \cap D \qquad (7.17)$$

VIII. Second order Necessary Conditions.

8.1 Second Order Derivative of E(Ω).

We shall now assume that Ω is an optimal solution to one of the previous problems and suppose that Ω is a smooth (enough) domain contained in D.We shall define the second order derivative of the cost function $E(\Omega)$ in a general way,then use it to derive the second order necessary conditions.

Under smoothness assumptions we have seen at the previous section that E(.) posesses a gradient at Ω: there exists $g=g(\Gamma)$ in $L^1(\Gamma)$, $\Gamma=\partial\Omega$, (g is the density gradient in the terminology of Zolésio[1976],[1980 a]) such that

$$dE(\Omega;\mathcal{V}) := \lim_{t>0,\, t\to 0} \left(E(\Omega_t(\mathcal{V})) - E(\Omega) \right)/t$$

$$= \int_\Gamma g(\Gamma)\ \mathcal{V}.n\ d\mu \qquad (8.1)$$

We define the boundary shape derivative of g (Zolésio J.P.[1988]),if it exists, as being the following element in $\mathfrak{D}'(\Gamma)$:

$$g'(\Gamma;\mathcal{V}) = [\ \partial/\partial t\{\ g(T_t(\Gamma))oT_t\ \}\ ]_{t=0}\ -[\ \mathrm{grad}_\Gamma g(\Gamma)].\mathcal{V} \tag{8.2}$$

where T_t is the flow mapping associated to the autonomeous vector field $\mathcal{V}$.

Following M.Delfour-J.P.Zolésio [1991] let us define the second order derivative.Given two autonomeous vector fields $\mathcal{V}$ and $\mathcal{W}$:

$$d^2E(\Omega;\mathcal{V},\mathcal{W}) = \lim_{t>0,t\to 0}\ (\ dE(\Omega_t(\mathcal{W});\mathcal{V})-dE(\Omega;\mathcal{V}))/t \tag{8.3}$$

We have the following expression for that second derivative.

Lemma 8.1.

$$d^2E(\Omega;\mathcal{V},\mathcal{W}) = \int_\Gamma g'(\Gamma;\mathcal{W})\ \mathcal{V}.n\ d\mu\ +\ \int_\Gamma (\ \mathrm{grad}_\Gamma g(\Gamma).\mathcal{W} + g(\Gamma)\ \mathrm{div}\mathcal{W})\ \mathcal{V}.n\ d\mu$$
$$+\int_\Gamma g < D\mathcal{V}.\mathcal{W}\ -D\mathcal{W}.\mathcal{V}\ ,\ n > d\mu \tag{8.4}$$

We use (8.4) for specific fields,namely A and B be two real functions defined on D such that:

A and B are both constants on each connected componants of $\partial\Omega$

$$\partial^2/\partial n^2\ A = \partial^2/\partial n^2\ B = 0 \text{ on } \partial\Omega, \quad \text{where } \partial^2/\partial n^2\ A = <D^2A.n,n>\ . \tag{8.5}$$

Let us define on Γ the two real functions v and w by:

$$v=\partial/\partial n\ A \text{ on } \Gamma,\ \ w=\partial/\partial n\ B \text{ on } \Gamma\ .$$

Corollary 8.2.

H being the mean curvature on $\partial\Omega$,$H=\mathrm{div}_\Gamma n$ we have:

$$d^2E(\Omega;\ \mathrm{grad}A,\mathrm{grad}B)\ =\ \int_\Gamma g'(\Gamma;\mathrm{grad}\ B)\quad v\ d\mu + \int_\Gamma\ H\ g(\Gamma)\ v\ w\ \ d\mu \tag{8.6}$$

Proof. We use the fact that $(\Delta A)_{|\Gamma} = \Delta_\Gamma(A_{|\Gamma}) + H\partial/\partial_n A + \partial^2/\partial_n 2\ \ A\ = H\ v$ (8.7)

(where Δ_Γ is the Laplace-Beltrami operator on $\partial\Omega$,see Zolésio J.P.[1979],[1980 a]) and the by part integration on $\partial\Omega$ in (8.4).

8.2 Second Order Necessary Condition for Problems $(\mathcal{D}_o)$,$(\mathcal{DC}_o)$,$(\mathcal{D}_o{}^\alpha)$, $(\mathcal{DC}_o{}^\alpha)$.

We know ,see J.P.Zolésio [1980 b],that the derivative of E at any smooth domain Ω is given by:

- when f is $L^2(D)$

$$dE(\Omega;V)= \int_{\partial\Omega} (1/2\ |\mathrm{grad}y|^2\text{-}fy)\mathcal{V}.n\ d\mu\ -\ \int_{\partial\Omega} \partial_n y\ \mathrm{grad}y.\mathcal{V}\ d\mu$$

– when f is in $H^{-1}(D)$ with its support in $\mathbf{C}$, and $\mathbf{C}\cap\partial\Omega$ is empty.

$$dE(\Omega;V)= \int_{\partial\Omega}\ 1/2\ |\mathrm{grad}y|^2\ \ d\mu\ -\ \int_{\partial\Omega} \partial_n y\ \mathrm{grad}y.\mathcal{V}\ d\mu$$

but as y is constant on $\partial\Omega$ we get grad $y.\mathcal{V} = \partial_n y\ \ \mathcal{V}.n$ and we get ,with the convention that f is zero on $\partial\Omega$ in the second situation :

$$dE(\Omega;V)= \int_{\partial\Omega} (-1/2\ |\mathrm{grad}y|^2\text{-}fy)\mathcal{V}.n\ d\mu \tag{8.8}$$

then the density gradient of E is $g(\Gamma)= -1/2\ |\text{grad}y|^2 - fy$ where y is the solution of the Dirichlet problem $-\Delta y=f$ in Ω, $y=0$ on $\partial\Omega$ .With (8.6) we just need to compute the boundary shape derivative $g'(\Gamma;\text{grad } B)$ in order to get $d^2E(\Omega;.,.)$.The density gradient g is in fact the restriction to $\partial\Omega$ of the function $\mathfrak{g}=-1/2\ |\text{grad}y|^2-fy$ defined on Ω and for which we can compute the shape derivative $\mathfrak{g}'(\Omega;W)= -\text{ grad } y.\text{grad } y'(\Omega;W) - f\, y'(\Omega;W)$.But the trace on $\partial\Omega$ of the shape derivative of $\mathfrak{g}$ and the boundary shape derivative $g'(\Gamma;W)$ (g being the restriction to $\partial\Omega$ of $\mathfrak{g}$) are related by:

$$g'(\Gamma;W) = \mathfrak{g}'(\Omega;W)_{|\partial\Omega} + \partial_n \mathfrak{g}\ W.n \tag{8.9}$$

The shape derivative y' is solution of the following Dirichlet BVP,see J.P.Zolésio[1979],[1980 a]:

$$-\Delta y' = 0 \text{ in } \Omega\,, \quad y' = -\partial_n y \text{ on } \partial\Omega \tag{8.10}$$

Lemma 8.3.

$$g'(\Gamma;W) = -\partial_n y\ (\ \partial_n y'(\Omega;W)_{|\partial\Omega} - \langle\partial^2/\partial n^2\ y.n,n\rangle\ W.n) \tag{8.11}$$

Proof. It follows from (8.9),(8.10) and classical arguments.

Proposition 8.4.

Let Ω be an optimal solution to one of the problems $(\mathfrak{D}_o)$,$(\mathfrak{DC}_o)$.Assume that Ω is a smooth open domain in D and that y is smooth enough.Then we have:

For all v and w smoothly defined on $\partial\Omega$, $d^2E(\Omega;\ \text{grad}A,\text{grad}B) = 0$

Proof. As Ω is optimal from (7.14) we get $\partial_n y = 0$ on $\partial\Omega$ so that $y'=0$.Then $g'(\Gamma;W) =0$ and $g(\Gamma) =0$, and the result derives from (8.6).

When the measure of Ω is imposed to be equal to α, i.e.when we consider the problems $(\mathfrak{D}_o{}^{\alpha})$ and $(\mathfrak{DC}_o{}^{\alpha})$, the second order derivative is not zero for the first order necessary condition (7.17),with $\sigma=0$,$G=0$ and $y=0$ on $\partial\Omega$ leads to $(\partial_n y)^2 = c^2$ on $\partial\Omega$. Then y' is not zero and ,from (8.7), we get

$\langle\partial^2/\partial n^2\ y.n,n\rangle = -H\partial_n y$ when f is equal to zero in a neighborhood of $\partial\Omega$. So that (8.11) can be written as

$$g'(\Gamma;W) = -\partial_n y\ \partial_n y'(\Omega;W)_{|\partial\Omega} - H\ c^2\ W.n \tag{8.12}$$

and we get the following second order necessary condition.

Proposition 8.5.

Let Ω be an optimal solution of problem $(\mathfrak{DC}_o{}^{\alpha})$ with G=0.Assume that Ω is smooth.then we have: $(\partial_n y)^2=c^2$ (constante) on $\partial\Omega$ and for all smooth real fuction v defined on $\partial\Omega$ with $\int_\Gamma v\, d\mu = 0$ we have:

$$\int_\Gamma \partial_n y\ \partial_n y'(\Omega;\text{grad}A)_{|\partial\Omega}\ v\ d\mu + \int_\Gamma 1/2\, H\, c^2\, v^2\ d\mu \geq 0. \tag{8.13}$$

IX. Regularity Result.

Assume that Ω is an optimal solution to problem ($\mathcal{D}_0^{\alpha}$). Then from (7.11), as $\delta=0$ (i.e. we have no boundary term in that problem), the necessary first order optimality condition can be written as follows:

for any admissible field $\mathcal{V}$ we have :

$$< -1/2\ \text{grad}(X_\Omega\ |\text{grad}y|^2) + \text{div}(X_\Omega\ (\text{grad}y.*\text{grad}y)) + (fy-G)\ \text{grad}X_\Omega,\ \mathcal{V} > =0 \qquad (9.1)$$

As $\mathcal{V}$ can be chosen free divergence over D we obtain classiquely from (9.1) that the gradient $\mathcal{G}(\Omega)$ is equal to gradp for some p. In fact we can here give more precise informations on that 'preassure term p' which appears to be proportional to the characteristic function X_Ω of the optimal domain.

Proposition 9.1

Let Ω be an optimal solution to problem ($\mathcal{D}_0^{\alpha}$). Then we have : there exist a real number β such that

$$(\Delta y + f)\ \text{grad}y = (\ \beta + G - yf)\ \text{grad}X_\Omega \qquad (9.2)$$

Proof.

The gradient $\mathcal{G}(\Omega) = -1/2\ \text{grad}(X_\Omega\ |\text{grad}y|^2) + \text{div}(X_\Omega\ (\text{grad}y.*\text{grad}y)) + (fy-G)\ \text{grad}X_\Omega$ can be written , as grady=0 a.e. in $D\backslash\Omega$ (so that X_Ω grady = grady a.e. in D),

$$\mathcal{G}(\Omega) = -1/2\ \text{grad}(\ |\text{grad}y|^2) + \text{div}(\ (\text{grad}y.*\text{grad}y)) + (fy-G)\ \text{grad}X_\Omega \qquad (9.3)$$

but $-1/2\ \text{grad}(\ |\text{grad}y|^2) + \text{div}(\ (\text{grad}y.*\text{grad}y)\ ) = \Delta y\ \text{grad}y$

Now,following the lemma 9.2 it is enough for the field $\mathcal{V}$ to be an admissible field to verify the following condition

$$\int_{\partial\Omega} \mathcal{V}.n\ d\mu = \int_\Omega \text{div}\mathcal{V}\ dx = 0 \qquad (9.4)$$

and the orthogonality condition (9.1) to all fields verifying that condition (9.4) leads to (9.3) where b is obtained as follows: let V_0 be a vector field over D such that $\int_\Omega \text{div}(V_0)\ dx \neq 0$, then for any field V, the field $V - \gamma\ Vo$ verifies the condition (9.4) when $\gamma = (\int_\Omega \text{div}(V_0)\ dx)^{-1}\ (\int_\Omega \text{div}(V)\ dx)$

The optimality condition $< \mathcal{G}(\Omega),\ V - \gamma\ Vo >= 0$ is equivalent to

$$< \mathcal{G}(\Omega),\ V > = \gamma < \mathcal{G}(\Omega)\ ,\ V_0 >$$

but $\gamma = (\int_\Omega \text{div}(V_0)\ dx)^{-1}\ < -\ \text{grad}X_\Omega,\ V >$, so that we get :

$< \mathcal{G}(\Omega),\ V > = < \mathcal{G}(\Omega)\ ,\ V_0 >\ (\int_\Omega \text{div}(V_0)\ dx)^{-1}\ < -\ \text{grad}X_\Omega,\ V >$ for any field V,then the condition (9.) holds with

$$\beta = -< \mathcal{G}(\Omega)\ ,\ V_0 >\ (\int_\Omega \text{div}(V_0)\ dx)^{-1}\ .$$

Lemma 9.2

Let Ω be a measurable subset in D and V be given in $W^{1,\infty}(D)$ such that $\int_\Omega \text{div}V\ dx = 0$ .Then there exits $\tau>0$ and V in $L^\infty(0,\tau,W^{1,\infty}(D))$ such that $\int_{\Omega_t(V)} \text{div}\mathcal{V}(t)\ dx = 0$ for all t $0 \le t \le \tau$.

References.

Alt H.W.,Caffarelli L.A [1980].Existence and Regularity for a Minimum Problem with Free Boundary. *Preprint of Courant Institut, N.Y.*

Deny J, Lions J.L. [1953]. Les espaces du type de Beppo Levi. *Ann. Inst. Fourier (Grenoble) 5,305-370*

A. Fasano [1990] Lecture at the Nato Advance Study "Shape Optimization and Free Boundary Problems", Montréal.

Federer H., Ziemer W.P.[1972]. The Lebesgue Set of a Function Whose Distribution Derivative are p-th Power Summable. *Indiana Univ.Math.J.22,139-158*

Stampacchia G. [1960] Seminaire de Mathematiques superieures. *Presse de l'Universite de Montréal.*

Témam R.[1983] Problèmes mathematiques en plasticite. *Gauthiers Villars ,Paris.*

De Giorgi E., Colombini F., Piccinini L.C.[1972] Frontiera Orientate di Misura Minima e Questioni Collegate. *Scuola Normale Superiore, Pisa.*

Delfour M.C.,Zolésio J.P.[1991] To appear in *SIAM J. on Control and Optim.*

Sokolowski J., Zolésio J.P.[1990] . Introduction to Shape Optimization.Shape Sensitivity Analysis. *Springer Verlag.*

Zolésio J.P.[1976] Un Résultat d'Existence de Vitesse Convergente en Optimisation de Domaine.*C.R.Acad.Sc.,Paris 283 ,Ser. A,Pp. 855-859*

Zolésio J.P.[1980a] The Material Derivative (or Speed Method) for Shape Optimization. *In "Optimization of Distributed Parameter Systems", J.Céa,E.Haug eds.,Sljthoff aand Noordhoff,pp.1089-1154..*

Zolésio J.P.[1980b] Shape variational Formulation for Free Boundary Value Problems. *In "Optimization of Distributed Parameter Systems", J.Céa,E.Haug eds.,Sljthoff aand Noordhoff,pp.1055-1194..*

Zolésio J.P.[1979a] Optimisation de domaine par déformation. *These de Doctorat d'état. Nice , France.*

Zolésio J.P.[1988] Shape Derivaive and Shape Acceleration in Boundary Value Problems.'*Lecture notes in Control and Information Sc.,no 104,pp.153-191,A.Bermudez ed.,Springer Verlag,Heidelberg,Berlin,New-York,Tokyo.*

Zolésio J.P.[1979b] Probleme a frontiere libre issu de la physique des plasmas . C.Acad. Sc. Paris ,1979

Lecture Notes in Control and Information Sciences

Edited by M. Thoma and A. Wyner

Vol. 137: S. L. Shah, G. Dumont (Eds.)
Adaptive Control Strategies for Industrial Use
Proceedings of a Workshop
Kananaskis, Canada, 1988
VI, 360 pages. 1989

Vol. 138: D. C. McFarlane, K. Glover
Robust Controller Design Using Normalized Coprime Factor Plant Descriptions
X, 206 pages. 1990

Vol. 139: V. Hayward, O. Khatib (Eds.)
Experimental Robotics I
The First International Symposium
Montreal, June 19-21, 1989
XVII, 613 pages. 1990

Vol. 140: Z. Gajic, D. Petkovski, X. Shen
Singularly Perturbed and Weakly Coupled Linear Control Systems
A Recursive Approach
VII, 202 pages. 1990

Vol. 141: S. Gutman
Root Clustering in Parameter Space
VIII, 153 pages. 1990

Vol. 142: A. N. Gündeş, C. A. Desoer
Algebraic Theory of Linear Feedback Systems with Full and Decentralized Compensators
V, 176 pages. 1990

Vol. 143: H.-J. Sebastian, K. Tammer (Eds.)
System Modelling and Optimization
Proceedings of the 14th IFIP Conference
Leipzig, GDR, July 3-7, 1989
X, 960 pages. 1990

Vol. 144: A. Bensoussan, J. L. Lions (Eds.)
Analysis and Optimization of Systems
Proceedings of the 9th International Conference, Antibes, June 12-15, 1990
XII, 992 pages. 1990

Vol. 145: M. B. Subrahmanyam
Optimal Control with a Worst-Case Performance Criterion and Applications
X, 133 pages. 1990

Vol. 146: D. Mustafa, K. Glover
Minimum Entropy H_∞ Control
IX, 144 pages. 1990

Vol. 147: J. P. Zolésio (Ed.)
Stabilization of Flexible Structures
Third Working Conference,
Montpellier, France, January 1989
V, 327 pages, 1990

Vol. 148: In preparation

Vol. 149: K. H. Hoffmann, W. Krabs (Eds.)
Optimal Control of Partial Differential Equations
Proceedings of the IFIP WG 7.2 International Conference, Irsee, April 9-12, 1990
VI, 245 pages. 1991

Vol. 150: L. C. G. J. M. Habets
Robust Stabilization in the Gap-topology
IX, 126 pages. 1991

Vol. 151: J. M. Skowronski, H. Flashner, R. S. Guttalu (Eds.)
Mechanics and Control
Proceedings of the 3rd Workshop on Control Mechanics in Honor of the 65th Birthday of George Leitmann
January 22-24, 1990, University of Southern California
IV, 497 pages. 1991

Vol. 152: J. D. Aplevich
Implicit Linear Systems
XI, 176 pages. 1991

Vol. 153: O. Hájek
Control Theory in the Plane
X, 272 pages. 1991

Vol. 154: A. Kurzhanski, I. Lasiecka (Eds.)
Modelling and Inverse Problems of Control for Distributed Parameter Systems
Proceedings of IFIP (W.G. 7.2)-IIASA Conference
Laxenburg, Austria, July 24-28, 1989
VII, 179 pages. 1991

Vol. 155: M. Bouvet, G. Bienvenu (Eds.)
High-Resolution Methods in Underwater Acoustics
V, 249 pages. 1991

Lecture Notes in Control and Information Sciences

Edited by M. Thoma and A. Wyner

Vol. 156: R. P. Hämäläinen, H. K. Ehtamo (Eds.)
Differential Games –
Developments in Modelling and Computation
Proceedings of the Fourth International Symposium on Differential Games and Applications
August 9-10, 1990, Helsinki University of Technology, Finland
XIII, 292 pages. 1991

Vol. 157: R. P. Hämäläinen, H. K. Ehtamo (Eds.)
Dynamic Games in Economic Analysis
Proceedings of the Fourth International Symposium on Differential Games and Applications
August 9-10, 1990, Helsinki University of Technology, Finland
XIII, 311 pages. 1991

Vol. 158: K. Warwick, M. Kárný, A. Halousková (Eds.)
Advanced Methods in Adaptive Control for Industrial Applications
X, 331 pages. 1991

Vol. 159: X. Li, J. Yong (Eds.)
Control Theory of Distributed Parameter Systems and Applications
Proceedings of the IFIP WG 7.2 Working Conference, Shanghai, China, May 6–9, 1990
VIII, 219 pages. 1991

Vol. 160: P. V. Kokotović (Ed.)
Foundations of Adaptive Control
IX, 525 pages. 1991

Vol. 161: L. Gerencsér, P. E. Caines (Eds.)
Topics in Stochastic Systems:
Modelling, Estimation and Adaptive Control
IV, 401 pages, 1991

Vol. 162: C. Canudas de Wit (Ed.)
Advanced Robot Control
IX, 314 pages, 1991

Vol. 163: V. L. Mehrmann
The Autonomous Linear Quadratic Control Problem
Theory and Numerical Solution
VI, 177 pages, 1991

Vol. 164: I. Lasiecka, R. Triggiani
Differential and Algebraic Riccati Equations with Application to Boundary/Point Control Problems: Continuous Theory and Approximation Theory
XI, 160 pages, 1991

Vol. 165: G. Jacob, F. Lamnabhi-Lagarrigue (Eds.)
Algebraic Computing in Control
Proceedings of the First European Conference
Paris, March 13–15, 1991
IX, 385 pages. 1991

Vol. 166: L. L. M. van der Wegen
Local Disturbance Decoupling with Stability for Nonlinear Systems
V, 135 pages, 1991

Vol. 167: M. Rao
Integrated System for Intelligent Control
VIII, 133 pages. 1992

Vol. 168: P. Dorato, L. Fortuna, G. Muscato
Robust Control for Unstructured Perturbations – An Introduction
VI, 117 pages. 1992

Vol. 169: V. M. Kuntzevich, M. Lychak
Guaranteed Estimates, Adaptation and Robustness in Control Systems
IV, 209 pages. 1992

Vol. 170: J. M. Skowronski, H. Flashner, R. S. Guttalu (Eds.)
Mechanics and Control
Proceedings of the 4th Workshop on Control Mechanics, January 21–23, 1991
University of Southern California, USA
IV, 301 pages. 1992

Vol. 171: P. Stefanidis, A. P. Papliński, M. J. Gibbard
Numerical Operations with Polynomial Matrices
Application to Multi-Variable Dynamic Compensator Design
VIII, 205 pages. 1992

Vol. 172: H. Tolle, E. Ersü
Neurocontrol
Learning Control Systems
Inspired by Neuronal Architectures and Human Problem Solving Strategies
X, 211 pages. 1992

Vol. 173: W. Krabs
On Moment Theory and Controllability of One-Dimensional Vibrating Systems and Heating Processes
VII, 174 pages. 1992

Lecture Notes in Control and Information Sciences

Edited by M. Thoma and A. Wyner

Vol. 174: A.J.M. Beulens, H.-J. Sebastian (Eds.)
Optimization-Based Computer-Aided
Modelling and Design
Proceedings of the First Working Conference
of the IFIP TC 7.6 Working Group,
The Hague, The Netherlands, 1991
VIII, 270 pages, 1992

Vol. 175: E. Rogers, D.H. Owens
Stability Analysis for Linear Repetitive Processes
VII, 197 pages, 1992

Vol. 176: B.L. Rozovskii, R.B. Sowers (Eds.)
Stochastic Partial Differential Equations
and Their Applications
Proceedings of IFIP WG 7/1 International Conference
University of North Carolina at Charlotte, NC
June 6 - 8, 1991
VIII, 251 pages, 1992

Vol. 177: I. Karatzas, D. Ocone (Eds.)
Applied Stochastic Analysis
Proceedings of a US-French Workshop,
Rutgers University, New Brunswick, N.J.
April 29 - May 2, 1991
X, 311 pages, 1992

Vol. 178: J.P. Zolésio (Ed.)
Boundary Control and Boundary Variation
Proceedings of IFIP WG 7.2 Conference,
Sophia Antipolis, France
October 15 - 17, 1990
VIII, 392 pages 1992